ENGINEERING TOOLS, TECHNIQUES AND TABLES

DEVELOPMENT OF COMPLEX ELECTROMAGNETIC PROBLEMS USING FDTD SUBGRIDDING IN HYBRID COMPUTATIONAL TECHNIQUES

ENGINEERING TOOLS, TECHNIQUES AND TABLES

Additional books in this series can be found on Nova's website under the Series tab.

Additional e-books in this series can be found on Nova's website under the e-book tab.

ENGINEERING TOOLS, TECHNIQUES AND TABLES

DEVELOPMENT OF COMPLEX ELECTROMAGNETIC PROBLEMS USING FDTD SUBGRIDDING IN HYBRID COMPUTATIONAL TECHNIQUES

KHAIRAN N. RAMLI,
RAED A. ABD-ALHAMEED
AND
PETER S. EXCELL
EDITORS

nova publishers
New York

For permission to use material from this book please contact us:
Telephone 631-231-7269; Fax 631-231-8175
Web Site: http://www.novapublishers.com

NOTICE TO THE READER

Additional color graphics may be available in the e-book version of this book.

Library of Congress Cataloging-in-Publication Data

Development of complex electromagnetic problems using FDTD subgridding in hybrid computational techniques/editors, K. N. Ramli, R. A. Abd-Alhameed and P. S. Excell (Electromagnetics and Radio Frequency Engineering School of Engineering, Design and Technology, Bradford University, UK).
pages cm. -- (Engineering tools, techniques, and tables)
Includes bibliographical references and index.
ISBN 978-1-61122-013-1 (hardcover)
1. Electromagnetism--Mathematical models. 2. Finite differences. 3. Time-domain analysis. 4. Numerical grid generation (Numerical analysis) I. Abd-Alhameed, R. A. (Raed A.)
QC760.4.M37D48 2014
537.01'515353--dc23

2014029133

Published by Nova Science Publishers, Inc. † New York

CONTENTS

PREFACE

Complex electromagnetic problems using new hybridised computational technique combining the frequency domain Method of Moments (MoM), Finite-Difference Time-Domain (FDTD) and a subgridded Finite-Difference Time-Domain (SGFDTD) method are studied and discussed in detail. The techniques are desirable to predict electromagnetic absorption in inhomogeneous, anisotropic and lossy dielectric materials irradiated by geometrically intricate sources.

In Method of Moments modelling, the surface kernel solution is derived for 1-D, 2-D and 3-D. The electric surface patch integral formulation is solved by independent linear basis function methods in the circumferential and axial directions of the antenna wires. A similar orthogonal basis function is used on the end surface and appropriate attachments with the wire surface are employed to satisfy the requirements of current continuity. The surface current distributions on structures which may include closely spaced parallel wires are investigated and analysed. The results are found to be stable and showed good agreement with less comprehensive earlier work by others. Moreover, Galerkin technique is employed to predict the effect of a chiral bianisotropic substrate of a rectangular microstrip resonator in the spectral domain.

The work also investigates the interaction between overhead high voltage transmission lines and underground utility pipelines using the FDTD technique for the whole structure, combined with a subgridding method at points of interest. The induced fields above the pipeline are computed. The Perfectly Matched Layer (PML) concept has been utilized to circumvent open-region geometries. The establishment of edge elements has greatly improved the performance of this method and the computational burden due to huge numbers of time steps, in the order of tens of millions, has been eased to tens of thousands by employing quasi-static methods.

The book also illustrates the principle of the equivalent surface boundary employed close to the antenna for MoM-FDTD-SGFDTD hybridisation. It shows the benefit of using hybrid methods due to their ability to analyse a system of multiple discrete regions by employing the principle of equivalent sources to excite the coupling surfaces. A human body model is developed to interact with a short range RFID antenna. The near and far field radiation pattern are explored for which the cumulative distribution function of antenna radiation efficiency is presented. The field distributions show reasonable and stable results at 900 MHz. This work helps deeper investigation of the phenomena in the interaction between electromagnetic fields and human tissues.

FDTD method is further used at powerline and mobile communication frequency to model the interaction between the EM wave and electrically small objects. The method is also employed for lumped-element biological cell model in which Floquet periodic boundary conditions are imposed to introduce a periodic replication effect of the biological cell. Furthermore, the hybrid MoM-FDTD technique permits different parts of an antenna to be computed. The Equivalence Principle describes the fields between the different parts accurately. The method facilitates precise computation of antenna in proximity with lossless and lossy dielectric media.

In: Development of Complex Electromagnetic … ISBN: 978-1-61122-013-1
Editors: K. N. Ramli et al.

Chapter 1

NUMERICAL SOLUTION OF MAXWELL EQUATIONS USING FDTD AND MOM

K. N. Ramli[1], R. A. Abd-Alhameed[2] and P. S. Excell[3]
[1]Faculty of Electrical and Electronics Engineering,
Universiti Tun Hussein Onn Malaysia,
Parit Raja, Batu Pahat, Johor, Malaysia
[2]Mobile and Satellite Communications Research Centre,
Bradford University, Bradford, UK
[3]Institute for Arts, Science and Technology,
Glyndwr University,Wrexham, UK

ABSTRACT

This chapter briefly describes the historical background of the numerical solution for Maxwell equations using FDTD and MoM techniques in solving complex electromagnetic scattering problems. It also presents the state of the art of the original contributions of the analysis methods towards further developments of the hybrid computational technquiesin Electromagnetics.

1.1. Background and Motivations

Over the past years, a number of numerical and analytical approaches to Maxwell's time-dependent curl equations were broadly used with the increases in computer memory capacity and relentless advances in computational imitating efficiency.

Consequently, the demand for efficient field modelling tools in electromagnetic scattering problems is ceaselessly expanding.

In general, computational electromagnetic techniques have been applied to vast areas including the study of the radiation, scattering and penetration of electromagnetic wave with 3-D objects, in problems related to telecommunication, electromagnetic compatibility (EMC), microwave devices, waveguide structures and medical diagnosis.

Apparently, many considerations need to be taken into account when deciding to choose the most suitable numerical technique in order to solve a specific problem.

Some of the main factors include the geometry of the scattering and radiating objects, computer requirements, the domain of interest whether time-domain or frequency-domain, and the absorbing boundary conditions (ABC). The material characteristics and its geometry play important roles in choosing the approach used to mathematically model the properties of the electromagnetic interaction. In biomedical application for instance, the object is inhomogeneous, anisotropy, penetrable body which cannot be easily included in the formulations [1]. However, the ability of finite-difference time-domain (FDTD) algorithm to simulate and investigate the inhomogeneous, anisotropy media makes it very useful and effective technique thus far. Typically, there are two main categories of numerical algorithms approaches applied by researches. They are namely the frequency-domain integral formulation using the method of moments (MoM) [2-8], and the time-domain differential formulation by means of the FDTD method [9-14]. Integro-differential equation method is generally global in nature so that the initial and boundary conditions must be imposed as the algorithm continues. In contrast, differential-equation technique is typically local in nature so that the initial and boundary conditions are both directly included in the computational algorithm. The former technique usually postulates extensive analytical pre-processing whereas the later demands negligible analytical pre-processing. The methods mentioned above can be applied to certain specific geometries of concern.

Consequently, the differential-equation formulations are increasingly well-known due to the fact that it can solve any type of geometries in the problem space of the computational domain. In the past, MoM has been applied for aircraft field analysis [15], dual-reflector antenna and feeding horn [16] and microstrip antennas [17]. Moreover, FDTD has been employed for patch antennas [18], sloped interfaces [19-20] and SAR calculation [21].

1.2. HISTORY OF FDTD

First introduced by Yee in 1966 [22], FDTD is widely used to solve electromagnetic scattering problems due to its muscular characteristics including:

- Simplicity: The second-order accurate central finite-difference approximations for spatial and temporal derivatives of the electric and magnetic vector field components is directly used to solve Maxwell's equations explicitly in the absence of linear algebra.
- Fidelity: Wideband and narrowband applications can be easily implemented by applying different type of time pulse shape such as Gaussian and sinusoidal wave respectively.
- Robustness: Numerical dispersions in FDTD computations can be enclosed to model very large variety of electromagnetic scattering problems accurately. Furthermore, the FDTD algorithm can be easily implemented on parallel computers for faster simulation time.
- Effectiveness: Problems involving nonlinear media can be inherently alleviated in a straightforward manner in the time-domain compared with those in the frequency-domain technique.
- Versatility: It can intrinsically be used to model inhomogeneous, anisotropy materials such as biological tissues, geophysical strata and shielding metal structures.

Generally, the algorithm used by Yee was described by the electric field component which was spatially and temporally offset from the magnetic field component to acquire the update equations. These equations were used in a leap-frog manner to propagate the electric and magnetic fields ahead in time. The equations provide the present fields in terms of the past fields all over the computational domain. After Yee’s publication, the approach was widely used with different endeavour [23-27].

The boundaries of the computational domain in FDTD need to be carefully treated when simulating problems in open regions. Spurious reflections will generally occur from the termination of the grid. The problem can be solved by means of the well known method called the absorbing boundary condition (ABC). It is generally meant to absorb any outgoing propagating waves without ideally producing spurious reflections. The ABCs was first proposed in 1971 by Merewether [28] to solve the open region difficulties. The development chronicle to magnify the practicability study of the technique was continued in the literature by [29-33] which were based of nonmaterial type.

In contrast, Berenger presented a new idea in 1994 called the perfectly matched layer (PML) ABC which was based on material category [34]. The state of the art of Berenger's PML contributes to notably better precision when compared to the other ABCs in the written works [35, 36] for broad assortment of applications.

The main handicap of FDTD lies in the truth that only consistent grids can be used. Accordingly, the geometry resemblance in FDTD is restricted to staircase-shaped boundaries which lead to a large number of computer memory requirements and the CPU time particularly when dealing with curvature geometries with fine features [37]. The total number of cells in the computational domain grows significantly due to a global fine mesh. Another FDTD weakness is the presence of error due to numerical dispersion [38, 39].

In this case, many scientists were prompted to examine the subgridding scheme as an approach to parry the problem. A variety of methods have been proposed to boost the efficiency of FDTD technique such as non-uniform meshing [40], sub-cellular technique [41], non-orthogonal meshing [42], alternative direction implicit (ADI) method [43-45], higher-order technique [46, 47], hybrid method [48-52] and subgridding method [53-59]. In FDTD subgridding technique, the smaller size components in a structure is filled with fine grids and the remaining of the space is represented by coarse grids. The fields on the boundary between coarse and fine grids are basically unknown in nature. They are predicted by using spatial and temporal interpolations.

The regions of the coarse and fine grids are computed by the FDTD method and are kept in time step which satisfy the Courant stability condition. Consequently, the stable subgridding algorithm can refine the mesh locally. Hence, the accuracy of the solution can be improved without increasing the computational efforts significantly.

1.3. History of MoM

The MoM is basically a general procedure for solving linear equations. The "moments" in its name is due to the process of taking moments by multiplying the suitable weighting functions and integrating. In other words, it is essentially the technique of weighted residuals applicable for solving both the differential and integral equations. The advantages of MoM are accuracy, versatility and the potential to compute the near and far zone parameters. Furthermore, the method proved its ability to solve real complex antenna geometry in both frequency and time domain. The use of MoM and related matrix methods has become widespread in electromagnetic areas since the published paper of Richmond in 1965 [60] by generating a system of linear equations for the unknown current density and enforcing the boundary conditions at discrete points in the scattering body. Afterwards, he developed a point-matching solution for scattering by conducting bodies of arbitrary shape [61]. In 1967, Harrington documented the mathematical concept of MoM by which the functional equations of the field theory were reduced to matrix equations [62]. Later, he published a book on MoM which was a step forward towards the development of the numerical techniques [63].

The prime drawback of using MoM lies beneath its rectangular and triangular basis function. Their usage to examine the problem of electromagnetic scattering by dielectric objects with high dielectric constant leads to spurious charges. This problem was alleviated by means of solenoidal basis function [64, 65]. Wilton and Govind [66] made an effort to circumvent the error currents and anomaly behaviour of the solution near the edges by means of triangle expansion functions with suitable singular pulses at the edges.

The MoM has been favourably employed in variety of electromagnetic applications [67-77] such as scattering problems [67-70], synthesis of slotted waveguide array antenna [71], field analysis in circular-loop antenna [72, 73], the solution for patch antenna using volume integral equation (VIE) [74], VHF propagation modelling [75], microwave tomography system [76], and surface current distribution in normal mode helical antenna [77].

1.4. State of Art of the Original Contribution

The dominant focus of the research work is the modelling and analysis of the complex electromagnetic problems by means of MoM, FDTD and subgridded FDTD (SGFDTD) scheme to be employed in several applications. In order to achieve the research goal, the basic idea is to use the hybridisationmethod in which the tools for electromagnetic field modelling problems can be designed with more accuracy and efficiency. The surface kernel solution of MoM technique is derived for 1-D, 2-D and 3-D. The method is used to predict the surface current distributions on structures with closely spaced parallel wires, such as dipoles, loops and helical antennas.In addition, the Galerkin procedure is imposed to determine the effect of a chiral bianisotropic substrate of a rectangular microstrip resonatorin the spectral domain. Next, the present work is devoted to mathematical modelling and implementing SGFDTD in 2-D Cartesian coordinate keeping minimum reflection at the boundary. This method is applied to the interaction between overhead transmission lines to the underground pipeline for validation purposes. The SGFDTD formulation is then embedded inside the hybrid MoM-FDTD method.

The full code with the adaptation of subgridding inside the hybrid MoM-FDTD design problems is written in Fortran as a platform. The hybridisation of MoM-FDTD-SGFDTD code is used to analyse and investigate the applications in electromagnetic problems for validation such as the interaction between EM fields to the human body. FDTD method is used to model the interaction between the EM wave and electrically small objects in free space, lossless and lossy media at powerline frequency and mobile communication frequency. Quasi-static approach is introduced to parry the computational burden due to huge number of time steps. In addition, the method is also employed to model biological cells incorporating lumped-element biological cell model. Floquet periodic boundary conditions are imposed to imitate the effect of periodic replication of the cell. It is noteworthy that aheterogeneous hybrid computational method is also discussed in the book. The combination of MoM and FDTD allow different parts of an antenna to be simulated accurately. The fields between these two regions are intimately described by the implementation of the Equivalence Principle. The current crosses the boundary of the two regions may precisely be computed by incorporating the extension of the equivalence principle. The method paves the way for precise computation of antenna in proximity with lossless and lossy dielectric media.

1.5. Overview of the Book

Chapter 1 postulates historical background and literature survey of FDTD and MoM techniques used to solve electromagnetic scattering issues. It should be noted that a more detailed review of existing literature is reported at the beginning of each chapter with separate references at the end.

Chapter 2 unfolds the theoretical concept of FDTD principles including the derivation of the magnetic and electric field update equations, parameters that control the stability and accuracy, plane wave source modelling concept by applying the equivalent surface, and finally the implementation of Berenger's PML absorbing boundary condition.

Chapter 3 presents the surface kernel solution of the MoM. The formulation begins from 1-D, and then extended to 2-D and 3-D antenna problems. The surface current distributions on structures with closely spaced parallel wires, such as dipoles, loops and helical antennas, are computed by using the method of moments with a general surface patch formulation. The modelling method employed a two-dimensional electric surface patch integral equation formulation solved by independent piecewise-linear basis function methods in the circumferential and axial directions of the wire.

Chapter 4 explains the modelling and analysis of quasi-static FDTD subgridding technique in two-dimensional approach. The interaction between overhead transmission lines and underground pipeline at power-line frequency is modelled. FDTD technique is used for the whole structure spatial problem combined with subgrid method at the pipeline. The soil in the common corridor has been designed as arbitrarily inhomogeneous.

Chapter 5 exposes the effect of a chiral bianisotropic substrate of a rectangular microstrip resonator using the integral equation formulation based on Galerkin procedure for moment method in the spectral domain. It is aimed to study the effect of the chirality of a bianisotropic substrate upon the resonant frequency and the half-power bandwidth. In addition, the effect of a magnetic anisotropy via an asymptotic approach is also discussed.

Chapter 6 describes the hybridisation MoM-FDTD-SGFDTD computational method. The modelling on multiple-region hybrid techniques with frequency-domain MoM and time-domain FDTD and subgridding are suggested and investigated. The method is validated for near field and far field applications particularly on the interaction between electromagnetic fields and human body in which the RFID antenna is located and moved at several positions in front and back of inhomogeneous human body model.

Chapter 7 portrays the implementation of frequency scaling approach in order to reduce the number of FDTD time steps. The reflection on the interface layers has been reduced in lossless and lossy penetrating media. The accuracy of the FDTD scaling approach with the models of homogenous and layered spheres in free space, lossless and lossy media is verified. This work facilitates to the validity of using the scaled-frequency FDTD method to obtain induced electric fields at extremely low frequency (50 Hz) and at mobile communication frequencies.

Chapter 8 reveals microdosimetricmodeling of bioelectromagnetic interactions at cellular level. The approach employs FDTD method, combined with an arbitrarily-oriented implementation of the Hodgkin-Huxley cell-membrane model, the Floquet periodic boundary condition, the frequency-scaling technique and the lumped-element model for membrane behavior. The work supports deeper understanding of the phenomena in the interaction between EM fields and biological systems at various levels of spatial definition.

Chapter 9 discloses a heterogeneous hybrid computational method. Different methods are employed at different parts of antenna simulation to permit the most suitable method to be used for each element. The method employs FDTD and MoMto determine the electric and magnetic fields in two regionsinterfaced by Equivalence Principle. The work assists in the prediction of antenna behavior in proximity with lossy dielectric volumes.

REFERENCES

[1] A. Taflove and S. C. *Hagness, Computational electrodynamics: The finite-difference time-domain method,* 3rd ed., Boston, MA: Artech House, 2005.

[2] K. Tap, P. Pathak and R. Burkholder, "Complex source beam - moment method procedure for accelerating numerical integral equation solutions of radiation and scattering problems," *IEEE Transactions on Antennas and Propagation*, vol. 62, no. 4, pp. 2052 - 2062, 2014.

[3] D. J. Ludick, R. Maaskant, D. B. Davidson, U. Jakobus, R. Mittra, and D. De Villiers, "Efficient analysis of large aperiodic antenna arrays using the domain Green's function method," *IEEE Transactions on Antennas and Propagation*, vol. 62, no. 4, part 1, pp. 1579 - 1588, 2014.

[4] X.-M. Pan and X.-Q. Sheng, "Improved algebraic preconditioning for MoM solutions of large-scale electromagnetic problems," *IEEE Antennas and Wireless Propagation Letters*, vol. 13, pp. 106-109, 2014.

[5] M. A. Echeverri Bautista, M. A. Francavilla, F. Vipiana, and G. Vecchi, "A hierarchical fast solver for EFIE-MoM analysis of multiscale structures at very low frequencies," *IEEE Transactions on Antennas and Propagation*, vol. 62, no. 3, pp. 1523-1528, 2014.

[6] M. Shafieipour, I. Jeffrey, J. Aronsson, and V. I. Okhmatovski, "On the equivalence of RWG method of moments and the locally corrected nyström method for solving the electric field integral equation," *IEEE Transactions on Antennas and Propagation*, vol. 62, no. 2, pp. 772-782, 2014.

[7] N. Hendijani, J. Cheng and R. J. Adams, "Combined field integral equation using a constraint-based Helmholtz decomposition," *IEEE Transactions on Antennas and Propagation*, vol. 62, no. 3, pp. 1500-1503, 2014.

[8] J. Alvarez, L. D. Angulo, M. R. Cabello, A. Rubio Bretones, and S. G. Garcia, "An analysis of the leap-frog discontinuous Galerkin method for Maxwell's equations," *IEEE Transactions on Microwave Theory and Techniques*, vol. 62, no. 2, pp. 197-207, 2014.

[9] R. Xiong, B. Chen, Y. Yi, H.-L. Chen, and H.-F. Zhao, "Two-step method for the FDTD analysis of long apertures having depth," *IEEE Antennas and Wireless Propagation Letters*, vol. 12, pp. 39-42, 2013.

[10] A. C. Lesina, A. Vaccari and A. Bozzoli, "A modified RC-FDTD algorithm for plasmonics in Drude dispersive media nanostructures," *6th European Conference on Antennas and Propagation (EUCAP)*, pp. 687-690, 2012.

[11] G. M. Noetscher, Y. Xu and S. N. Makarov, "Accuracy of point source models with coincident phase centers in a cubic FDTD grid for arbitrary source orientation," *IEEE Antennas and Propagation Society International Symposium (APSURSI)*, pp. 1-2, 2012.

[12] V. Nayyeri, M. Soleimani and M. Dehmollaian, "Implementation of a PEMC boundary condition in the 2-D FDTD technique," *IEEE Antennas and Propagation Society International Symposium (APSURSI)*, pp. 1-2, 2012.

[13] M. Weiwei, S. Dong and W. Xianliang, "UPML-FDTD parallel computing on GPU," *International Conference on Microwave and Millimeter Wave Technology (ICMMT)*, vol. 4, pp. 1-4, 2012.

[14] C. Guiffaut, A. Reineix and B. Pecqueux, "New oblique thin wire formalism in the FDTD method with multiwire junctions," *IEEE Transactions on Antennas and Propagation*, vol. 60, no. 3, pp. 1458-1466, 2012.

[15] A. Schroder, G. A. Rasek, H.-D. Bruns, Z. Reznicek, J. Kucera, S. E. Loos, and C. Schuster, "Analysis of high intensity radiated field coupling into aircraft using the method of moments," *IEEE Transactions on Electromagnetic Compatibility*, vol. 56, no. 1, pp. 113-122, 2014.

[16] S. R. Zang and J. R. Bergmann, "Analysis of omnidirectional dual-reflector antenna and feeding horn using method of moments," *IEEE Transactions on Antennas and Propagation*, vol. 62, no. 3, pp. 1534-1538, 2014.

[17] M. Amir, S. Bedra, S. Benkouda, and T. Fortaki, "Bacterial foraging optimisation and method of moments for modelling and optimisation of microstrip antennas," *IET Microwaves, Antennas and Propagation*, vol. 8, no. 4, pp. 295-300, 2014.

[18] W. G. Whittow, C. C. Njoku and Y. C. Vardaxoglou, "Fine scale simulations of patch antennas with heterogeneous substrates," *USNC-URSI Radio Science Meeting (Joint with AP-S Symposium)*, pp. 223, 2013.

[19] D.-C. Liu and H.-C. Chang, "The second-order condition of FDTD method at sloped dielectric interfaces by averaging the permittivities along surface normal," *IEEE Transactions on Antennas and Propagation*, vol. 60, no. 11, pp. 5259-5267, 2012.

[20] D. M. Shyroki, "Modeling of sloped interfaces on a Yee grid," *IEEE Transactions on Antennas and Propagation*, vol. 59, no. 9, pp. 3290-3295, 2011.

[21] S. K. Kamepally, B. P. Kumar and C. S. Paidimarry, "FDTD estimation for accurate specific absorption rate in a tumor," *Annual International Conference on Emerging Research Areas and International Conference on Microelectronics, Communications and Renewable Energy (AICERA/ICMiCR)*, pp. 1-5, 2013.

[22] K. S. Yee, "Numerical solution of initial boundary value problems involving Maxwell's equations in isotropic media," *IEEE Transactions on Antennas and Propagation*, vol. AP-14, pp. 302-307, 1966.

[23] A. Taflove and M. E. Brodwin, "Numerical solution of steady-state electromagnetic scattering problems using the time-dependent maxwell's equations," *IEEE Transactions on Microwave Theory and Techniques*, vol. 23, pp. 623-630, 1975.

[24] A. Taflove and M. E. Brodwin, "Computation of the electromagnetic fields and induced temperatures within a model of the microwave-irradiated human eye," *IEEE Transactions on Microwave Theory and Techniques*, vol. 23, pp. 888-896, 1975.
[25] R. Holland, "THREDE: A free-field EMP coupling and scattering code," *IEEE Transactions on Nuclear Science*, vol. NS-24, pp. 2416-2421, 1977.
[26] K. S. Kunz and K. M. Lee, "A three-dimensional finite-difference solution of the external response of an aircraft to a complex transient EM environment: Part I-The method and its implementation," *IEEE Transactions on Electromagnetic Compatibility*, vol. EMC-20, pp. 328-333, 1978.
[27] A. Taflove, "Application of the finite-difference time-domain method to sinusoidal steady-state electromagnetic-penetration problems," *IEEE Transactions on Electromagnetic Compatibility*, vol. EMC-22, pp. 191-202, 1980.
[28] D. E. Merewether, "Transient currents induced on a metallic body of revolution by an electromagnetic pulse," *IEEE Transactions on Electromagnetic Compatibility*, vol. EMC-13, pp. 41-44, 1971.
[29] B. Engquist and A. Majda, "Absorbing boundary conditions for numerical simulation of waves," *Applied Mathematical Sciences*, vol. 74, pp. 1765-1766, 1977.
[30] G. Mur, "Absorbing boundary conditions for the finite-difference approximation of the time-domain electromagnetic-field equations," *IEEE Transactions on Electromagnetic Compatibility*, vol. EMC-23, pp. 377-382, 1981.
[31] Z. P. Liao, H. L. Wong, B. P. Yang, and Y. F. Yuan, "A transmitting boundary for transient wave analysis," *Scientia Sinica (Series A)*, vol. XXVII, pp. 1063-1076, 1984.
[32] R. G. Keys, "Absorbing boundary conditions for acoustic media," *Geophysics*, vol. 50, pp. 892-902, 1985.
[33] R. L. Higdon, "Absorbing boundary conditions for difference approximations to the multi-dimensional wave equation," *Mathematics of Computation*, vol. 47, pp. 437-459, 1986.
[34] J. P. Berenger, "A perfectly matched layer for the absorption of electromagnetic waves," *Journal of Computational Physics*, vol. 114, pp. 185-200, 1994.
[35] D. T. Prescott and N. V. Shuley, "Reflection analysis of FDTD boundary conditions - Part I: Time-space absorbing boundaries," *IEEE*

Transactions on Microwave Theory and Techniques, vol. 45, pp. 1162-1170, 1997.

[36] D. T. Prescott and N. V. Shuley, "Reflection analysis of FDTD boundary conditions - Part II: Berenger's PML absorbing layers," *IEEE Transactions on Microwave Theory and Techniques*, vol. 45, pp. 1171-1178, 1997.

[37] A. Akyurtlu, D. H. Werner, V. Veremey, D. J. Steich, and K. Aydin, "Staircasing errors in FDTD at an air-dielectric interface," *IEEE Microwave and Guided Wave Letters*, vol. 9, pp. 444-446, 1999.

[38] J. B. Schneider and C. L. Wagner, "FDTD dispersion revisited: Faster-than-light propagation," *IEEE Microwave and Guided Wave Letters*, vol. 9, pp. 54-56, 1999.

[39] J. Represa, C. Pereira, M. Panizo, and F. Tadeo, "A simple demonstration of numerical dispersion under FDTD," *IEEE Transactions on Education*, vol. 40, pp. 98-102, 1997.

[40] R. Wu, G. Benqing and Y. Shiming, "An improved method in FDTD simulation to reduce reflection from non-uniform mesh," *3rd International Conference on Microwave and Millimeter Wave Technology Proceedings*, pp. 646-649, 2002.

[41] S. Kapoor, "Sub-cellular technique for finite-difference time-domain method," *IEEE Transactions on Microwave Theory and Techniques*, vol. 45, pp. 673-677, 1997.

[42] P. Thoma and T. Weiland, "Numerical stability of finite difference time domain methods," *IEEE Transactions on Magnetics*, vol. 34, pp. 2740-2743, 1998.

[43] T. Namiki, "A new FDTD algorithm based on alternating-direction implicit method," *IEEE Transactions on Microwave Theory and Techniques*, vol. 47, pp. 2003-2007, 1999.

[44] F. H. Zheng, Z. Z. Chen and J. Z. Zhang, "Toward the development of a three-dimensional unconditionally stable finite-difference time-domain method," *IEEE Transactions on Microwave Theory and Techniques*, vol. 48, pp. 1550-1558, 2000.

[45] B. K. Huang, Y. S. Jiang and W. B. Wang, "A hybrid 2-D ADI-FDTD subgridding scheme," *IEEE International Symposium on Antennas, Propagation and EM Theory Proceedings*, pp. 707-710, 2003.

[46] M. F. Hadi and M. P. May, "A modified FDTD (2,4) scheme for modeling electrically large structures with high-phase accuracy," *IEEE Transactions on Antennas and Propagation*, vol. 45, pp. 254-264, 1997.

[47] Z. H. Shao, Z. X. Shen, Q. Y. He, and G. W. Wei, "A generalized higher order finite-difference time-domain method and its application in guided-wave problems," *IEEE Transactions on Microwave Theory and Techniques*, vol. 51, pp. 856-861, 2003.

[48] G. Cerri, P. Russo, A. Schiavoni, G. Tribellini, and P. Bielli, "MoM-FDTD hybrid technique for analysing scattering problems," *Electronic Letters*, vol. 34, pp. 438-440, 1998.

[49] A. Monorchio and R. Mittra, "A hybrid finite-element finite-difference time-domain (FE/FDTD) technique for solving complex electromagnetic problems," *IEEE Microwave and Guided Wave Letters*, vol. 8, pp. 93-95, 1998.

[50] N. V. Venkatarayalu, Y. B. Gan, R. Lee, and L. W. Li, "Antenna modeling using stable hybrid FETD-FDTD method," *Proceedings of 2007 International Symposium on Antennas and Propagation,* Honolulu, US, pp. 3736-3739, 2007.

[51] S. S. Zivanovic, K. S. Yee and K. K. Mei, "A subgridding method for the time-domain finite-difference method to solve Maxwell's equations," *IEEE Transactions on Microwave Theory and Techniques*, vol. 39, pp. 471-479, 1991.

[52] P. Thoma and T. Weiland, "A consistent subgridding scheme for the finite difference time domain method," *International Journal of Numerical Modelling-Electronic Networks Devices and Fields*, vol. 9, pp. 359-374, 1996.

[53] S. Chaillou, J. Wiart and W. Tabbara, "A subgridding scheme based on mesh nesting for the FDTD method," *Microwave and Optical Technology Letters*, vol. 22, pp. 211-214, 1999.

[54] S. M. Wang, F. L. Teixeira, R. Lee, and J. F. Lee, "Optimization of subgridding schemes for FDTD," *IEEE Microwave and Wireless Components Letters*, vol. 12, pp. 223-225, 2002.

[55] K. Xiao, D. J. Pommerenke and J. L. Drewniak, "A three-dimensional FDTD subgridding algorithm based on interpolation of current density," *IEEE International Symposium on Electromagnetic Compatibility*, vol. 1, pp. 118-123, 2004.

[56] K. N. Ramli, R. A. Abd-Alhameed and P. S. Excell, "Modelling and analysis of complex EM fields problems using hybrid procedure," *26th International Conference on CAD/CAM, Robotics and Factories of the Future 2011,* Kuala Lumpur, Malaysia, pp. 232-238, 2011.

[57] K. N. Ramli, R. A. Abd-Alhameed, C. H. See, P. S. Excell, and J. M. Noras, "Hybrid computational scheme for antenna-human body

interaction," *Progress In Electromagnetics Research*, vol. 133, pp. 117-136, 2013.

[58] K. N. Ramli and R. A. Abd-Alhameed, "HVAC electromagnetic field computation using SGFDTD method," *IEEE Asia-Pacific Conference on Applied Electromagnetics (APACE),* Melaka, Malaysia, pp. 36-39, 2012.

[59] K. N. Ramli, R. A. Abd-Alhameed, C. H. See, A. Atojoko, J. M. Noras, and P. S. Excell, "Electromagnetic field computation for power transmission lines using quasi-static sub-gridding finite-difference time-domain approach," *Electromagnetics*, vol. 34, no. 1, pp. 47-65, 2014.

[60] J. H. Richmond, "Digital computer solutions of the rigorous equations for scattering problems," *Proceedings of the IEEE*, vol. 53, pp. 796- 804, 1965.

[61] J. H. Richmond, "A wire-grid model for scattering by conducting bodies," *IEEE Transactions on Antennas and Propagation*, vol. 14, pp. 782-786, 1966.

[62] R. F. Harrington, "Matrix methods for field problems," *Proceedings of the IEEE*, vol. 55, pp. 136-149, 1967.

[63] R. F. Harrington, *Field computation by moment methods*. New York: The Macmillan Co., 1968.

[64] L. Mendes and S. Antenor, "A tridimensional solenoidal basis function to analyze 3D scattering of homogeneous dielectrics using the method of moments," *SBMO/IEEE MTT-S International Microwave and Optoelectronics Conference Proceedings*, vol. 2, pp. 897-900, 1995.

[65] S. Antenor and L. Mendes, "Method of moments with solenoidal basis functions: an approach to analyze the scattering by inhomogeneous dielectrics," *Third International Conference on Computation in Electromagnetics (Conf. Publ. No. 420)*, pp. 318-321, 1996.

[66] D. Wilton and S. Govind, "Incorporation of edge conditions in moment method solutions," *IEEE Transactions on Antennas and Propagation*, vol. 25, pp. 845-850, 1977.

[67] N. C. Albertsen, J. E. Hansen and N. E. Jensen, "Computation of radiation from wire antennas on conducting bodies," *IEEE Transactions on Antennas and Propagation*, vol. 22, pp. 200-206, 1974.

[68] E. H. Newman and D. M. Pozar, "Electromagnetic modeling of composite wire and surface geometries," *IEEE Transactions on Antennas and Propagation*, vol. 26, pp. 784-789, 1978.

[69] M. E. Peters and E. H. Newman, "Method of moments analysis of anisotropic artificial media composed of dielectric wire objects," *IEEE*

Transactions on Microwave Theory and Techniques, vol. 43, pp. 2023-2027, 1995.

[70] S. M. Rao, D. R. Wilton and A. W. Glisson, "Electromagnetic scattering by surfaces of arbitrary shape," *IEEE Transactions on Antennas and Propagation*, vol. 30, pp. 409-418, 1982.

[71] A. J. Sangster and A. H. I. McCormick, "Theoretical design/synthesis of slotted waveguide arrays," *IEE Proceedings Microwaves, Antennas and Propagation*, vol. 136, pp. 39-46, 1989.

[72] L. W. Li, C. P. Lim and M. K. Leong, "Method of moments analysis of electrically large circular-loop antennas: non-uniform currents," *IEEE Antennas and Propagation Society International Symposium*, vol. 4, pp. 2542-2545, 1999.

[73] L. W. Li, M. S. Yeo and M. K. Leong, "Method of moments analysis of EM fields in a multilayered spheroid radiated by a thin circular loop antenna," *IEEE Transactions on Antennas and Propagation*, vol. 52, pp. 2391-2402, 2004.

[74] S. N. Makarov, S. D. Kulkarni, A. G. Marut, and L. C. Kempel, "Method of moments solution for a printed patch/slot antenna on a thin finite dielectric substrate using the volume integral equation," *IEEE Transactions on Antennas and Propagation*, vol. 54, pp. 1174-1184, 2006.

[75] J. T. Johnson, R. T. Shin, J. C. Eidson, L. Tsang, and J. A. Kong, "A method of moments model for VHF propagation," *Antennas and Propagation Society International Symposium*, vol. 1, pp. 362-365, 1997.

[76] J. L. Hu, Z. P. Wu, H. McCann, L. E. Davis, and C. G. Xie, "Quasi-three-dimensional method of moments for analyzing electromagnetic wave scattering in microwave tomography systems," *IEEE Sensors Journal*, vol. 5, pp. 216-223, 2005.

[77] R. A. Abd-Alhameed, K. N. Ramli and P. S. Excell, "The complete surface-current distribution in a normal-mode helical antenna," *IEEE Antennas and Propagation Magazine*, vol. 54, no. 1, pp. 95-108, 2012.

In: Development of Complex Electromagnetic … ISBN: 978-1-61122-013-1
Editors: K. N. Ramli et al.

Chapter 2

FDTD TECHNIQUE FOR FIELD TRUNCATION

K. N. Ramli[1], R. A. Abd-Alhameed[2] and P. S. Excell[3]
[1]Faculty of Electrical and Electronics Engineering,
Universiti Tun Hussein Onn Malaysia,
Parit Raja, Batu Pahat, Johor, Malaysia
[2]Mobile and Satellite Communications Research Centre,
Bradford University, Bradford, UK
[3]Institute for Arts, Science and Technology,
Glyndwr University, Wrexham, UK

This chapter explains the fundamental concepts of the FDTD computational method. It includes the derivation of FDTD updating equations for electric and magnetic field components, the accuracy and stability which necessitate the FDTD computational method. The solution of Maxwell's equations is included by means of finite differences. The general idea of the ABC to the specific Berenger's PML in 2-D and 3-D are also described.

2.1. INTRODUCTION

Over the past few years, finite-difference time-domain (FDTD) method [1-11] have become increasingly prevalent in the computational electromagnetic problems due to its simplicity, efficiency, robustness and versatility scheme for highly complex configuration in the computational domain. Generally, FDTD technique is the most well-known numerical method for the solution of problems in electromagnetic simulation ranging from RF to optical frequencies. It is considered to be one of the most powerful numerical techniques for solving partial differential equations of any kind. In addition, it can be utilized to solve the spatial as well as the temporal distributions of electric and magnetic fields in various media. In principle, FDTD is a method that divides the solution domain into finite discrete points and then replaces the partial differential equation with a set of difference equations. It has successfully been applied to many problems of propagation, radiation and scattering of electromagnetic waves such as antenna, radar, wireless communication system, high speed electronic, photonic, radiography, x-ray crystallography, bio-electromagnetic and geophysical imaging. A good measure of its success lies in the fact that thousand of papers on the subject have been published in journals and international symposium, apart from the books and tutorials devoted to it [11]. Moreover, much specific and general purpose commercial software is available on the market which further extends its appeal globally. Furthermore, three books are used as the main references to the recent FDTD research written by Taflove and Hagness [11], Taflove [12] and Kunz and Luebbers [13].

2.2. FDTD Updating Stencils

A pioneering way of describing the electromagnetic phenomena was introduced by James Clerk Maxwell in 1865 [14]. Later in 1873, he published an article called "Treatise on Electricity and Magnetism" in which the discoveries of Coulomb, Oersted, Ampere and Faraday were united into four refined constructed mathematical equations known as Maxwell's equations. The differential time domain Maxwell's equations in a linear medium are given by:

$$\frac{\partial B}{\partial t} = -\nabla \times E - J_m \tag{2.1}$$

$$\frac{\partial D}{\partial t} = \nabla \times H - J_e \tag{2.2}$$

B is the magnetic flux density in Wb/m^2, D is the electric flux density in C/m^2, E is the electric field in V/m and H is the magnetic field in A/m. J_m is the magnetic current density in V/m^2 and is defined to relate any magnetic loss to the field. J_e is the electric current density in A/m^2 and is defined to relate any electric loss to the field. J_m and J_e are respectively given by:

$$J_m = \sigma^* \times H \tag{2.3}$$

$$J_e = \sigma \times E \tag{2.4}$$

σ^* is magnetic resistivity in Ω/m and σ is the electric conductivity in S/m. In materials with field-independent, direction-independent and frequency-independent electric and magnetic properties, the following proportions apply:

$$B = \mu \times H \tag{2.5}$$

$$D = \varepsilon \times E \tag{2.6}$$

μ is the magnetic permeability in H/m and ε is the electric permittivity in F/m. Inserting (2.3) and (2.5) to (2.1) and dividing by μ gives:

$$\frac{\partial H}{\partial t} = \frac{1}{\mu}\left(-\nabla \times E - \sigma^* H\right) \tag{2.7}$$

$\sigma^* H$ is the magnetic losses which may exist inside the medium. Inserting (2.4) and (2.6) to (2.2) and dividing by ε gives:

$$\frac{\partial E}{\partial t} = \frac{1}{\varepsilon}\left(\nabla \times H - \sigma E\right) \tag{2.8}$$

σE is the electric losses which may exist inside the medium. In Cartesian coordinates, equations (2.7) and (2.8) yield the following six scalar equations:

$$\frac{\partial H_x}{\partial t} = \frac{1}{\mu}\left(\frac{\partial E_y}{\partial z} - \frac{\partial E_z}{\partial y} - \sigma^* H_x\right) \tag{2.9}$$

$$\frac{\partial H_y}{\partial t} = \frac{1}{\mu}\left(\frac{\partial E_z}{\partial x} - \frac{\partial E_x}{\partial z} - \sigma^* H_y\right) \tag{2.10}$$

$$\frac{\partial H_z}{\partial t} = \frac{1}{\mu}\left(\frac{\partial E_x}{\partial y} - \frac{\partial E_y}{\partial x} - \sigma^* H_z\right) \tag{2.11}$$

$$\frac{\partial E_x}{\partial t} = \frac{1}{\varepsilon}\left(\frac{\partial H_z}{\partial y} - \frac{\partial H_y}{\partial z} - \sigma E_x\right) \tag{2.12}$$

$$\frac{\partial E_y}{\partial t} = \frac{1}{\varepsilon}\left(\frac{\partial H_x}{\partial z} - \frac{\partial H_z}{\partial x} - \sigma E_y\right) \tag{2.13}$$

$$\frac{\partial E_z}{\partial t} = \frac{1}{\varepsilon}\left(\frac{\partial H_y}{\partial x} - \frac{\partial H_x}{\partial y} - \sigma E_z\right) \tag{2.14}$$

A space point in a rectangular grid is defined from Yee's original notation [15] as:

$$(i, j, k) = (i\Delta x, j\Delta y, k\Delta z) \tag{2.15}$$

Let F denote any function of discrete space and time given by:

$$F(i\Delta x, j\Delta y, k\Delta z, n\Delta t) \equiv F^n_{i,j,k} \tag{2.16}$$

Δx, Δy and Δz are the grid space increments in *x, y* and *z* directions respectively, and Δt is the time increment. Using a central finite-difference approximation, space and time derivatives of F can be written as:

$$\frac{\partial F}{\partial x} = \frac{F^{n}_{i+1/2,j,k} - F^{n}_{i-1/2,j,k}}{\Delta x} + O(\Delta x)^2 \tag{2.17}$$

$$\frac{\partial F}{\partial t} = \frac{F^{n+1/2}_{i,j,k} - F^{n-1/2}_{i,j,k}}{\Delta t} + O(\Delta t)^2 \tag{2.18}$$

In equation (2.17), $O(\Delta x)^2$ is the error term that represents all the remaining terms in a Taylor series expansion. It is known as a central finite difference scheme in space with second-order accuracy. Similarly, (2.18) is second-order accurate in time. Applying Yee's finite-difference scheme to (2.9) gives:

$$\frac{H_x\big|^{n+1/2}_{i,j,k} - H_x\big|^{n-1/2}_{i,j,k}}{\Delta t} = \frac{1}{\mu_{i,j,k}}\left(\begin{array}{l} \dfrac{E_y\big|^{n}_{i,j,k+1/2} - E_y\big|^{n}_{i,j,k-1/2}}{\Delta z} - \\ \dfrac{E_z\big|^{n}_{i,j+1/2,k} - E_z\big|^{n}_{i,j-1/2,k}}{\Delta y} - \sigma^{*}_{i,j,k} H_x\big|^{n}_{i,j,k} \end{array} \right) \tag{2.19}$$

The $H_x\big|^{n}_{i,j,k}$ field component in (2.19) is evaluated at time step n. However, the value of $H_x\big|^{n}_{i,j,k}$ at time step n is not available and hence the following interpolated approximation is used:

$$H_x\big|^{n}_{i,j,k} = \frac{H_x\big|^{n+1/2}_{i,j,k} + H_x\big|^{n-1/2}_{i,j,k}}{2} \tag{2.20}$$

By substituting equation (2.20) in (2.19), leaving $H_x\big|^{n+1/2}_{i,j,k}$ on the left hand side and passing the all remaining terms to the right, assuming cubical FDTD cells are used, the finite difference updating equation for the magnetic and electric field components can be derived as:

$$H_x\Big|_{i,j,k}^{n+1/2} = D_{a,H_x}\Big|_{i,j,k} H_x\Big|_{i,j,k}^{n-1/2} + D_{b,H_x}\Big|_{i,j,k} \begin{pmatrix} E_y\Big|_{i,j,k+1/2}^{n} - E_y\Big|_{i,j,k-1/2}^{n} \\ + E_z\Big|_{i,j-1/2,k}^{n} - E_z\Big|_{i,j+1/2,k}^{n} \end{pmatrix} \tag{2.21}$$

$$H_y\Big|_{i,j,k}^{n+1/2} = D_{a,H_y}\Big|_{i,j,k} H_y\Big|_{i,j,k}^{n-1/2} + D_{b,H_y}\Big|_{i,j,k} \begin{pmatrix} E_z\Big|_{i+1/2,j,k}^{n} - E_z\Big|_{i-1/2,j,k}^{n} \\ + E_x\Big|_{i,j,k-1/2}^{n} - E_x\Big|_{i,j,k+1/2}^{n} \end{pmatrix} \tag{2.22}$$

$$H_z\Big|_{i,j,k}^{n+1/2} = D_{a,H_z}\Big|_{i,j,k} H_z\Big|_{i,j,k}^{n-1/2} + D_{b,H_z}\Big|_{i,j,k} \begin{pmatrix} E_x\Big|_{i,j+1/2,k}^{n} - E_x\Big|_{i,j-1/2,k}^{n} \\ + E_y\Big|_{i-1/2,j,k}^{n} - E_y\Big|_{i+1/2,j,k}^{n} \end{pmatrix} \tag{2.23}$$

$$E_x\Big|_{i,j,k}^{n+1} = C_{a,E_x}\Big|_{i,j,k} E_x\Big|_{i,j,k}^{n} + C_{b,E_x}\Big|_{i,j,k} \begin{pmatrix} H_z\Big|_{i,j+1/2,k}^{n+1/2} - H_z\Big|_{i,j-1/2,k}^{n+1/2} \\ + H_y\Big|_{i,j,k-1/2}^{n+1/2} - H_y\Big|_{i,j,k+1/2}^{n+1/2} \end{pmatrix} \tag{2.24}$$

$$E_y\Big|_{i,j,k}^{n+1} = C_{a,E_y}\Big|_{i,j,k} E_y\Big|_{i,j,k}^{n} + C_{b,E_y}\Big|_{i,j,k} \begin{pmatrix} H_x\Big|_{i,j,k+1/2}^{n+1/2} - H_x\Big|_{i,j,k-1/2}^{n+1/2} \\ + H_z\Big|_{i-1/2,j,k}^{n+1/2} - H_z\Big|_{i+1/2,j,k}^{n+1/2} \end{pmatrix} \tag{2.25}$$

$$E_z\Big|_{i,j,k}^{n+1} = C_{a,E_z}\Big|_{i,j,k} E_z\Big|_{i,j,k}^{n} + C_{b,E_z}\Big|_{i,j,k} \begin{pmatrix} H_y\Big|_{i+1/2,j,k}^{n+1/2} - H_y\Big|_{i-1/2,j,k}^{n+1/2} \\ + H_x\Big|_{i,j-1/2,k}^{n+1/2} - H_x\Big|_{i,j+1/2,k}^{n+1/2} \end{pmatrix} \tag{2.26}$$

It can be seen that the coefficients on the left hand side are referred to as Yee's updating coefficients. The electric field coefficients are given by:

$$C_a\Big|_{i,j,k} = \left(1 - \frac{\sigma_{i,j,k}\Delta t}{2\varepsilon_{i,j,k}}\right) \Bigg/ \left(1 + \frac{\sigma_{i,j,k}\Delta t}{2\varepsilon_{i,j,k}}\right) \tag{2.27}$$

$$C_{b_p}\Big|_{i,j,k} = \left(\frac{\Delta t}{\varepsilon_{i,j,k}\Delta_p}\right) \Bigg/ \left(1+\frac{\sigma_{i,j,k}\Delta t}{2\varepsilon_{i,j,k}}\right) \tag{2.28}$$

The magnetic updating coefficients can be written as:

$$D_a\Big|_{i,j,k} = \left(1-\frac{\sigma^*_{i,j,k}\Delta t}{2\mu_{i,j,k}}\right) \Bigg/ \left(1+\frac{\sigma^*_{i,j,k}\Delta t}{2\mu_{i,j,k}}\right) \tag{2.29}$$

$$D_{b_p}\Big|_{i,j,k} = \left(\frac{\Delta t}{\mu_{i,j,k}\Delta_p}\right) \Bigg/ \left(1+\frac{\sigma^*_{i,j,k}\Delta t}{2\mu_{i,j,k}}\right) \tag{2.30}$$

The subscript p can be x, y or z and Δ_p is the cell size in the p-direction. Assuming the structure under investigation contains different types of material such as dielectric or magnetic, electric and magnetic field updating coefficients can be easily calculated from equations (2.27) to (2.30) before the FDTD time stepping algorithm starts.

The orientation of the fields in Figure 2.1 is known as the FDTD lattice or Yee cell. The magnetic and electric fields are located on the faces and the edges of the cube respectively. Each electric field vector component is surrounded by four circulating magnetic field vector components and vice versa. Both the electric and magnetic field vector components are located half a cell from each other. In addition, this arrangement permits easy implementations of the central finite difference approximations and the integral form of the Faraday's law and the Ampere's law.

The system of difference equations is solved at the nodes. Figure 2.2 illustrates the typical relationship between field components within a quarter of a cell and on a plane distinctly helpful when handling boundary conditions of a closed region.

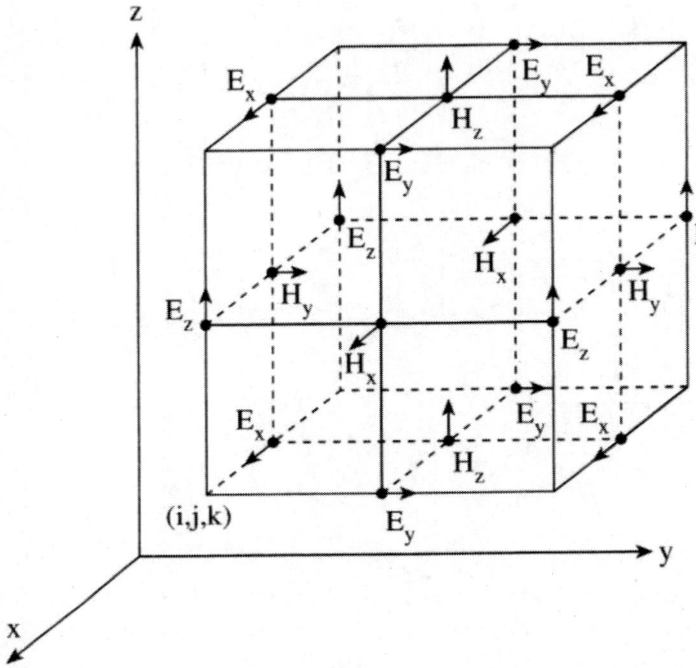

Figure 2.1. The electric and magnetic field components distribution on the FDTD lattice [16].

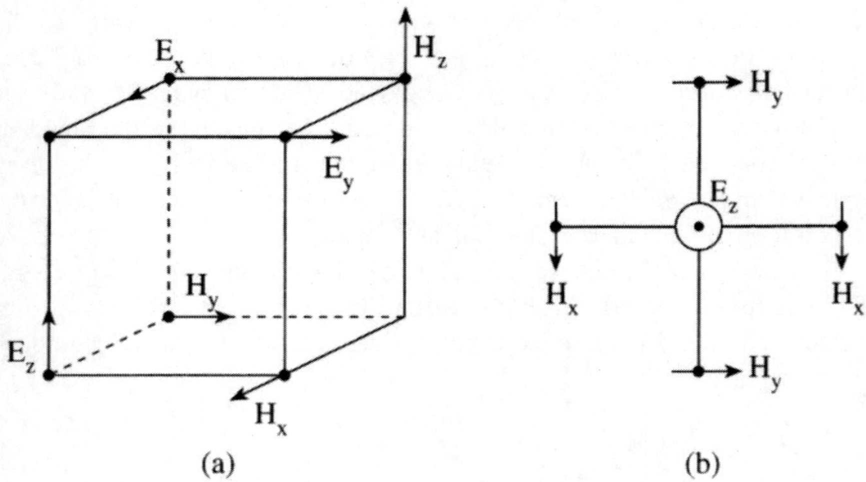

Figure 2.2. Relationship between field components: (a) within a quarter of a unit cell, (b) on a plane [16].

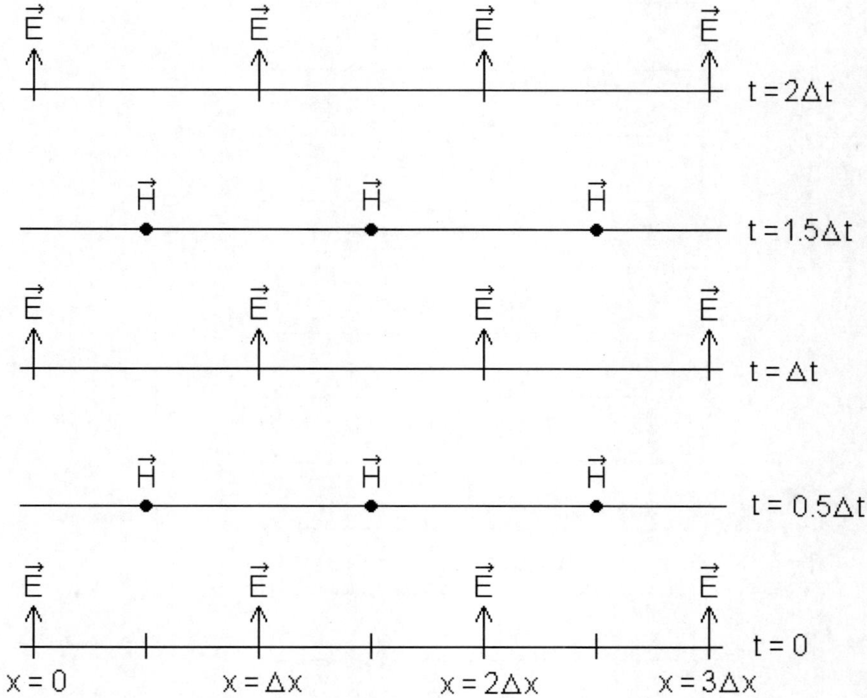

Figure 2.3. Space-time chart of the Yee algorithm in a leapfrog arrangement [11].

The field quantities are solved with a leapfrog arrangement where a half time step separates the solutions of the electric and magnetic fields as shown in Figure 2.3. The time-stepping of the FDTD algorithm is continued until the desired late time pulse response or steady state behaviour is reached. Figure 2.4 depicts the time-stepping FDTD algorithms flow chart. Apparently, a suitable size of the time step should be chosen properly to avoid the late time instability of the algorithms, after determining the spatial resolution based on the geometrical features and the operating frequency. It can be seen that a leapfrog arrangement between the electric and magnetic field vector components is used to implement the time step of the FDTD algorithm. The grids of the electric and magnetic fields are displaced half a cell between them indicating that the computer must work through them in turn.

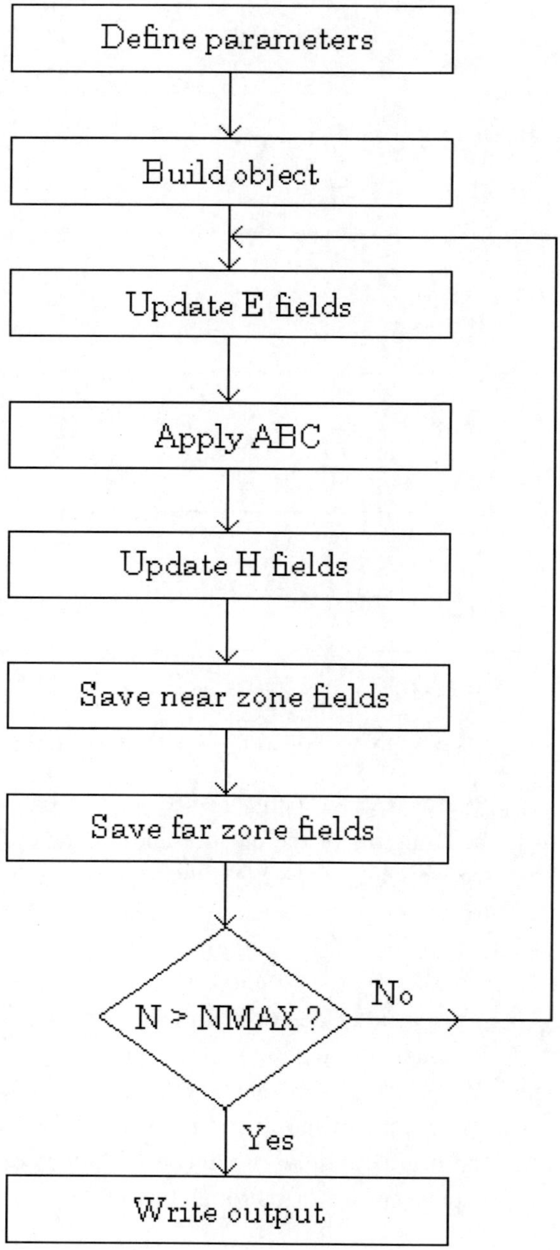

Figure 2.4. The flow chart of the time stepping FDTD algorithm [13].

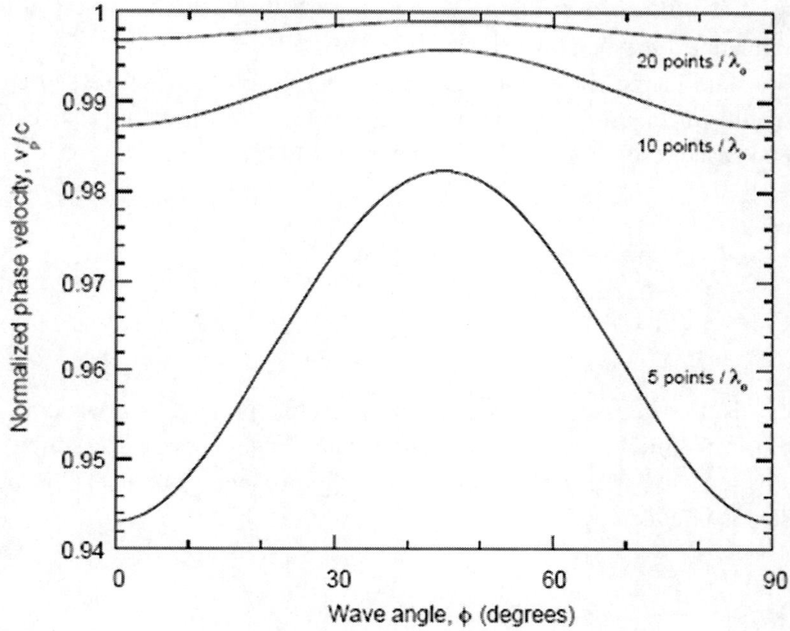

Figure 2.5. Variation of the numerical phase velocity with wave propagation angle in two-dimensional FDTD grid [11].

2.3. NUMERICAL DISPERSION

The accuracy and stability of FDTD method is controlled by two significant parameters namely the space resolution Δx, Δy, Δz and the time resolution Δt. Waves of different frequencies will propagate at slightly different speeds through the grid due to the approximations inherent in FDTD. As a result, the grid dispersion error will occur in the FDTD. The difference in the propagation speed also depends on the direction of propagation relative to the grid. Hence, the grid dispersion error must be reduced to an acceptable level to increase the accuracy and stability of the results. To ensure the FDTD numerical accuracy, the spatial increment h used in the simulation is required to be small enough compared to the wavelength. Usually, h is less than or equal to $\lambda_{min}/10$, where λ_{min} is the shortest wavelength at the frequency of computation. The numerical grid produces a certain amount of numerical dispersion error, when the grid size and the angle of propagating wave are increased as illustrated in Figure 2.5. Numerical dispersion is observed to be approximately 1% when the cell size is exactly $\lambda/10$.

The time increment Δt must satisfy the CFL criterion [11, 13] to ensure the stability of the computation process, which sets the relation between the time and space resolution for three-dimensional FDTD given by:

$$v_{\max}\Delta t \le \frac{1}{c\sqrt{\frac{1}{(\Delta x)^2}+\frac{1}{(\Delta y)^2}+\frac{1}{(\Delta z)^2}}} \tag{2.31}$$

v_{max} is the maximum phase velocity of the propagating wave in any material in the model. For the three-dimensional cubic cell space lattice with $\Delta x = \Delta y = \Delta z = h$ in a d-dimensional spatially homogeneous FDTD grid, the time step is bounded by:

$$\Delta t \le \frac{h}{v_{\max}\sqrt{d}} \tag{2.32}$$

d is the number of the space dimensions (d = 1, 2 or 3). The time step should be chosen as large as possible to minimise the burden on the computer run time requirement. Smaller time step will give more accurate result with longer simulation run time. The time resolution value of $h/(2v_{max})$ is always conceived in many FDTD codes to compromise the accuracy and stability of FDTD simulation whilst the CFL criterion is satisfied.

2.4. Incident Wave Source Conditions

An electromagnetic wave excitation will be introduced into the FDTD lattice for modelling the scattering problem. This can be achieved by using the electric and magnetic field vector components to realize the wave source compared to the total number of field components in the space lattice. In addition, the required computation time and resources to simulate the source is relatively small compared to the required ordinary time-stepping of the fields. As a result, the maximum algorithm efficiency is achieved. The compact electromagnetic wave sources can be classified into two general classes namely the hard source and the total/scattered field formulation for the plane wave excitation in one and three dimensional lattices.

2.4.1. Hard Source Excitation

The hard source is set up by assigning a desired time function to specific electric or magnetic field components in the FDTD lattice. In 1-D TM grid, the hard source on E_z could be established at the grid source point i_s to generate a continuous sinusoidal wave of frequency f_o which is switched on at $n = 0$. There are three common hard sources discussed here. The first type of the hard source can be expressed as [11]:

$$E_z\Big|_{i_s}^{n} = E_o \sin\left(2\pi f_o n\Delta t\right) \tag{2.33}$$

The second hard source provides a wideband Gaussian pulse with finite DC content. The pulse is centred at the time step n_o and has a $1/e$ characteristic decay of n_{decay} time steps. It is simply given by:

$$E_z\Big|_{i_s}^{n} = E_o e^{-\left(\left(n-n_o\right)/n_{decay}\right)^2} \tag{2.34}$$

Equation (2.34) has a nonzero value at $n = 0$. n_o should take a value of at least $3n_{decay}$ if a smooth transition from zero into the Gaussian pulse is needed. The third hard source provides zero DC content, bandpass Gaussian pulse with Fourier spectrum symmetrical about f_o. The pulse is centred at the time step n_o and has a $1/e$ characteristic decay of n_{decay} time steps. It is expressed as:

$$E_z\Big|_{i_s}^{n} = E_0 e^{-\left(\left(n-n_0\right)/n_{decay}\right)^2} \sin\left[2\pi f_0\left(n-n_0\right)\Delta t\right] \tag{2.35}$$

2.4.2. Plane Wave Source Excitation

The plane wave source excitation model was used by Yee in 1966 [15]. An arbitrary incident plane wave is used for the excitation purposes. The incident plane wave is modelled and approximated within the FDTD space lattice. Two methods of excitation of the plane wave on the space lattice is discussed here, namely the initial condition technique and the total/scattered field formulation technique. The initial condition method was used by Yee in the implementation of the FDTD method to represent an incident plane wave with special applications in the RCS simulations. The values of electric and magnetic field vector components at the zero time step of the incident plane wave throughout the problem space are preset in sign and magnitude. As a result, this will give the desired polarisation of the incident plane wave. However, this method has a drawback due to the fact that it is a non-compact wave source. It requires a large number of additional free space cells to contain initial conditions of long duration pulses or continuous sinusoids for oblique incident angles of the propagating wave. This will unfortunately increase the size of the problem space which one must avoid. Hence, the technique is limited to special usage only. The most popular method used in many FDTD software is the total/scattered field formulation method [17, 18] due to the fact that it allows the modelling of the FDTD with long-duration pulsed or sinusoidal illuminations for arbitrary plane wave propagation directions.

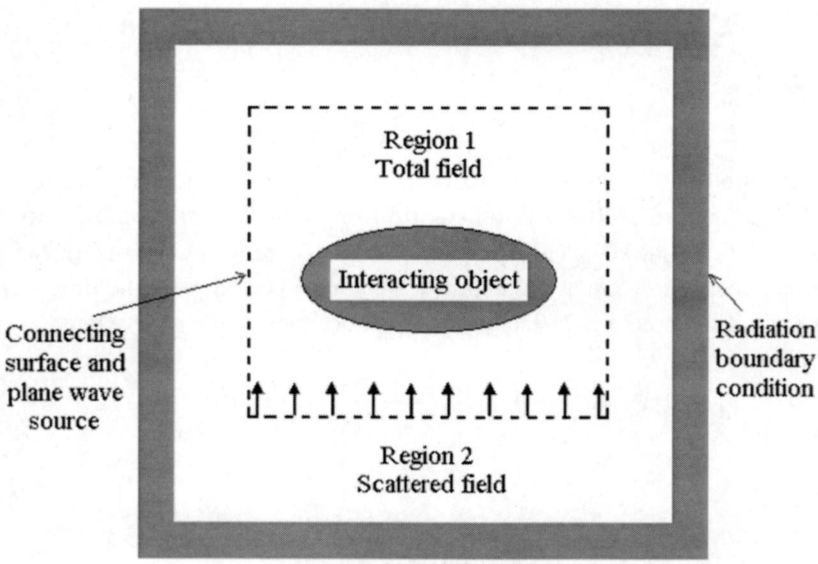

Figure 2.6. Total and scattered field zoning for a generic scattering case [11].

Figure 2.6 presents the total and scattered field zoning for a generic scattering case. Basically, it shows two distinct regions of interest namely the total field and scattered field region. Both of the regions are separated by a non-physical virtual surface applied numerically with a special treatment to include the incident wave excitation and to split the problem space into the total field and scattered field regions. As a result, the scattered field vector values can be computed in the scattered field region without the presence of the incident field. Furthermore, the arbitrary incident plane wave with different oblique incidence angles using incident-field array (IFA) excitation scheme [11] can be modelled efficiently. The IFA is generally an FDTD-based lookup table from which the incident-field values are covered on the FDTD lattice in the propagating direction.

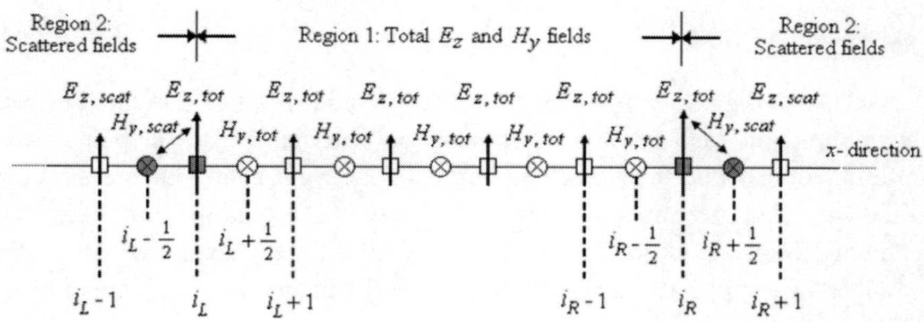

Figure 2.7. Total and scattered field components for 1-D FDTD grid.

2.4.2.1. Total/Scattered Field Formulation in 1-D

The basic principle of the total/scattered formulation example is best illustrated by simple example of one-dimensional transverse magnetic (TM) field case, as shown in Figure 2.7 [11]. The formulation divides the *x*-directed array of E_z and H_y components into two regions, region 1 (total fields) and region 2 (scattered fields) by a virtual surface. At this surface interface between the two regions, there exists a special set of *E* and *H* components as depicted by the grey-coloured components. These four field components are E_z at i_L and i_R and H_y at $(i_L\text{-}^1/_2)$ and $(i_R\text{+}^1/_2)$. Subscripts *L* and *R* represent left and right respectively. According to Yee algorithm [11], these may have different types, total or scattered, respectively. The incident fields, E_{inc} and H_{inc} are known and may be calculated at the outer interface points. They are added to the regular updating equations. For example, E_z at i_L, the left interface surface, is considered to be a total field component and by applying the basic FDTD algorithm, it is updated as:

$$E_z\Big|_{tot,i_L}^{n+1} = E_z\Big|_{tot,i_L}^{n} + \frac{\Delta t}{\varepsilon_o \Delta}\left(H_y\Big|_{tot,i_L+1/2}^{n+1/2} - H_y\Big|_{scat,i_L-1/2}^{n+1/2}\right) \tag{2.36}$$

Subscripts *tot* and *scat* stand for total and scattered fields respectively. It is clear that the above equation is inconsistent, so $H_{y,tot}$ at $(i_L\text{-}^1/_2)$ must be subtracted from $H_{y,tot}$ at $(i_L+^1/_2)$ to advance the value of $E_{z,tot}$ at i_L. To correct this updating equation, the vector function $H_{y,inc}$ at $(i_L\text{-}^1/_2)$ is added as an excitation wave source of the FDTD algorithm. The boundary E_z updating equation is given by:

$$E_z\Big|_{tot,i_L}^{n+1} = E_z\Big|_{tot,i_L}^{n} + \frac{\Delta t}{\varepsilon_o \Delta}\left(H_y\Big|_{tot,i_L+1/2}^{n+1/2} - H_y\Big|_{scat,i_L-1/2}^{n+1/2}\right) - \frac{\Delta t}{\varepsilon_o \Delta} H_y\Big|_{inc,i_L-1/2}^{n+1/2} \tag{2.37}$$

Subscript *inc* stands for incident field. The added term is assumed to be a known function such as sinusoidal wave or Gaussian pulse for plane wave representation, while the rest of the terms of the right hand side are assumed stored in computer memory from the previous updating time step. By following proper modifications of equation (2.37), the updating equations for the other three special magnetic and electric boundary field components are expressed as follows:

$$H_y\Big|_{scat,i_L-1/2}^{n+1/2} = H_y\Big|_{scat,i_L-1/2}^{n-1/2} + \frac{\Delta t}{\mu_o \Delta}\left(E_z\Big|_{tot,i_L}^{n} - E_z\Big|_{scat,i_L-1}^{n}\right) - \frac{\Delta t}{\mu_o \Delta} E_z\Big|_{inc,i_L}^{n} \tag{2.38}$$

$$E_z\Big|_{tot,i_R}^{n+1} = E_z\Big|_{tot,i_R}^{n} + \frac{\Delta t}{\varepsilon_o \Delta}\left(H_y\Big|_{scat,i_R+1/2}^{n+1/2} - H_y\Big|_{tot,i_R-1/2}^{n+1/2}\right) + \frac{\Delta t}{\varepsilon_o \Delta} H_y\Big|_{inc,i_R+1/2}^{n+1/2} \tag{2.39}$$

$$H_y\Big|_{scat,i_R+1/2}^{n+1/2} = H_y\Big|_{scat,i_R+1/2}^{n-1/2} + \frac{\Delta t}{\mu_o \Delta}\left(E_z\Big|_{scat,i_R+1}^{n} - E_z\Big|_{tot,i_R}^{n}\right) + \frac{\Delta t}{\mu_o \Delta} E_z\Big|_{inc,i_R}^{n} \tag{2.40}$$

Equations (2.38)-(2.40) are the one-dimensional total/scattered field FDTD formulations to generate the equivalent effects of the plane wave at one surface interface. The same principle can be extended to 2-D and 3-D scattering problems.

2.4.2.2. Total/Scattered Field Formulation in 3-D

In general, the equivalence principle surface implementation in FDTD code is employed with electric and magnetic field nodes are located at half a cell apart between each other. In three-dimensional case, the interface of the total and the scattered field zone is constituted by six flat planes to form a closed rectangular box as depicted in Figure 2.8. Furthermore, each face of the rectangular box consists of two tangential electric and magnetic fields in a closed surface. The notation used by Taflove in [11] is used in order to simplify the equations. Equations from (2.21) to (2.26) are used as the basis of the updating equations. A cubic-cell space lattice is assumed for simplicity.

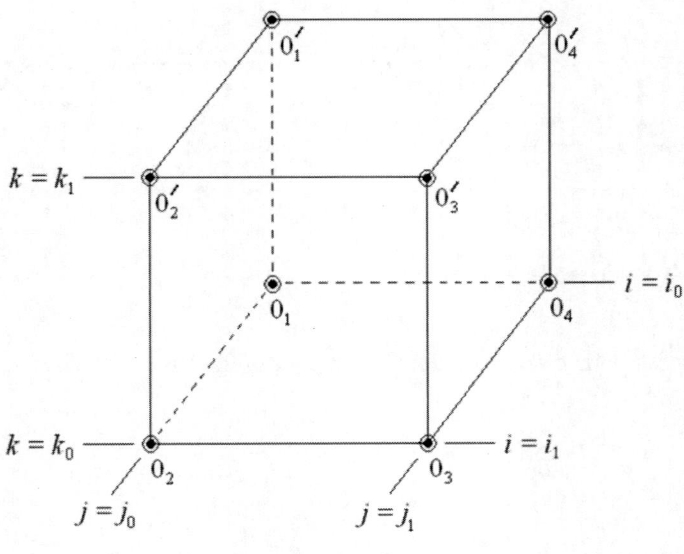

Figure 2.8. Coordinate origins for the calculation of incident field.

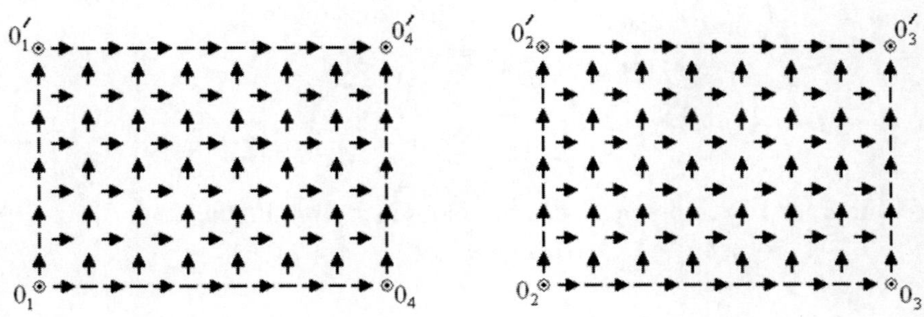

Figure 2.9. Location of $E_y(\rightarrow)$ and $E_z(\uparrow)$ components in planes $i = i_o$ and $i = i_1$.

The E_y and E_z components can be determined from Figure 2.9. E_y ($i = i_o$; $j = j_o + ½, \ldots, j_1 - ½$; $k = k_o, \ldots, k_1$) is given by:

$$E_y\Big|_{i_0,j,k}^{n+1} = \{E_y\Big|_{i_0,j,k}^{n+1}\}_{(2.25)} + C_{b,E_y}\Big|_{i_0,j,k} H_{z,inc}\Big|_{i_0-1/2,j,k}^{n+1/2} \tag{2.41}$$

E_z ($i = i_o$; $j = j_o, \ldots, j_1$; $k = k_o + ½, \ldots, k_1 - ½$) is given by:

$$E_z\Big|_{i_0,j,k}^{n+1} = \{E_z\Big|_{i_0,j,k}^{n+1}\}_{(2.26)} - C_{b,E_z}\Big|_{i_0,j,k} H_{y,inc}\Big|_{i_0-1/2,j,k}^{n+1/2} \tag{2.42}$$

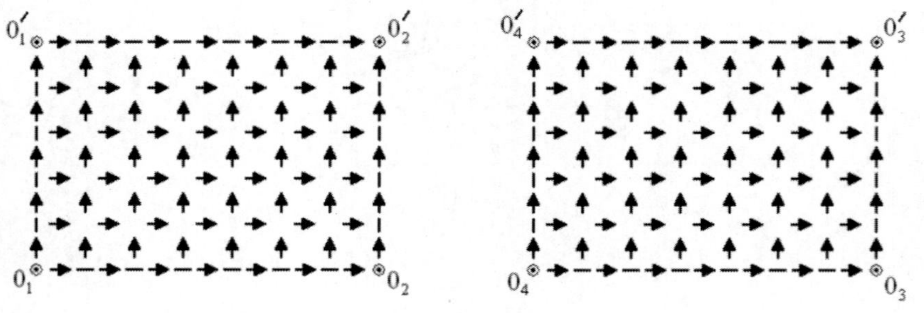

Figure 2.10. Location of $E_x(\rightarrow)$ and $E_z(\uparrow)$ components in planes $j = j_o$ and $j = j_1$.

E_y ($i = i_1$; $j = j_o + ½, \ldots, j_1 - ½$; $k = k_o, \ldots, k_1$) is given by:

$$E_y\Big|_{i_1,j,k}^{n+1} = \{E_y\Big|_{i_1,j,k}^{n+1}\}_{(2.25)} - C_{b,E_y}\Big|_{i_1,j,k} H_{z,inc}\Big|_{i_1+1/2,j,k}^{n+1/2} \tag{2.43}$$

E_z ($i = i_1$; $j = j_o, \ldots, j_1$; $k = k_o + ½, \ldots, k_1 - ½$) is given by:

$$E_z\Big|_{i_1,j,k}^{n+1} = \{E_z\Big|_{i_1,j,k}^{n+1}\}_{(2.26)} + C_{b,E_z}\Big|_{i_1,j,k} H_{y,inc}\Big|_{i_1+1/2,j,k}^{n+1/2} \tag{2.44}$$

The E_x and E_z components can be determined from Figure 2.10. E_x ($i = i_o + ½, \ldots, i_1 - ½$; $j = j_o$; $k = k_o, \ldots, k_1$) is given by:

$$E_x\Big|_{i,j_0,k}^{n+1} = \{E_x\Big|_{i,j_0,k}^{n+1}\}_{(2.24)} - C_{b,E_x}\Big|_{i,j_0,k} H_{z,inc}\Big|_{i,j_0-1/2,k}^{n+1/2} \tag{2.45}$$

E_z ($i = i_o, \ldots, i_1; j = j_o; k = k_o + ½, \ldots, k_1 - ½$) is given by:

$$E_z\Big|_{i,j_0,k}^{n+1} = \{E_z\Big|_{i,j_0,k}^{n+1}\}_{(2.26)} + C_{b,E_z}\Big|_{i,j_0,k} H_{x,inc}\Big|_{i,j_0-1/2,k}^{n+1/2} \quad (2.46)$$

E_x ($i = i_o + ½, \ldots, i_1 - ½; j = j_1; k = k_o, \ldots, k_1$) is given by:

$$E_x\Big|_{i,j_1,k}^{n+1} = \{E_x\Big|_{i,j_1,k}^{n+1}\}_{(2.24)} + C_{b,E_x}\Big|_{i,j_1,k} H_{z,inc}\Big|_{i,j_1+1/2,k}^{n+1/2} \quad (2.47)$$

E_z ($i = i_o, \ldots, i_1; j = j_1; k = k_o + ½, \ldots, k_1 - ½$) is given by:

$$E_z\Big|_{i,j_1,k}^{n+1} = \{E_z\Big|_{i,j_1,k}^{n+1}\}_{(2.26)} - C_{b,E_z}\Big|_{i,j_1,k} H_{x,inc}\Big|_{i,j_1+1/2,k}^{n+1/2} \quad (2.48)$$

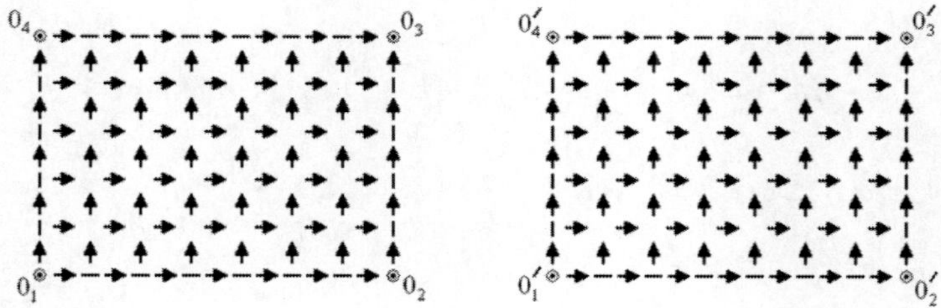

Figure 2.11. Location of $E_x(\rightarrow)$ and $E_y(\uparrow)$ components in planes $k = k_o$ and $k = k_1$.

The E_x and E_y components can be determined from Figure 2.11. E_x ($i = i_o +$ ½, …, i_1 - ½; $j = j_o, \ldots, j_1$; $k = k_o$) is given by:

$$E_x\Big|_{i,j,k_0}^{n+1} = \{E_x\Big|_{i,j,k_0}^{n+1}\}_{(2.24)} + C_{b,E_x}\Big|_{i,j,k_0} H_{y,inc}\Big|_{i,j,k_0-1/2}^{n+1/2} \quad (2.49)$$

E_y ($i = i_o, \ldots, i_1; j = j_o + ½, \ldots, j_1 - ½; k = k_o$) is given by:

$$E_y\Big|_{i,j,k_0}^{n+1} = \{E_y\Big|_{i,j,k_0}^{n+1}\}_{(2.25)} - C_{b,E_y}\Big|_{i,j,k_0} H_{x,inc}\Big|_{i,j,k_0-1/2}^{n+1/2} \quad (2.50)$$

E_x ($i = i_o + ½, \ldots, i_1 - ½; j = j_o, \ldots, j_1; k = k_1$) is given by:

$$E_x\Big|_{i,j,k_1}^{n+1} = \{E_x\Big|_{i,j,k_1}^{n+1}\}_{(2.24)} - C_{b,E_x}\Big|_{i,j,k_1} H_{y,inc}\Big|_{i,j,k_1+1/2}^{n+1/2} \tag{2.51}$$

E_y ($i = i_o, \ldots, i_1; j = j_o + ½, \ldots, j_1 - ½; k = k_1$) is given by:

$$E_y\Big|_{i,j,k_1}^{n+1} = \{E_y\Big|_{i,j,k_1}^{n+1}\}_{(2.25)} + C_{b,E_y}\Big|_{i,j,k_1} H_{x,inc}\Big|_{i,j,k_1+1/2}^{n+1/2} \tag{2.52}$$

The consistency condition for the magnetic field components located 0.5Δ outside of each face of the total/scattered-field interface are given by analogy. The H_y and H_z components can be determined from Figure 2.9. H_y ($i = i_o - ½; j = j_o, \ldots, j_1; k = k_o + ½, \ldots, k_1 - ½$) is given by:

$$H_y\Big|_{i_0-1/2,j,k}^{n+1/2} = \{H_y\Big|_{i_0-1/2,j,k}^{n+1/2}\}_{(2.22)} - D_{b,H_y}\Big|_{i_0-1/2,j,k} E_{z,inc}\Big|_{i_0,j,k}^{n} \tag{2.53}$$

H_z ($i = i_o - ½; j = j_o + ½, \ldots, j_1 - ½; k = k_o, \ldots, k_1$) is given by:

$$H_z\Big|_{i_0-1/2,j,k}^{n+1/2} = \{H_z\Big|_{i_0-1/2,j,k}^{n+1/2}\}_{(2.23)} + D_{b,H_z}\Big|_{i_0-1/2,j,k} E_{y,inc}\Big|_{i_0,j,k}^{n} \tag{2.54}$$

H_y ($i = i_1 + ½; j = j_o, \ldots, j_1; k = k_o + ½, \ldots, k_1 - ½$) is given by:

$$H_y\Big|_{i_1+1/2,j,k}^{n+1/2} = \{H_y\Big|_{i_1+1/2,j,k}^{n+1/2}\}_{(2.22)} + D_{b,H_y}\Big|_{i_1+1/2,j,k} E_{z,inc}\Big|_{i_1,j,k}^{n} \tag{2.55}$$

H_z ($i = i_1 + ½; j = j_o + ½, \ldots, j_1 - ½; k = k_o, \ldots, k_1$) is given by:

$$H_z\Big|_{i_1+1/2,j,k}^{n+1/2} = \{H_z\Big|_{i_1+1/2,j,k}^{n+1/2}\}_{(2.23)} - D_{b,H_z}\Big|_{i_1+1/2,j,k} E_{y,inc}\Big|_{i_1,j,k}^{n} \tag{2.56}$$

The H_x and H_z components can be determined from Figure 2.10. H_x ($i = i_o, \ldots, i_1; j = j_o - ½; k = k_o + ½, \ldots, k_1 - ½$) is given by:

$$H_x\Big|_{i,j_0-1/2,k}^{n+1/2} = \{H_x\Big|_{i,j_0-1/2,k}^{n+1/2}\}_{(2.21)} + D_{b,H_x}\Big|_{i,j_0-1/2,k} E_{z,inc}\Big|_{i,j_0,k}^{n} \tag{2.57}$$

H_z ($i = i_o + ½, \ldots, i_1 - ½; j = j_o - ½; k = k_o, \ldots, k_1$) is given by:

$$H_z\Big|_{i,j_0-1/2,k}^{n+1/2} = \{H_z\Big|_{i,j_0-1/2,k}^{n+1/2}\}_{(2.23)} - D_{b,H_z}\Big|_{i,j_0-1/2,k} E_{x,inc}\Big|_{i,j_0,k}^{n} \tag{2.58}$$

H_x ($i = i_o, \ldots, i_1; j = j_1+½; k = k_o + ½, \ldots, k_1 - ½$) is given by:

$$H_x\Big|_{i,j_1+1/2,k}^{n+1/2} = \{H_x\Big|_{i,j_1+1/2,k}^{n+1/2}\}_{(2.21)} - D_{b,H_x}\Big|_{i,j_1+1/2,k} E_{z,inc}\Big|_{i,j_1,k}^{n} \tag{2.59}$$

H_z ($i = i_o + ½, \ldots, i_1 - ½; j = j_1+½; k = k_o, \ldots, k_1$) is given by:

$$H_z\Big|_{i,j_1+1/2,k}^{n+1/2} = \{H_z\Big|_{i,j_1+1/2,k}^{n+1/2}\}_{(2.23)} + D_{b,H_z}\Big|_{i,j_1+1/2,k} E_{x,inc}\Big|_{i,j_1,k}^{n} \tag{2.60}$$

The H_x and H_y components can be determined from Figure 2.11. H_x ($i = i_o, \ldots, i_1; j = j_o + ½, \ldots, j_1 - ½; k = k_o - ½$) is given by:

$$H_x\Big|_{i,j,k_0-1/2}^{n+1/2} = \{H_x\Big|_{i,j,k_0-1/2}^{n+1/2}\}_{(2.21)} - D_{b,H_x}\Big|_{i,j,k_0-1/2} E_{y,inc}\Big|_{i,j,k_0}^{n} \tag{2.61}$$

H_y ($i = i_o + ½, \ldots, i_1 - ½; j = j_o, \ldots, j_1; k = k_o - ½$) is given by:

$$H_y\Big|_{i,j,k_0-1/2}^{n+1/2} = \{H_y\Big|_{i,j,k_0-1/2}^{n+1/2}\}_{(2.22)} + D_{b,H_y}\Big|_{i,j,k_0-1/2} E_{x,inc}\Big|_{i,j,k_0}^{n} \tag{2.62}$$

H_x ($i = i_o, \ldots, i_1; j = j_o + ½, \ldots, j_1 - ½; k = k_1 + ½$) is given by:

$$H_x\Big|_{i,j,k_1+1/2}^{n+1/2} = \{H_x\Big|_{i,j,k_1+1/2}^{n+1/2}\}_{(2.21)} + D_{b,H_x}\Big|_{i,j,k_1+1/2} E_{y,inc}\Big|_{i,j,k_1}^{n} \tag{2.63}$$

H_y ($i = i_o + ½, \ldots, i_1 - ½; j = j_o, \ldots, j_1; k = k_1 + ½$) is given by:

$$H_y\Big|_{i,j,k_1+1/2}^{n+1/2} = \{H_y\Big|_{i,j,k_1+1/2}^{n+1/2}\}_{(2.22)} - D_{b,H_y}\Big|_{i,j,k_1+1/2} E_{x,inc}\Big|_{i,j,k_1}^{n} \tag{2.64}$$

2.5. Absorbing Boundary Conditions (ABC)

ABC is generally one of the most critical elements in FDTD simulation and analysis for several reasons. It is introduced at the outer boundary of the FDTD lattice to limit the problem space into bounded regions and hence the electromagnetic wave interaction problems can be solved easily. It is meant to simulate the extension of the FDTD mesh to infinity. In other words, it is used to truncate the computational region at the edges of the finite FDTD mesh. The key to the success of an ABC is that it should be able to absorb the waves travelling outward from the FDTD lattice with extremely low reflection at the boundary.

There have been numerous ABC approaches [11] to effectively approximate an infinite mesh. Mur [17] used analytical ABC which provides the effective reflection at the boundary between -35 to -40 dB. Liao et al. [19] implemented both spatial and temporal extrapolation of the electromagnetic fields by means of Newton backward-difference polynomial. The result was 20 dB less reflective than the second order Mur. Nevertheless, the method was delicate to the frequency and direction of propagation of the incident wave [20]. Therefore, the ABC boundary must be located at large distances from the radiator and scatterer to be more efficient and effective. Mei et al. [21] used error cancellation procedure to improve the performance of the local ABC. However, the method had a considerable residual error due to the assumption of zero degree of the incidence wave angle. One of the most flexible and efficient methods is the perfectly matched layer (PML) developed by Berenger [22]. Though PML is perfect in theory, spurious reflection is present in the actual FDTD computation [23, 24]. The basic idea in [22] is to produce an artificial lossy medium by complex electric permittivity and magnetic permeability surrounding the FDTD problem space so that the wave will be absorbed before it hits the boundary. In addition, this technique permitted the boundary reflections below -80 dB. The PML boundary can now be placed close to the scatterer and radiator. Berenger's PML is now considered as the predominant and standard used in the most FDTD imitations [25, 26].

2.5.1. 3-D Berenger's PML

In 3-D, all six Cartesian field vector components are separated to realize the following twelve modified version of Maxwell's equations [22]:

$$\mu_o \frac{\partial H_{xy}}{\partial t} + \sigma_y^* H_{xy} = -\frac{\partial (E_{zx} + E_{zy})}{\partial y} \tag{2.65}$$

$$\mu_o \frac{\partial H_{xz}}{\partial t} + \sigma_z^* H_{xz} = \frac{\partial (E_{yx} + E_{yz})}{\partial z} \tag{2.66}$$

$$\mu_o \frac{\partial H_{yz}}{\partial t} + \sigma_z^* H_{yz} = -\frac{\partial (E_{xy} + E_{xz})}{\partial z} \tag{2.67}$$

$$\mu_o \frac{\partial H_{yx}}{\partial t} + \sigma_x^* H_{yx} = \frac{\partial (E_{zx} + E_{zy})}{\partial x} \tag{2.68}$$

$$\mu_o \frac{\partial H_{zx}}{\partial t} + \sigma_x^* H_{zx} = -\frac{\partial (E_{yx} + E_{yz})}{\partial x} \tag{2.69}$$

$$\mu_o \frac{\partial H_{zy}}{\partial t} + \sigma_y^* H_{zy} = \frac{\partial (E_{xy} + E_{xz})}{\partial y} \tag{2.70}$$

$$\varepsilon_o \frac{\partial E_{xy}}{\partial t} + \sigma_y E_{xy} = \frac{\partial (H_{zx} + H_{zy})}{\partial y} \tag{2.71}$$

$$\varepsilon_o \frac{\partial E_{xz}}{\partial t} + \sigma_z E_{xz} = -\frac{\partial (H_{yx} + H_{yz})}{\partial z} \tag{2.72}$$

$$\varepsilon_o \frac{\partial E_{yz}}{\partial t} + \sigma_z E_{yz} = \frac{\partial (H_{xy} + H_{xz})}{\partial z} \tag{2.73}$$

$$\varepsilon_o \frac{\partial E_{yx}}{\partial t} + \sigma_x E_{yx} = -\frac{\partial (H_{zx} + H_{zy})}{\partial x} \tag{2.74}$$

$$\varepsilon_o \frac{\partial E_{zx}}{\partial t} + \sigma_x E_{zx} = \frac{\partial (H_{yx} + H_{yz})}{\partial x} \tag{2.75}$$

$$\varepsilon_o \frac{\partial E_{zy}}{\partial t} + \sigma_y E_{zy} = -\frac{\partial (H_{xy} + H_{xz})}{\partial y} \tag{2.76}$$

PML matching conditions and lattice structures are basically similar to the TM case.

2.5.2. 3-D PML Numerical Implementations

The standard time-stepping of the Yee algorithm cannot be used since the attenuation of the outgoing source waves in the PML layer is very fast. Consequently, an explicit exponential time step algorithm is suggested in the PML layer [27]. From equations (2.70) and (2.71), the updating H_{zy} and E_{xy} in the PML medium can both be expanded by the expressions:

$$H_{zy}\Big|_{i,j,k}^{n+1/2} = e^{-\left(\frac{\sigma^*_{y_{i,j,k}}\Delta t}{\mu_{i,j,k}}\right)} H_{zy}\Big|_{i,j,k}^{n-1/2} + \frac{1 - e^{-\left(\frac{\sigma^*_{y_{i,j,k}}\Delta t}{\mu_{i,j,k}}\right)}}{\sigma^*_{y_{i,j,k}}\Delta_p} \begin{pmatrix} E_{xy}\Big|_{i,j+1/2,k}^{n} - E_{xy}\Big|_{i,j-1/2,k}^{n} \\ + E_{xz}\Big|_{i,j+1/2,k}^{n} - E_{xz}\Big|_{i,j-1/2,k}^{n} \end{pmatrix} \tag{2.77}$$

$$E_{xy}\Big|_{i,j,k}^{n+1} = e^{-\left(\frac{\sigma_{y_{i,j,k}}\Delta t}{\varepsilon_{i,j,k}}\right)} E_{xy}\Big|_{i,j,k}^{n} + \frac{1 - e^{-\left(\frac{\sigma_{y_{i,j,k}}\Delta t}{\varepsilon_{i,j,k}}\right)}}{\sigma_{y_{i,j,k}}\Delta_p} \begin{pmatrix} H_{zx}\Big|_{i,j+1/2,k}^{n+1/2} - H_{zx}\Big|_{i,j-1/2,k}^{n+1/2} \\ + H_{zy}\Big|_{i,j+1/2,k}^{n+1/2} - H_{zy}\Big|_{i,j-1/2,k}^{n+1/2} \end{pmatrix} \tag{2.78}$$

The subscript p can be x, y or z and Δ_p is the cell size in the p-direction. In addition, the updating equation for the other ten vector field components can be derived by using the same technique. The thickness of the PML layer can be reduced to obtain the optimum profile of the conductivity by means of geometric progression [28, 29]:

$$\sigma(\rho) = \sigma_o g^{\frac{\rho}{\Delta_p}} = -\frac{\varepsilon_o c}{2\Delta_p} \frac{\ln g}{\left(g^N - 1\right)} \ln R(0) g^{\frac{\rho}{\Delta_p}} \tag{2.79}$$

N is the number of cells in the PML layer and g is the geometric progression grading factor. This is an empirical factor in a function of the number of cells in the scatterer region. Equation (2.79) shows that the conductivity increases by the factor g from one cell to the other.

CONCLUSION

The fundamental concepts of the FDTD computational method were briefly explained in this chapter. Some of them deal with the solution of Maxwell's equations by means of finite differences, the derivation of FDTD updating equations for electric and magnetic field vector components, and the explanation on the accuracy and stability which necessitate the FDTD computational method. In addition, the excitation source inside the FDTD lattice was also explained. The general idea of the ABC to the specific Berenger's PML in 2-D and 3-D were well treated. Last but not least, Fortran source code was written successfully in 2-D FDTD with subgridding (SGFDTD) and 3-D MoM-FDTD-SGFDTD hybrid technique for different applications in electromagnetic scattering problems.

REFERENCES

[1] W. G. Whittow, C. C. Njoku and Y.C. Vardaxoglou, "Fine scale simulations of patch antennas with heterogeneous substrates," *USNC-URSI Radio Science Meeting (Joint with AP-S Symposium)*, pp. 223, 2013.

[2] R. Xiong, B. Chen, Y. Yi, H. -L. Chen and H. -F. Zhao, "Two-step method for the FDTD analysis of long apertures having depth," *IEEE Antennas and Wireless Propagation Letters*, vol. 12, pp. 39-42, 2013.

[3] S. K. Kamepally, B. P. Kumar and C. S. Paidimarry, "FDTD estimation for accurate specific absorption rate in a tumor," *Annual International Conference on Emerging Research Areas and International Conference on Microelectronics, Communications and Renewable Energy (AICERA/ICMiCR)*, pp. 1-5, 2013.

[4] C. Guiffaut, A. Reineix and B. Pecqueux, "New oblique thin wire formalism in the FDTD method with multiwire junctions," *IEEE*

Transactions on Antennas and Propagation, vol. 60, no. 3, pp. 1458-1466, 2012.

[5] A. C. Lesina, A. Vaccari and A. Bozzoli, "A modified RC-FDTD algorithm for plasmonics in Drude dispersive media nanostructures," *6th European Conference on Antennas and Propagation (EUCAP)*, pp. 687-690, 2012.

[6] G. M. Noetscher, Y. Xu and S. N. Makarov, "Accuracy of point source models with coincident phase centers in a cubic FDTD grid for arbitrary source orientation," *IEEE Antennas and Propagation Society International Symposium (APSURSI)*, pp. 1-2, 2012.

[7] V. Nayyeri, M. Soleimani and M. Dehmollaian, "Implementation of a PEMC boundary condition in the 2-D FDTD technique," *IEEE Antennas and Propagation Society International Symposium (APSURSI)*, pp. 1-2, 2012.

[8] D. -C. Liu and H. -C. Chang, "The second-order condition of FDTD method at sloped dielectric interfaces by averaging the permittivities along surface normal," *IEEE Transactions on Antennas and Propagation*, vol. 60, no. 11, pp. 5259-5267, 2012.

[9] M. Weiwei, S. Dong and W. Xianliang, "UPML-FDTD parallel computing on GPU," *International Conference on Microwave and Millimeter Wave Technology (ICMMT)*, vol. 4, pp. 1-4, 2012.

[10] D. M. Shyroki, "Modeling of sloped interfaces on a Yee grid," *IEEE Transactions on Antennas and Propagation*, vol. 59, no. 9, pp. 3290-3295, 2011.

[11] A. Taflove and S. C. Hagness, Computational electrodynamics: The finite-difference time-domain method, 3rd ed., Boston, MA: Artech House, 2005.

[12] A. Taflove, Advances in computational electrodynamics: The finite-difference time-domain method, Dedham, MA: Artech House, 1998.

[13] K. S. Kunz and R. J. Luebbers, The finite difference time domain method for electromagnetics, Boca Raton, FL: CRC Press, 1993.

[14] J. D. Jackson, Classical electrodynamics, 3rd ed., New York: John Wiley & Sons Inc., 1998.

[15] K. S. Yee, "Numerical solution of initial boundary value problems involving Maxwell's equations in isotropic media," *IEEE Transactions on Antennas and Propagation*, vol. AP-14, pp. 302-307, 1966.

[16] M. N. O. Sadiku, Numerical techniques in electromagnetics, Boca Raton: CRC Press, 2001.

[17] G. Mur, "Absorbing boundary conditions for the finite-difference approximation of the time-domain electromagnetic-field equations," *IEEE Transactions on Electromagnetic Compatibility*, vol. EMC-23, pp. 377-382, 1981.

[18] K. Umashankar and A. Taflove, "A novel method to analyze electromagnetic scattering of complex objects," *IEEE Transactions on Electromagnetic Compatibility*, vol. EMC-24, pp. 397-405, 1982.

[19] Z. P. Liao, H. L. Wong, B. P. Yang, and Y. F. Yuan, "A transmitting boundary for transient wave analysis," *Scientia Sinica (Series A)*, vol. XXVII, pp. 1063-1076, 1984.

[20] J. Y. Fang, "Absorbing boundary conditions applied to model wave propagation in microwave integrated-circuits," *IEEE Transactions on Microwave Theory and Techniques*, vol. 42, pp. 1506-1513, 1994.

[21] K. K. Mei and J. Y. Fang, "Superabsorption - A method to improve absorbing boundary conditions," *IEEE Transactions on Antennas and Propagation*, vol. 40, pp. 1001-1010, 1992.

[22] J. P. Berenger, "A perfectly matched layer for the absorption of electromagnetic waves," *Journal of Computational Physics*, vol. 114, pp. 185-200, 1994.

[23] A. F. Oskooi, An investigation of the perfectly matched layer for inhomogeneous media. M. Sc. Thesis, Cambridge: Massachusetts Institute of Technology, 2008.

[24] J. P. Berenger, "Numerical reflection from FDTD-PMLs: a comparison of the split PML with the unsplit and CFS PMLs," *IEEE Transactions on Antennas and Propagation*, vol. 50, pp. 258-265, 2002.

[25] D. T. Prescott and N. V. Shuley, "Reflection analysis of FDTD boundary conditions - Part I: Time-space absorbing boundaries," *IEEE Transactions on Microwave Theory and Techniques*, vol. 45, pp. 1162-1170, 1997.

[26] D. T. Prescott and N. V. Shuley, "Reflection analysis of FDTD boundary conditions - Part II: Berenger's PML absorbing layers," *IEEE Transactions on Microwave Theory and Techniques*, vol. 45, pp. 1171-1178, 1997.

[27] R. Holland, "Finite-difference time-domain (FDTD) analysis of magnetic diffusion," *IEEE Transactions on Electromagnetic Compatibility*, vol. 36, pp. 32-39, 1994.

[28] J. P. Berenger, "Perfectly matched layer for the FDTD solution of wave-structure interaction problems," *IEEE Transactions on Antennas and Propagation*, vol. 44, pp. 110-117, 1996.

[29] J. P. Berenger, "Improved PML for the FDTD solution of wave-structure interaction problems," *IEEE Transactions on Antennas and Propagation*, vol. 45, pp. 466-473, 1997.

In: Development of Complex Electromagnetic … ISBN: 978-1-61122-013-1
Editors: K. N. Ramli et al.

Chapter 3

SURFACE KERNEL SOLUTION OF THE METHOD OF MOMENTS

***K. N. Ramli*[1], *M. Lashab*[2], *K. H. Sayidmarie*[3] and *R. A. Abd-Alhameed*[4]**
[1]Faculty of Electrical and Electronics Engineering,
Universiti Tun Hussein Onn Malaysia,
Parit Raja, Batu Pahat, Johor, Malaysia
[2]Electronics Department, University of Skikda, Skikda, Algeria
[3]Department of Communication Engineering,
College of Electronic Engineering, University of Mosul, Mosul, Iraq
[4]Mobile and Satellite Communications Research Centre,
Bradford University, Bradford, UK

ABSTRACT

The classical methods known as asymptotic or high frequency, such as physical optics (PO) or geometrical theory of diffraction (GTD), can present an approximate or almost exact solution for electromagnetic scattering to some given antenna shape. The numerical methods known as the full wave lead generally to an exact solution for any given shape of antenna, these can be listed as finite element method (FEM), finite-difference time-domain method (FDTD) and moment method (MoM). The moment method is based on the digitalization of the electric field integral equation (EFIE) or the magnetic field integral equation (MFIE), obtained from the vector solution of Maxwell equation and for given boundary conditions. The moment method presented is ideally used for perfectly conducting surface and homogenous medium, for inhomogeneous medium leading to volume integral a hybrid solution MoM/FEM is preferred, also for large structures practically considered as planar a MoM/PO is preferred to reduce the computing time.

In this work, the moment method is used in spatial domain, applied to homogenous medium and perfectly conducting surfaces, one-dimensional and two-dimensional antennas, cylindrical square or circular antennas. An investigation of the surface current distribution in a Normal-Mode Helical Antenna (NMHA) is also reported.

This enables precise prediction of the performance of NMHAs, since traditional wire-antenna simulations ignore important details.

A moment-method formulation was developed using two geometrically-orthogonal basis functions to represent the total non-uniform surface current distribution over the wire of the helix. Extended basis functions were used to reliably treat the discontinuity of the current at the free ends. A surface kernel was used all over the antenna structure. The surface current distribution was computed for different antenna geometries such as dipoles, loops and helices. For helices the currents were investigated for different pitch distances and number of turns. It was found that the axially-directed component of the current distribution around the surface of the wire was highly non-uniform and that there was also a significant circumferential current flow due to inter-turn capacitance, both effects that are overlooked by standard filamentary current representations using an extended kernel. The impedance characteristic showed good agreement with the predictions of a standard filamentary-current code, in the case of applied uniform excitation along the local axis of the wire. However, the power loss computations of the present technique produce significantly different results compared to those well established methods when wires are closely spaced.

3.1. Introduction

Great research work has been carried out to improve the numerical method applied to electromagnetic scattering, especially for very complicated geometries or for stratified and inhomogeneous medium. The moment method is still one the most powerful method, which could give very exact solution for many antenna structures [1-3], but for some cases where the geometry of the antenna is complicated or its medium is inhomogeneous, in this case a hybrid solution is needed [4, 5]. The choice of the numerical solution depends on the antenna shape and the type of medium where the antenna is placed. The moment method (MoM) is used for very simple structures, where surface integral equations are used [6, 7].

The finite element method (FEM) and the finite-difference time-domain (FDTD) are used for very complicated geometries where volumetric integral equations are used.

The computing time is usually $O(N^2)$ for the moment method and $O(N^3)$ for the other numerical methods, where N is the number of unknowns. For very large structures (over 100 wavelength), the computing time and the memory storage became very big, a hybrid with approximate methods such as physical optics (PO) is desirable [8, 9].

The moment method was first introduced by Harrington in 1968 [10], since then many effort were made to improve both the precision and the computational time spent. A good choice of basis and testing function can improve the precision and the computing time, for instance sinusoidal functions known as entire domain, when used good results are obtained but the computing time is a bit longer than ordinary functions [11-15].

Several research work have been carried out to reduce the computing time of the classical moment method, such as the fast multipole (FMPM) [16] and improved moment method based on wavelets [17-21].

These two methods have the ability to reduce the computing time by simply reducing the impedance matrix coefficients.

3.2. MAXWELL'S EQUATIONS

Maxwell equations should be applied to any antenna or 3-D structure to find the electromagnetic scattering. For homogenous and isotropic medium fed by magnetic or electric field, the equations are defined as [22].

$$\nabla \times \vec{E} = -j\omega\mu\vec{H} \tag{3.1}$$

$$\nabla \times \vec{H} = j\omega\varepsilon\vec{E} + \vec{J} \tag{3.2}$$

$$\nabla.\vec{E} = \frac{\rho}{\varepsilon} \tag{3.3}$$

$$\nabla.\vec{H} = 0 \tag{3.4}$$

We define $\vec{E}$ and $\vec{H}$ respectively the electric and magnetic field, $\vec{J}$ is the current density and ρ the electric charge density.

The medium is considered homogenous and isotropic as shown in Figure 3.1, the permittivity ε and the permeability μ are considered constant with respect to curl and divergence action.

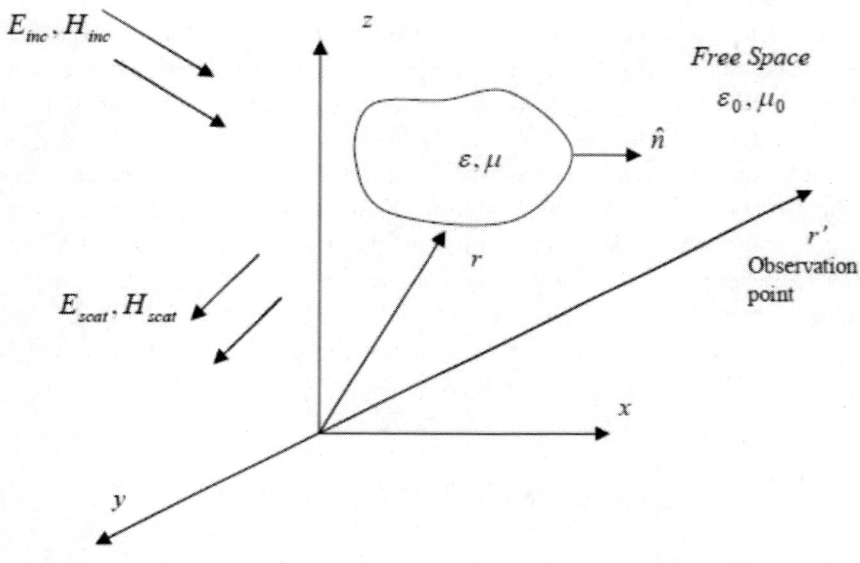

Figure 3.1. Homogenous and isotropic structure fed by electric or magnetic field.

3.2.1. Electric Field Integral Equations (EFIE)

Applying the boundary conditions upon the antenna or the object of arbitrary shape in Figure 3.1, having a perfectly conducting surface (PEC), the tangential electric field on the surface is null. This can be expressed as [23]:

$$\vec{E}_{inc} + \vec{E}_{scat} = 0 \tag{3.5}$$

where, $\vec{E}_{inc}$ is the incident electric field and $\vec{E}_{scat}$ is the scattered electric field. Using Maxwell's equations previously given, the scattered electric field can be expressed by the following equation.

$$\vec{E}_{scat} = -j\omega.\vec{A} - \nabla\Phi \tag{3.6}$$

where $\vec{A}$ is the magnetic potential vector, and Φ is the electric potential scalar. They are obtained after substitution of equation (3.6) into Maxwell's equations also by using the unit vector into the expression gauges de Lorentz [24], given by:

$$\vec{A} = \mu.\int_S J(r').G(r,r').dS' \tag{3.7}$$

$$\Phi = \frac{1}{\varepsilon}\int_S \sigma(r').G(r,r').dS' \tag{3.8}$$

where $G(r,r')$ is Green's function in the space domain, and σ is the charge density. They are expressed by the following relations:

$$G(r,r') = \frac{e^{-jk|r-r'|}}{4\pi.|r-r'|} \tag{3.9}$$

$$\sigma = -\frac{1}{j.\omega}.\nabla.J \tag{3.10}$$

where r is the vector describing the contour of the structure, r' is the observation vector and $k = \omega\sqrt{\varepsilon_0\mu_0}$ is the propagation constant in free space. By substituting (3.7) and (3.8) in (3.6), the following equation is derived:

$$\hat{n}\times\vec{E}_{inc} = \hat{n}\times(j\omega.\vec{A}+\nabla.\Phi) \tag{3.11}$$

$\hat{n}$ is the unit normal vector to the surface of the structure. Finally, the integral equation of the electric field can be given as:

$$\vec{E}_{inc}(r) = \frac{j}{\varepsilon_0 \omega} \int_S [k^2 J(r').G(r,r') + \nabla' J(r').\nabla G(r,r').] dS' \tag{3.12}$$

where ∇ and ∇' are respectively the gradient with respect to source point (x,y,z) and the divergence with respect to the observation point (x',y',z'). Equation (3.12) is defined as the electric field integral equation (EFIE). The temporal dependence $e^{j\omega t}$ is omitted here, but it is considered present in all this manuscript. The solution of the integral equation leads to the density of current in which the directivity and the gain of the antennas are obtained.

3.2.2. Magnetic Field Integral Equations (MFIE)

For perfectly conducting structure, the boundary conditions on the surface of the structure can expressed by the fact that the sum of the incident field and the reflected field lead to the density of current J_s. This is defined by the following equation:

$$\hat{n} \times H = \hat{n} \times (H_{inc} + H_{scat}) = J_S \tag{3.13}$$

The magnetic field integral equation (MFIE) is obtained in the same manner as previously done. The general expression is given as:

$$\hat{n} \times H_{inc}(r) = \frac{J_s(r)}{2} - \hat{n} \times \int_S J_s(r') \times \nabla' G(r,r').dS' \tag{3.14}$$

Here the integral is called principal value, where the case ($r = r'$) is excluded. This type of integral is called Cauchy integral [25].

3.3. MOMENT METHOD

3.3.1. Formulation

The moment method (MoM) is very well known technique for resolving linear equations [4]. For antenna analysis, the moment method converts the integral equation of the field into a matrix equation, by digitalizing this later, the matrix equation is given as [3, 22]:

$$\vec{g} = \vec{L}(\vec{f}) \tag{3.15}$$

where $\vec{g}$ is the excitation source, $\vec{L}$ is a linear integro-differential operator, and $\vec{f}$ is the unknown function, which is written as the sum of N independent functions $\vec{f}_n$, called the basis functions with unknowns amplitudes a_n.

$$\vec{f} = \sum_{n=1}^{N} a_n \vec{f}_n \tag{3.16}$$

By injecting equations (3.16) into (3.15) and owing to the fact that $\vec{L}$ is a linear operator, equation (3.15) can be written as:

$$\vec{g} = \sum_{n=1}^{N} a_n \vec{L}(\vec{f}_n) \tag{3.17}$$

The unknown amplitudes are determined by using asset of integral equation weighted by test functions w_n. These are inserted in equation (3.17) as an inner product defined as follows:

$$\langle \vec{w}_m, \vec{g} \rangle = \sum_{n=1}^{N} a_n \langle \vec{w}_m, \vec{L}(\vec{f}_n) \rangle \tag{3.18}$$

Equation (3.18) is written in a matrix form as given below:

$$[g_m] = [Z_{mn}][a_n] \tag{3.19}$$

Equation (3.19) often written in a similar manner to the Ohm's law, using the matrix so called impedance, defined by the following scalar product:

$$Z_{mn} = \left\langle \vec{w}_m, \vec{L}(\vec{f}_n) \right\rangle \tag{3.20}$$

Equation (3.20) is written in explicit form as:

$$Z_{mn} = \begin{bmatrix} \langle w_1, L(f_1) \rangle & \langle w_1, L(f_2) \rangle & \cdots \\ \langle w_2, L(f_1 \rangle & \langle w_2, L(f_2 \rangle & \cdots \\ \cdots \quad \cdots & \cdots \quad \cdots & \cdots \end{bmatrix} \tag{3.21}$$

Also the unknown amplitudes and the scalar product of the test function and excitation are given as:

$$[a_n] = \begin{bmatrix} a_1 \\ a_2 \\ \vdots \end{bmatrix}; \quad [g_m] = \begin{bmatrix} \langle w_1, g \rangle \\ \langle w_2, g \rangle \\ \vdots \end{bmatrix} \tag{3.22}$$

The impedance matrix is considered to be not singular or reversible, the unknowns are written as:

$$[a_n] = [Z_{mn}]^{-1} . [g_m] \tag{3.23}$$

The exact solution depends on the choice of the basis function and the test function. In this case, these have to be linear and independent.

3.3.2. Basis Functions

By choosing the test functions as Dirac delta $w_m(x) = \delta(x - x_m)$, the moment method is known as point matching method [26, 27]. In the case where the test functions and weighting function are similar $w_m(x) = f_m$, the moment method is known as Galerkin method. In this chapter, the basis and testing function are chosen as Dirac delta, triangular, rectangular and wavelets.

3.4. One-Dimensional Antennas

In this section, a one-dimensional (1-D) patch is under study as shown in Figure 3.2. The moment method is applied on this structure to find the current density and radar cross section (RCS). The patch is considered as very thin and fed by TE or TM wave depending on the polarization case.

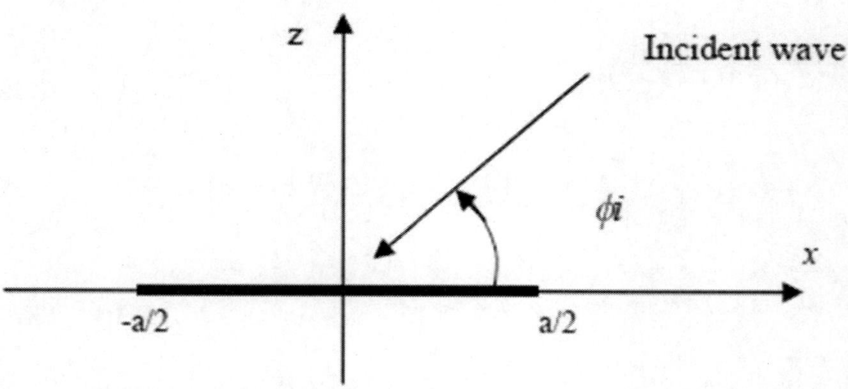

Figure 3.2. One-dimensional patch.

Since the path is 1-D, the integral equations of the electric field (TM case) and magnetic (TE case) respectively equation (3.12) and (3.14) are written in the following form.

$$E_Z^{inc}(x) = \frac{\omega\mu}{4}\int_{-a/2}^{a/2} J(x).H_0^{(2)}(k.|x - x'|).dx \tag{3.24}$$

$$-H_Z^{inc}(x) = \frac{J(x)}{2} + j\frac{k}{4}\int_{-a/2}^{a/2} J(x').H_1^{(2)}(k|x-x'|).dx' \tag{3.25}$$

where $H_0^{(2)}$ and $H_1^{(2)}$ are Hankel's functions zero and first order of the second type respectively, defined as given bellow:

$$H_0^{(2)}(k|r-r'|) = 4j.G(r,r') \tag{3.26}$$

$$H_1^{(2)} = \sqrt{\frac{2}{\pi kr}}\, e^{-j(kr-3\pi/4)} \tag{3.27}$$

The integral equations (3.24) and (3.25) are solved by applying the moment method as illustrated by equation (3.18). Equations given below are respectively for the TM and TE case.

$$\sum_{n=1}^{N} a_n \left\langle \delta(x-x_m), \frac{\omega\mu}{4}\int_{x_n}^{x_{n+1}} \Lambda(x-x_n).H_0^{(2)}(k|x-x'|).dx' \right\rangle =$$

$$\left\langle \delta(x-x_m), E_Z^{inc}(x) \right\rangle \tag{3.28}$$

$$\sum_{n=1}^{N} a_n \left\langle \delta(x-x_m), (\frac{\delta_{m,n}}{2} + j\frac{k}{4}\int_{x_n}^{x_{n+1}} \Lambda(x-x_n).H_1^{(2)}(k|x-x'|).dx' \right\rangle =$$

$$\left\langle \delta(x-x_m), -H_Z^{inc}(x) \right\rangle \tag{3.29}$$

where $\delta_{m,n}$ is the Kronecker delta defined by the following equation:

$$\delta_{m,n} = \begin{cases} 1 & m = n \\ 0 & m \neq n \end{cases}$$

The patch is cut into N segments as, $x_n = (n-1).\Delta_x$, with $\Delta_x = \frac{a}{N}$. The test functions are Dirac delta and the basis functions are chosen as triangular $\Lambda(x-x_n)$, defined as follows:

$$\Lambda(x-x_m)=\begin{cases}\dfrac{x-x_{m-1}}{\Delta_x}, & x_{m-1}<x<x_m\\ -\dfrac{x-x_{m+1}}{\Delta_x}, & x_m<x<x_{m+1}\\ 0, & elsewhere\end{cases}$$
(3.30)

The triangular function is represented graphically by the following Figure 3.3.

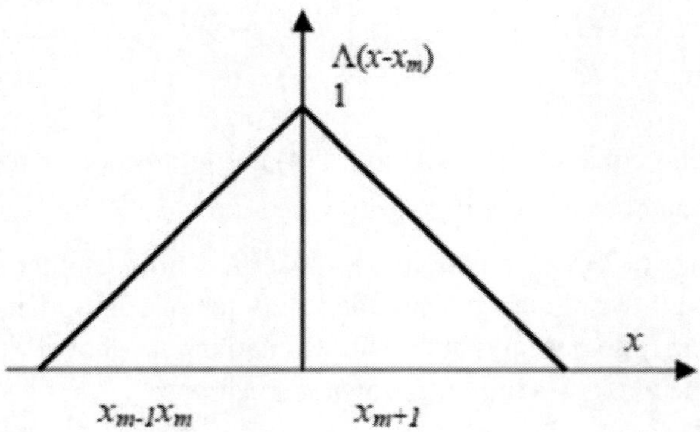

Figure 3.3. Triangular basis function.

Equations (3.28) and (3.29) can be written in a very simple form by applying the scalar product contained inside and taking into account the definition of the basis functions of equation (3.30).

$$\sum_{n=1}^{N} a_n \frac{\omega\mu}{4.\Delta_x}[\int_{x_{n-1}}^{x_n}(x-x_{n-1}).H_0^{(2)}(k|x_m-x|).dx$$

$$+\int_{x_n}^{x_{n+1}}(x_{n+1}-x).H_0^{(2)}(k|x_m-x|).dx]=E_Z^{inc}(x_m)$$
(3.31)

$$\sum_{n=1}^{N} a_n(\delta_{m,n}+j\frac{k}{4.\Delta_x}[\int_{x_{n-1}}^{x_n}(x-x_{n-1}).H_1^{(2)}(k|x_m-x|).dx$$

$$+\int_{x_n}^{x_{n+1}}(x_{n+1}-x).H_1^{(2)}(k|x_m-x|).dx].)=H_Z^{inc}(x_m) \tag{3.32}$$

The matrix elements are written in explicit form and given below by polarization case TM and TE respectively.

$$Z_{m,n}^{TM}=\frac{\omega\mu}{4.\Delta_x}\left[\int_{x_{n-1}}^{x_n}(x-x_{n-1})H_0^{(2)}(k|x_m-x|)dx+\int_{x_n}^{x_{n+1}}(x_{n+1}-x)H_0^{(2)}(k|x_m-x|)dx\right] \tag{3.33}$$

$$Z_{m,n}^{TE}=\delta_{m,n}$$
$$+j\frac{k}{4.\Delta_x}\left[\int_{x_{n-1}}^{x_n}(x-x_{x-1})H_1^{(2)}(|x_m-x|)dx+\int_{x_n}^{x_{n+1}}(x_{n+1}-x)H_1^{(2)}(k|x_m-x|)dx\right] \tag{3.34}$$

The integral equations (3.33) and (3.34) are written in Matlab code taking into account the presence of singularity when $x_n = x_m$. The singularity is due to discontinuity of Green function [28, 29], in all this chapter we have been trying to avoid this singularity using the technique of contournement related to the choice the integration path. The excitation as shown in Figure 3.2, projected in the x-axis leading to following equations:

$$g_m^{TM}=Eo.e^{jk.x_m\cos(\phi i)} \tag{3.35}$$

$$g_m^{TE}=Ho.e^{jk.x_m\cos(\phi i)} \tag{3.36}$$

Eo and *Ho* are corresponding unity quantities respectively to the incident electric and magnetic fields. After resolution of the impedance matrix (3.23), we obtain the density of current shown in Figure 3.4 and Figure 3.5.

It is remarkable in both cases that the current is divergent the boundary and the current is oscillating more and more for higher wavelength. The obtained results coincide exactly with the results obtained by Hatamzadeh [25].

It is worth well to present the radar cross section bi-static (RCS) of the 1-D patch. This quantity shows how easily this object is detected, the RCS depends on geometry of the object and angle of excitation. The RCS is defined in the literature by the following equation:

$$\sigma(\phi) = \lim_{r\to\infty} 2\pi r \frac{\left|E_{reflechis}\right|^2}{\left|E_{incident}\right|^2} \tag{3.37}$$

By taking the magnitude of the incident electric field $\left|E_{incident}\right|^2 = 1$, and $H_0^{(2)}(kr) \approx \sqrt{\frac{2j}{\pi.kr}} e^{-jk.r}$, for far field $r \to \infty$, ϕ is the observation angle.

By substituting in equation (3.31) and (3.32) respectively for the TM and TE case, the following relations are obtained [22]:

$$\sigma^{TM}(\phi) = \frac{k\eta^2}{4} \Delta_x \left| \sum_{i=1}^{N} a_n e^{jk.x_m \cos(\phi)} \right|^2 \tag{3.38}$$

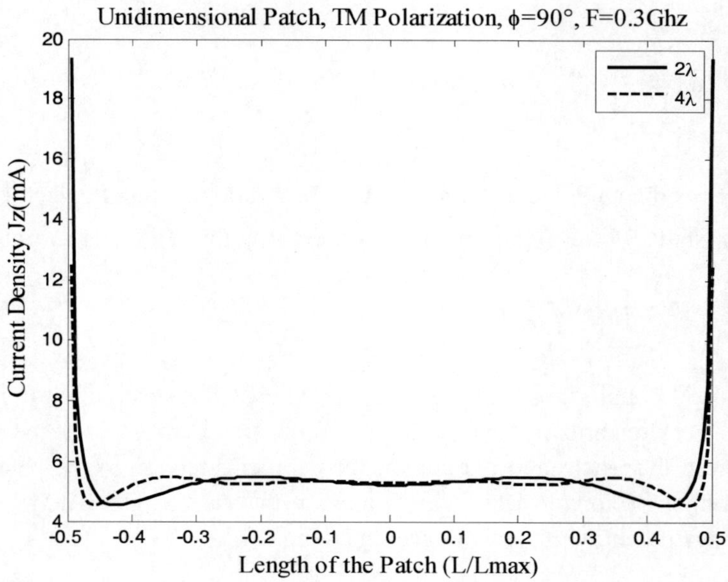

Figure 3.4. Current density of the path, with length 2λ and 4λ, incident angle $\varphi i = 90°$, TM polarization, $F = 0.3$ GHz.

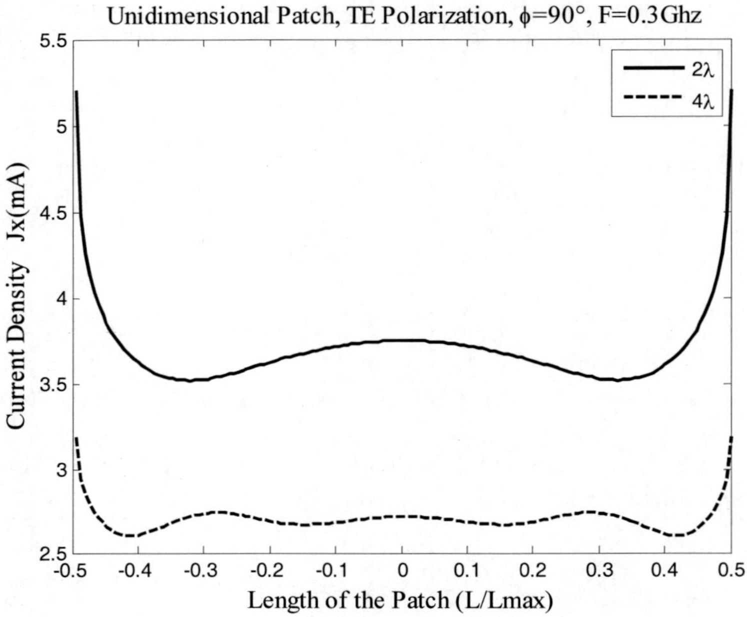

Figure 3.5. Current density of the path, with length 2*λ* and 4*λ*,incident angle *φi*= 90°, TE polarization, *F* = 0.3 GHz.

$$\sigma^{TE}(\phi) = \frac{k\eta^2}{4}\Delta_x \left| \sum_{i=1}^{N} a_n . \cos(\beta) e^{jk.x_m \cos(\phi)} \right|^2 \tag{3.39}$$

where β is the angle between the vector observation *r'* and the vector source *r*. The bi-static RCS is represented in decibel using the following formula:

$$\sigma_{dB} = 10.\log_{10}(\sigma(\phi)) \tag{3.40}$$

Figure 3.6 and Figure 3.7 show the bi-static RCS respectively for TM and TE case. It is remarkable that the RCS is more important for a perpendicular direction to the patch, and is minimal for parallel direction to the patch. It is also minimal for longer patch with respect to the wavelength. The results are in good agreement with the work presented in [25].

3.5. TWO-DIMENSIONAL ANTENNAS

In this section, the 2-D structure will be treated, such as the case of circular cylinder, square cylinder of infinitely long with homogenous, isotropic and perfectly conducting surface, as shown in Figure 3.8. The 2-D integral equations of both cases TM and TE, related to the circular cylinder can be extracted respectively from equations (3.12) and (3.14) according to the literature [8, 22]:

$$Ei_Z^{inc}(r) = \frac{\omega\mu}{4}\int_C J_z(r').H_0^{(2)}(k|\vec{r}-\vec{r}'|).dC' \quad (3.41)$$

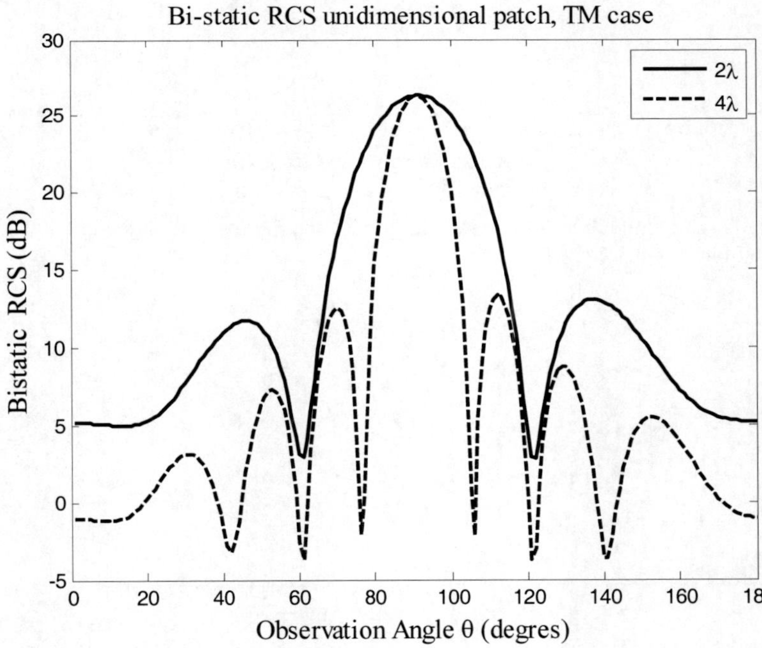

Figure 3.6. Bi-static RCS of the patch, length 2λ and 4λ, incident angle $\varphi i = 90°$, TM polarization, $F = 0.3$ GHz.

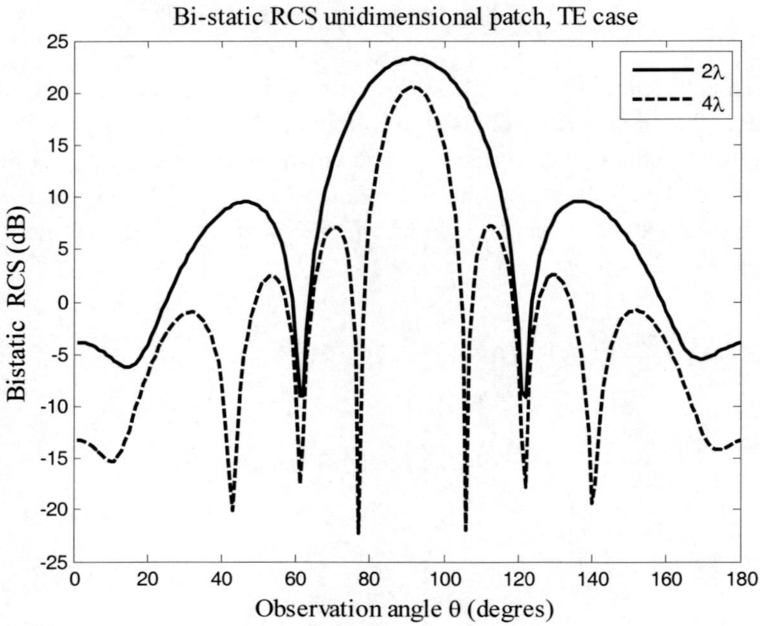

Figure 3.7. Bi-static RCS of the patch, length 2λ and 4λ, incident angle $\varphi i = 90°$, TE polarization, $F = 0.3$ GHz.

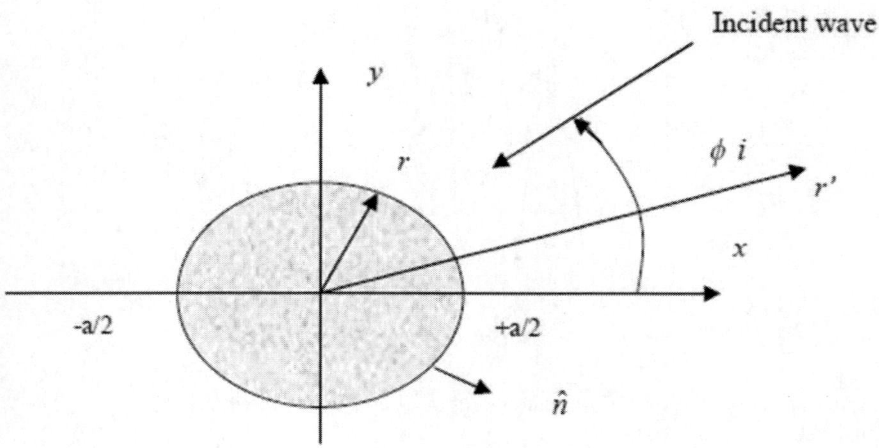

Figure 3.8. Perfectly conducting cylinder.

$$-H_Z^{inc}(r) = \frac{J_c(r)}{2} + j\frac{k}{4}\int_C J_c(r').H_1^{(2)}(k|\vec{r}-\vec{r}'|).\cos(\psi_{r,r'}).dC' \tag{3.42}$$

The index C insinuates that the integral should be done on the contour of the cylinder, also $\psi_{r,r'}$ is the angle between the vector $(\vec{r}-\vec{r}')$ and the normal vector to the surface $\hat{n}$. It is noteworthy that the current J_c, is a tangential current to the contour and not in the z-axis. Equations (3.41) and (3.42) can be written by just following the contour C, with changing variables from Cartesian coordinate to polar coordinate. The integral equations related to circular cylinder are given as:

$$E_Z^{inc}(\vec{r}(\theta)) = \frac{\omega\mu}{4}\int_{\theta=0}^{\theta=2\pi} J_Z(\theta').H_0^{(2)}(k|\vec{r}(\theta)-\vec{r}'(\theta')|).|\vec{r}'(\theta')|.d\theta' \tag{3.43}$$

$$-H_Z^{inc}(\theta) = \frac{J_\theta(\theta)}{2} + j\frac{k}{4}\int_{\theta=0}^{\theta=2\pi} J_\theta(\theta').H_1^{(2)}(k|\vec{r}-\vec{r}'|)\cos(\psi_{r,r'}).|\vec{r}'(\theta')|.d\theta' \tag{3.44}$$

The structure of Figure 3.8 presents an axial symmetry, by the fact that the 2-D problem is brought to a 1-D problem, where the only variable is the angle theta (θ). The integral equations (3.43) and (3.44) are digitalized by the moment method using equation (3.18). The following equations are respectively for the cases TM and TE.

$$\sum_{n=1}^{N} a_n \left\langle \delta(\theta-\theta_m), \frac{\omega\mu}{4}\int_{\theta_n}^{\theta_{n+1}} \Pi(\theta-\theta_n).H_0^{(2)}(k|R|).|\vec{r}'(\theta')|.d\theta' \right\rangle$$

$$= \left\langle \delta(\theta-\theta_m), E_Z^{inc}(\vec{r}) \right\rangle \tag{3.45}$$

$$\sum_{n=1}^{N} a_n \left\langle \delta(\theta-\theta_m), (\frac{\delta_{m,n}}{2} + j\frac{k}{4}\int_{\theta_n}^{\theta_{n+1}} \Pi(\theta-\theta_n).\cos(\psi_{r,r'}).H_1^{(2)}(k|R|).|\vec{r}'(\theta')|.d\theta' \right\rangle$$

$$= \left\langle \delta(\theta-\theta_m), -H_Z^{inc}(\vec{r}) \right\rangle \tag{3.46}$$

with,

$$R = \vec{r}(\theta) - \vec{r}'(\theta')\,,\quad \theta_n = (n-1).\Delta\theta\,,\quad \Delta\theta = \frac{2\pi}{N}$$

and

$$\theta_m = [(m-1) + \frac{1}{2}]\Delta\theta$$

The test functions are Dirac delta and the basis functions are rectangular type $\Pi(\theta - \theta_m)$, defined as follow:

$$\Pi(\theta-\theta_m) = \begin{cases} 1, & \theta_m - \frac{\Delta\theta}{2} < \theta < \theta_m + \frac{\Delta\theta}{2} \\ 0, & elsewhere. \end{cases}$$

(3.47)

The graphical presentation of the rectangular basis functions are given in Figure 3.9.

Equations (3.45) and (3.46) can be written more explicitly by performing their scalar product, giving the following equations:

$$Ei(\vec{r}(\theta_m)) = \frac{\omega\mu}{4}\sum_{i=1}^{N} a_n \int_{\theta_n}^{\theta_{n+1}} H_0^{(2)}(k|\vec{r}(\theta_m) - \vec{r}'(\theta')|).|\vec{r}'(\theta')|.d\theta' \tag{3.48}$$

$$-H_Z^{inc}(\theta_m) = \sum_{n=1}^{N} a_n \left[\frac{\delta_{m.n}}{2} + j\frac{k}{4}\int_{\theta_n}^{\theta_{n+1}} \cos(\psi_{r,r'}).H_1^{(2)}(kR).|r'(\theta')|.d\theta'\right] \tag{3.49}$$

The elements of the matrix respectively TM and TE are obtained by using equation (3.18) and their expressions are given below:

$$Z_{m,n}^{TM} = \frac{\omega\mu}{4}\int_{\theta_n}^{\theta_{n+1}} H_0^{(2)}(k|\vec{r}(\theta_m) - \vec{r}'(\theta')|).|\vec{r}'(\theta')|.d\theta' \tag{3.50}$$

$$Z_{m,n}^{TE} = \left[\frac{\delta_{m.n}}{2} + j\frac{k}{4}\int_{\theta_n}^{\theta_{n+1}} \cos(\psi_{r,r'}).H_1^{(2)}(kR).|r'(\theta')|.d\theta'\right] \tag{3.51}$$

In the same manner as done in the previous section, the excitation is projected in the (x,y) plane giving the following expressions:

$$g_m^{TM} = Eo.e^{jk(x_m.\cos(\phi i)+y_m.\sin(\phi i))} \tag{3.52}$$

$$g_m^{TE} = Ho.e^{jk(x_m.\cos(\phi i)+y_m.\sin(\phi i))} \tag{3.53}$$

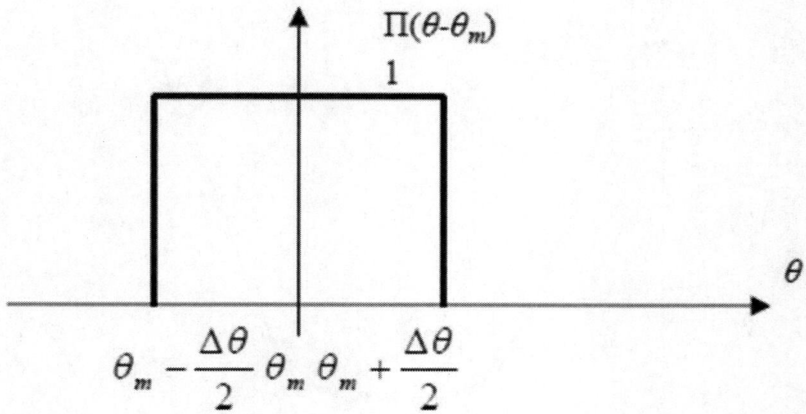

Figure 3.9. Rectangular basis functions.

Since the cylinder is circular, we have the following equations:

$x_m = |r(\theta_m)|.\cos(\theta_m)$ and $y_m = |r(\theta_m)|.\sin(\theta_m)$.

With $|r(\theta_m)| = \frac{a}{2}$, a is the diameter of the cylinder given in Figure 3.8, chosen equal to 1.5λ for the TM case and equal to 2λ for the TE case to compare the present results to the work curried out by Esteban [30]. The same circular cylinder was treated by the moment method in spectral domain. The current density is obtained after resolution of the matrix equation. The results are presented in Figure 3.10 for TM case and in Figure 3.11 for TE case. It can be noticed that the current density for both cases is greater in the direction of the feeding and minimum value obtained for the opposite side.

Following the exploration of the same results, the bi-static RCS for an infinitely long cylinder is presented in Figure 3.12 for the TM case, and in Figure 3.13 for TE case.

The radiation pattern given in Figure 3.14 shows the total pattern in both cases. However, it is clear that the maximum gain for TM case is facing to the feeding point, whereas the maximum gain for the TE case is in the opposite side of the feeding.

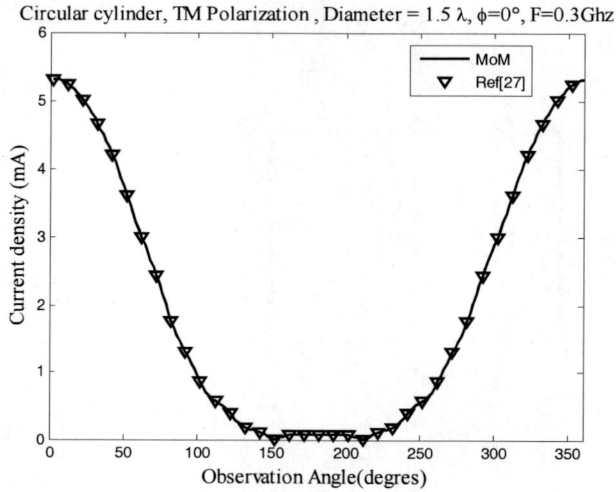

Figure 3.10. Current density of circular cylinder, cylinder diameter = 1.5λ, incident angle φi = 0°, TM polarization, F = 0.3 GHz.

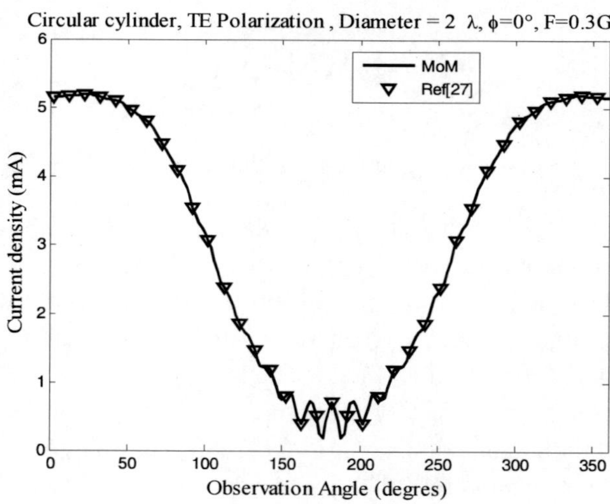

Figure 3.11. Current density of the circular cylinder, cylinder diameter = 2λ, incident angle φi = 0°, TE polarization, F = 0.3 GHz.

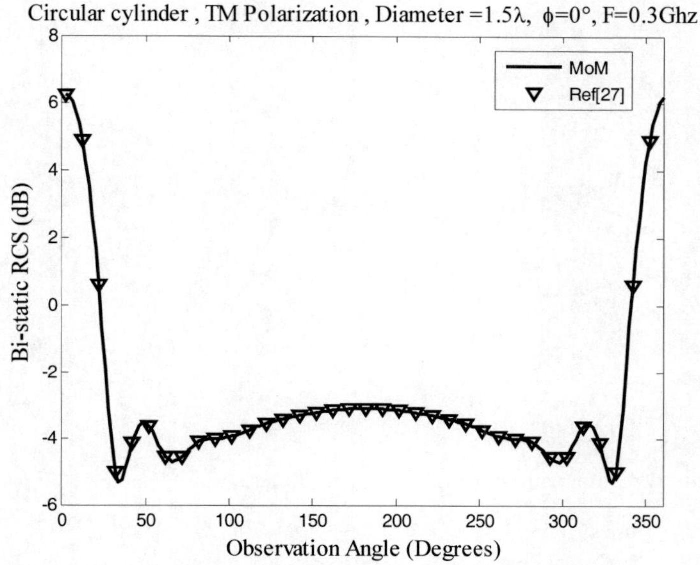

Figure 3.12. Bi-static RCS of the circular cylinder, cylinder diameter=1.5λ, incident angle $\varphi i = 0°$, TM polarization, $F = 0.3$ GHz.

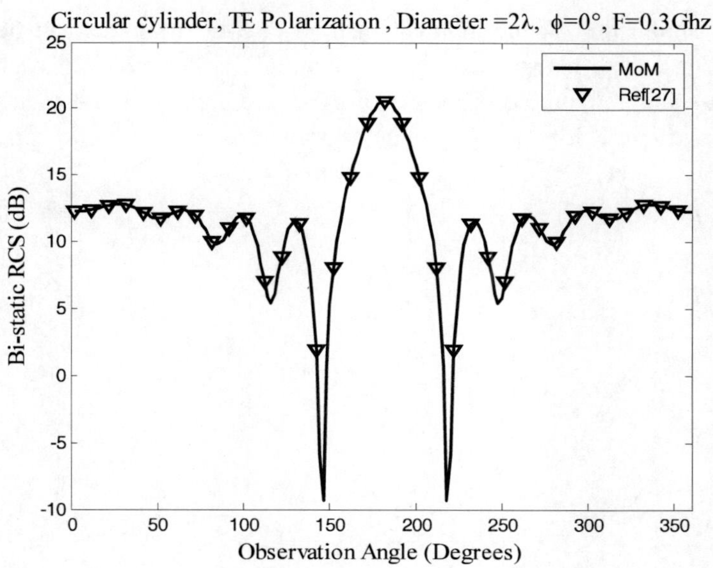

Figure 3.13. Bi-static RCS of the circular cylinder, cylinder diameter=2λ, incident angle $\varphi i = 0°$, TE polarization, $F = 0.3$ GHz.

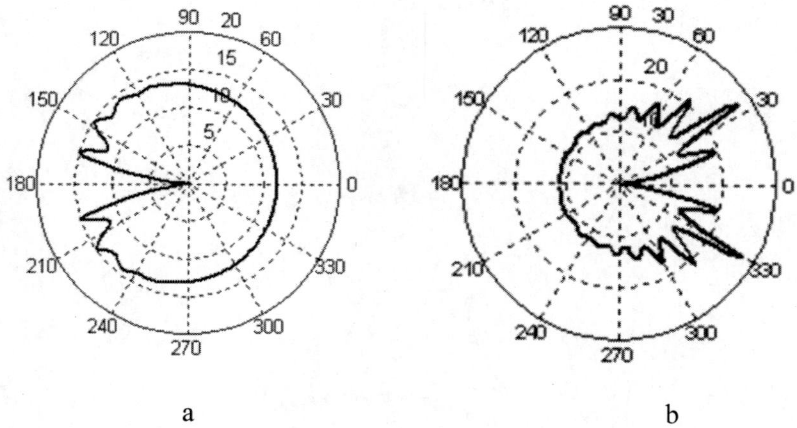

Figure 3.14. Radiation pattern for the circular cylinder, (a) TM case (b) TE case.

3.6. Three-Dimensional Antennas

3.6.1. Bodies of Revolution Structures

The application of the moment method in 3-D structures can be a very difficult task. The matrix of integral equations become very heavy and the unknowns are too much, especially for very complicated geometries, also when the structure does not present any symmetry.

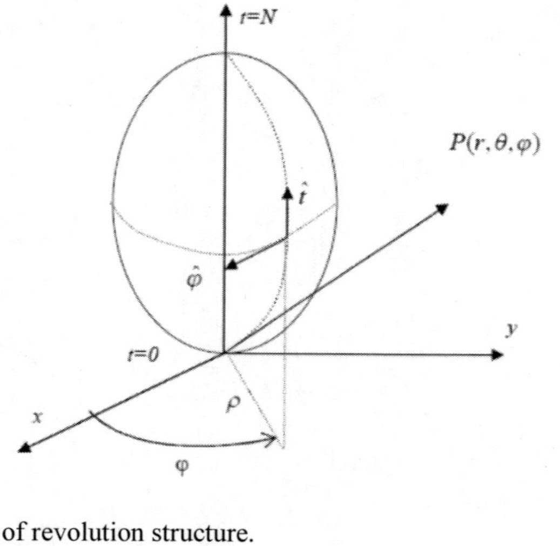

Figure 3.15. Body of revolution structure.

For simplicity, here in this section the moment method is applied for 3-D structures having symmetry of revolution known as bodies of revolution (BOR), as shown in Figure 3.15. Conical horn, parabolic reflector, circular waveguide and many others are examples of BOR structures. The solution of the integral equation of such structures can be brought to 2-D cases even though the structure is 3-D [27, 31]. The integral equation for 3-D structure and body of revolution, solved by the moment method is given below as the inner product [29]:

$$\left\langle E_{inc}, \vec{W} \right\rangle = \frac{j}{\varepsilon_0 \omega} \int_S \int_{S'} [k^2 (\vec{W}.\vec{J}).G(r,r') - (\nabla' \vec{J})(\nabla.\vec{W}) G(r,r')] dS'.dS \tag{3.54}$$

where $\vec{W}$ and $\vec{J}$ are respectively testing and basis functions as vector form, written in the tangent system of coordinate. The total current on the surface of the structure is written as [29]:

$$\vec{J}(t,\varphi) = \sum_{n=-\infty}^{+\infty} \left[J_t(t,\varphi) + J_\varphi(t,\varphi) \right] e^{jn\varphi} \tag{3.55}$$

where, J_t and J_φ are the two current density components, associated to the tangential unity vectors. The term $e^{jn\varphi}$ corresponds to the Fourier series expansion with n is the mode number.

3.6.2. Boundary Element Method (BEM)

Bodies of revolution antennas are generally treated by the boundary element method (BEM), where the magnetic field integral equation (MFIE) and the electric field integral equation (EFIE) are solved by the moment method using the surface equivalence principle [21, 32]. The principle enables the placement of electric current and magnetic current inside a closed volume, by just setting a discontinuity of the fields inside and outside the volume, as shown in Figure 3.16. The electric and magnetic currents are given as [33]:

$$J_s = \hat{n} \times (H_1^s - H_2^s) \tag{3.56}$$

$$M_s = -\hat{n} \times (E_1^s - E_2^s) \tag{3.57}$$

The fields inside the volume can be set to arbitrary values, since the equivalent currents are realized for any discontinuity. The internal fields are set to zero for simplicity of the problem, $H_2^s = 0$ and $E_2^s = 0$.

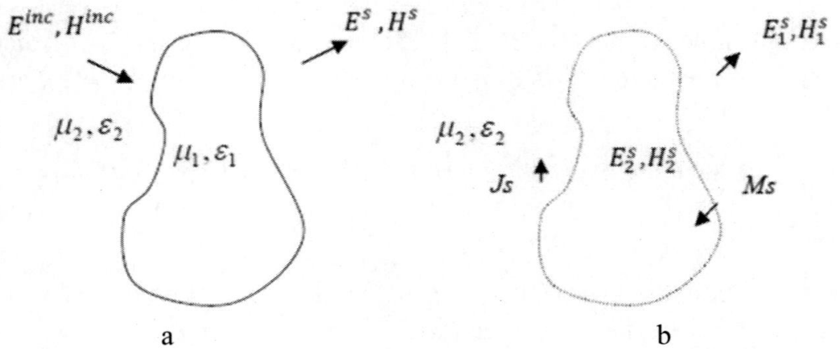

Figure 3.16. Surface equivalence principle, (a) Original problem (b) Replaced by surface of electric current and magnetic currents.

3.6.3. Parabolic Reflector Antenna

The reflector antenna of Figure 3.17, fed by a dipole antenna, is considered having symmetry of revolution, and thus it is a body of revolution, upon which we can apply the boundary element method and the surface equivalent principle.

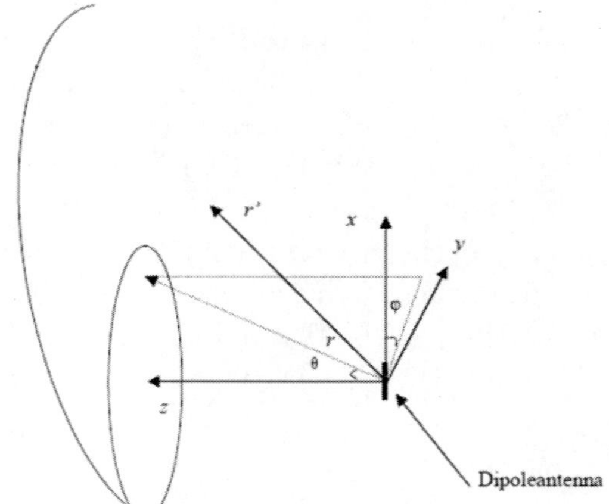

Figure 3.17. Parabolic reflector antenna.

To avoid heavy processing here we just consider the case of magnetic excitation. Equation (3.14) can be written as:

$$K(J(r)) = \frac{1}{2}J(r) - \hat{n}\times\int_S J(r')\times\nabla' G(r,r').ds' = \hat{n}\times H^i(r) \tag{3.58}$$

where, K is an integro-differential operator. Equation (3.58) is known as principle value integral where the current density is expanded as symmetry of revolution structure. The current components of equation (3.55) can written as wavelets expansion as in [33]. The idea behind choosing the wavelets as basis functions is mainly to reduce the computing time and to reduce the size of the impedance matrix.

$$J_t(t,\varphi) = \sum_{n=0}^{2^0-1} a_n^t.\phi_{j,n}^t(t,\varphi) + \sum_{m=0}^{j}\sum_{n=0}^{2^{m-1}} c_{m,n}^t \psi_{m,n}^t(t,\varphi) \tag{3.59}$$

$$J_\varphi(t,\varphi) = \sum_{n=0}^{2^0-1} a_n^\varphi.\phi_{j,n}^\varphi(t,\varphi) + \sum_{m=0}^{j}\sum_{n=0}^{2^{m-1}} c_{m,n}^\varphi \psi_{m,n}^\varphi(t,\varphi) \tag{3.60}$$

where $(\psi_{m,n}^t, \psi_{m,n}^\varphi)$ and $(\phi_{j,n}^t, \phi_{j,n}^\varphi)$ are respectively, the mother wavelet and the scaling wavelet. As matter of simplicity, the current expansion can be written in a simpler manner as:

$$J_t(t,\varphi) = \sum_{m=0}^{j}\sum_{n=0}^{2^m-1} c_{m,n}^t \psi_{m,n}^t(t,\varphi) \tag{3.61}$$

$$J_\varphi(t,\varphi) = \sum_{m=0}^{j}\sum_{n=0}^{2^m-1} c_{m,n}^\varphi \psi_{m,n}^\varphi(t,\varphi) \tag{3.62}$$

As application of the Galerkin technique, the wavelets are used as basis and testing functions. The following integral expresses the digitalized moment method as an inner product:

$$\left\langle \vec{W}, K(J(r)) \right\rangle = \left\langle \vec{W}, \hat{n} \times H^i(r) \right\rangle \tag{3.63}$$

where $\vec{W}$ is the basis and testing functions. Equation (3.63) is written in tangential coordinate $(\hat{u}_t, \hat{u}_\varphi, \hat{n})$ [34-36], giving two equations respectively in the $(\vec{t})$ axis and $(\vec{\varphi})$ axis:

$$\left[\frac{1}{2}\int_s \vec{W}^t.\vec{J}(r).ds + \int_s\int_s [(\hat{n}\times\vec{W}^t)\times\vec{J}(r')].\nabla G(r,r')ds'.ds\right]$$

$$= \int_s \vec{W}^t.(\hat{n}\times H^i)ds \tag{3.64}$$

$$\left[\frac{1}{2}\int_s \vec{W}^\varphi.\vec{J}(r).ds + \int_s\int_s [(\hat{n}\times\vec{W}^\varphi)\times\vec{J}(r')].\nabla G(r,r')ds'.ds\right]$$

$$= \int_s \vec{W}^\varphi.(\hat{n}\times H^i)ds \tag{3.65}$$

Knowing that,

$\hat{n}\times\vec{W}^\varphi = -W^\varphi\hat{t}$, and $\hat{n}\times\vec{W}^t = W^t\hat{\varphi}$

Also,

$$(\vec{W}^t\times\hat{n})\times\vec{J} = (W^t\hat{\varphi})\times(J_t\hat{t}' + J_\varphi\hat{\varphi}') \tag{3.66}$$

$$(\vec{W}^\varphi\times\hat{n})\times\vec{J} = -(W^\varphi\hat{t})\times(J_t\hat{t}' + J_\varphi\hat{\varphi}') \tag{3.67}$$

The excitation is written as:

$$n\times H^i = \begin{vmatrix} n & t & \varphi \\ 1 & 0 & 0 \\ 0 & H_t^i & H_\varphi^i \end{vmatrix} = -H_\varphi^i\hat{t} + H_t^i\hat{\varphi} \tag{3.68}$$

The inner product in equation (3.64) and (3.65) can be expressed as:

$$\vec{W}^t.\vec{J} = W^t\hat{t}.(J^t\hat{t}'+J^\varphi\hat{\varphi}') = W^t J^t \tag{3.69}$$

$$\vec{W}^\varphi.\vec{J} = W^\varphi\hat{t}.(J^t\hat{t}'+J^\varphi\hat{\varphi}') = W^\varphi J^\varphi \tag{3.70}$$

Concerning the gradient of Green's function, this can be written as:

$$\nabla G(r,r') = \hat{R}.(jk+\frac{1}{R}).G(r,r') \tag{3.71}$$

where

$$\hat{R} = \vec{R}/|R|,$$

$$R = |r-r'| = \sqrt{(\rho-\rho')^2+(z-z')^2+4\rho\rho'\sin^2(\frac{\varphi-\varphi'}{2})}$$

By replacing equations (3.69) and (3.70) into equations (3.64) and (3.65), and adding the summation in (3.61) and (3.62):

$$\sum_p\left[\int_0^T \frac{1}{2}W_q^t J_p^t.\rho.dt - \int_0^T\int_0^T (W_q^t J_p^t \hat{\varphi}\times\hat{t}'+W_q^t J_p^\varphi\hat{\varphi}\times\hat{\varphi}').I_{G1}.\rho'\,\rho.dt'.dt\right]$$

$$= \int_0^T W_q^t H_\varphi^i \hat{t}.\hat{t}.I_{G2}\rho.dt \tag{3.72}$$

$$\sum_p\left[\int_0^T \frac{1}{2}W_q^\varphi J_p^\varphi.\rho.dt + \int_0^T\int_0^T (W_q^\varphi J_p^t \hat{t}\times\hat{t}'+W_q^\varphi J_p^\varphi\hat{t}\times\hat{\varphi}').I_{G1}.\rho'\,\rho.dt'.dt\right]$$

$$= -\int_0^T W_q^\varphi H_t^i \hat{\varphi}.\hat{\varphi}.I_{G2}\rho.dt \tag{3.73}$$

$$I_{G1} = \int_0^{2\pi} \nabla G(r,r').e^{j\nu.\varphi'} d\varphi' \tag{3.74}$$

$$I_{G2} = \frac{1}{2\pi} \int_0^{2\pi} e^{-j\nu\varphi} d\varphi \tag{3.75}$$

where the index ν is the mode number and *T* is the maximum arc length of the antenna (for a sphere $T = \pi.r$, *r* is the radius). The indices (*p*,*q*) correspond respectively to the combinations (m_p, n_p) and (m_q, n_q). Equations (3.72) and (3.73) can be written as a matrix equation as [37]:

$$\begin{bmatrix} Z^{tt} & Z^{t\varphi} \\ Z^{\varphi.t} & Z^{\varphi\varphi} \end{bmatrix} \cdot \begin{bmatrix} c^t_{m,n} \\ c^\varphi_{m,n} \end{bmatrix} = \begin{bmatrix} H^1 \\ H^2 \end{bmatrix} \tag{3.76}$$

Each element of the matrix (3.76) is given based on equations (3.72) and (3.73):

$$Z^{tt}_{pq} = \int_t \frac{1}{2} . W^t_q J^t_p \rho . dt - \int_t \int_{t'} W^t_q J^t_p . \hat{\varphi} \times \hat{t}' . I_{G1} . \rho\rho' dt' dt \tag{3.77}$$

Then equation (3.77) is written as an inner product, using wavelets of different scale:

$$Z^{tt}_{pq} = \left\langle \psi_p, \left\langle \psi_q, \frac{1}{2} - T(t,t).\Omega(t,\xi) \right\rangle \right\rangle \tag{3.78}$$

where *T*(*t*,*t*) is the second term under the double integral, $\Omega(t,\varphi,\xi)$ is an operator of changing variables from (t,t') to (ξ,ξ'), ξ is a variable related the wavelet belonging to domain [0,1], finally ψ_p and ψ_q are wavelets of different rank. Equation (3.78) can be written in more explicit form as introduced by Wang [38] and Tretiakov [39].

$$Z_{pq}^{tt} = \int_0^1 \psi_q [\frac{1}{2}\psi_p - \int_0^1 \psi_p T(t,t)\ \Omega(t,\xi)\ d\xi.D(\xi)]D(\xi')d\xi' \tag{3.79}$$

With $D(\xi) = |dt/d\xi|$, in the same manner the other elements of the matrix can be expressed as:

$$Z_{pq}^{t\varphi} = -\int_t \int_{t'} W_q^t J_p^\varphi .\hat{\varphi} \times \hat{\varphi}'.I_{G1}.\rho\rho' dt' dt \tag{3.80}$$

$$Z_{pq}^{t\varphi} = -\langle \psi_q, \langle \psi_p, T(t,\varphi).\Omega(t,\xi) \rangle \rangle \tag{3.81}$$

and,

$$Z_{pq}^{\varphi.t} = \int_t \int_{t'} W_q^\varphi J_p^t .\hat{t} \times \hat{t}'.I_{G1}.\rho\rho' dt' dt \tag{3.82}$$

$$Z_{pq}^{\varphi.t} = \langle \psi_q, \langle \psi_p, T(\varphi,t).\Omega(t,\xi) \rangle \rangle \tag{3.83}$$

Also,

$$Z_{pq}^{\varphi\varphi} = \int_t \frac{1}{2}.W_q^\varphi J_p^\varphi \rho.dt + \int_t \int_{t'} W_q^\varphi J_p^\varphi .\hat{t} \times \hat{\varphi}'.I_{G1}.\rho\rho' dt' dt \tag{3.84}$$

$$Z_{pq}^{\varphi\varphi} = \left\langle \psi_p, \left\langle \psi_q, \frac{1}{2} + T(\varphi,\varphi).\Omega(t,\xi) \right\rangle \right\rangle \tag{3.85}$$

We express the excitation using the wavelets as described previously:

$$H_q^1 = \langle \psi_q, H^t I_{G2}.\Omega(t,\xi) \rangle \tag{3.86}$$

$$H_q^2 = -\langle \psi_q, H^\varphi I_{G2}.\Omega(t,\xi) \rangle \tag{3.87}$$

Once the matrix equation (3.76) is solved the unknowns $[c^{t}_{m,n}, c^{\varphi}_{m,n}]$ which are the components of the current density are obtained. To find the co-polarization and cross-polarization, a transformation from tangential coordinate to polar coordinate have to be done as in [40]. The wavelet employed is constructed from Haar orthogonal wavelet with vanishing moment N=7. The lowest resolution level is chosen as $2^j = 2^7 = 128$. Since 128 wavelets are involved, a system with 128 × 128 matrix elements is generated.The radiation pattern given in Figure 3.18 and Figure 3.19 are obtained for given ratio focal distance over length of diameter, with F/D = 0.5, the diameter = 10λ, the incident angle φ=45° and the frequency F = 30 GHz. The figures show a gain of almost 10 dB is obtained, and the co-polarization is maximum at the front side with its very narrow main lobe (20° width).

In reverse, the cross polarization is null at the front of the antenna and maximum at its sides. The radiation pattern given in Figure 3.20 and Figure 3.21 show the rectangular and polar radiation pattern respectively. with F/D = 0.5, the incident angle of the excitation is 45°, and the diameter = 5λ and the frequency F = 30 GHz. The figures show a gain of almost 15 dB is obtained with the main lobe of the co-polarization is larger than last one (40° width). The obtained results were compared to previously obtained results in the literature [16, 41] and they were in good agreement.

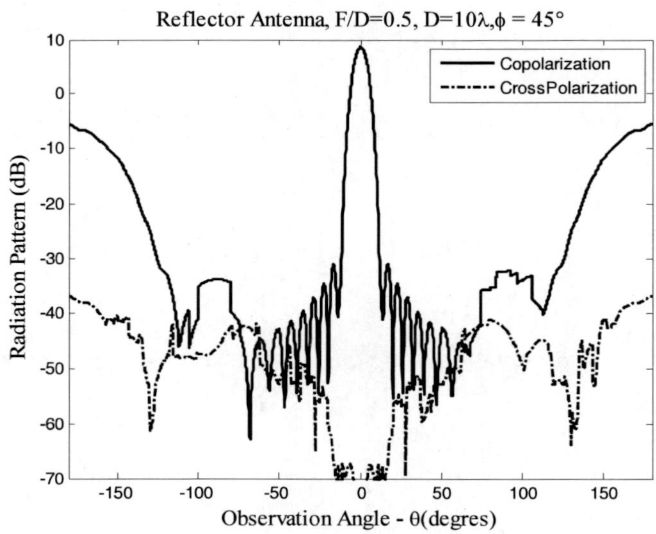

Figure 3.18. Rectangular radiation pattern of the reflector antenna, D=10λ at frequency F=30 GHz.

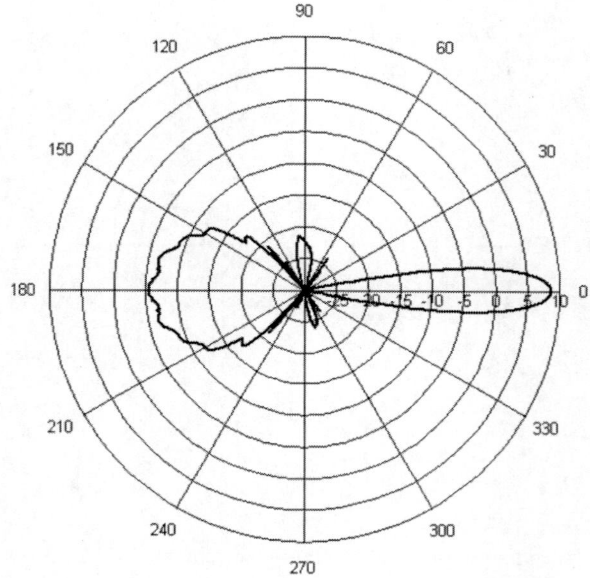

Figure 3.19. Polar radiation pattern of the reflector antenna, D=10λ at frequency F=30 GHz, co-polarization.

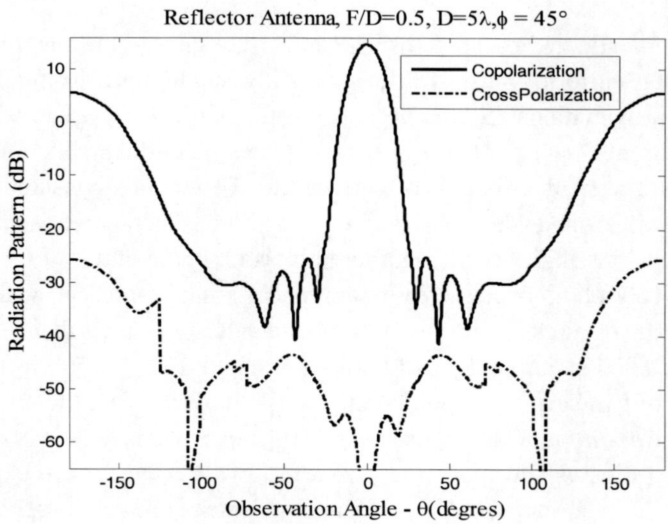

Figure 3.20. Rectangular radiation pattern of the reflector antenna, D=5λ at frequency F=30 GHz.

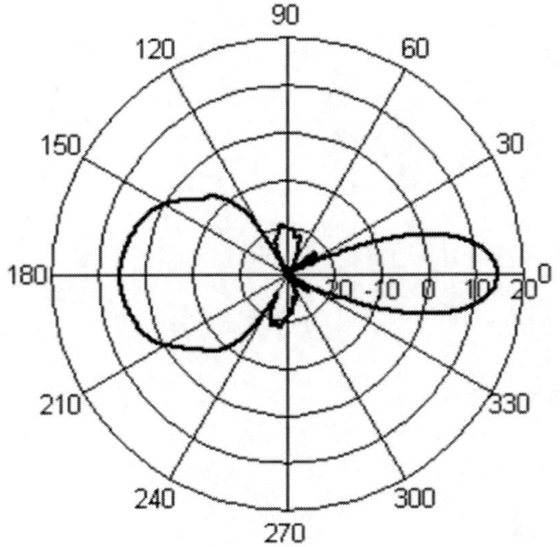

Figure 3.21. Polar radiation pattern of the reflector antenna, D=5λ at frequency F=30 GHz, co-polarization.

3.7. Normal-Mode Helical Antenna

When parallel wires are close together the surface current distribution becomes non-uniform. This effect has been investigated previously, subject to certain approximations. Smith [42, 43] and Olaofe [44] assumed that the average current flowing in a set of parallel wires was equal, which means that the cross-sectional distribution of surface current remains constant along the wires. Tulyathan and Newman [45] used a more general treatment but still neglected the possibility of a circumferential component in the surface current: it is intuitively obvious that such a component must be present when there is significant displacement current flow in the inter-wire capacitance. A more general detailed solution by Abd-Alhameed and Excell [46] included the modelling of two surface current components at any point on the wire surface, subject to certain geometry constraints. And more recently [47-49] in which two parallel dipoles and loop antennas were investigated for the existence of the non-uniform surface currents, in which the antenna power losses were fully covered in [49]. In addition, most of the methods used for analysis of wire antennas of arbitrary shape (including the possibility of closely parallel wires) assume a uniform surface current distribution across the cross section, (e.g., Djordjevic et al. [50], Burke and Poggio [51] and Richmond [52]).

Hence, surface resistive losses and reactive effects that may be augmented by the non-uniform surface current will not be correctly predicted.

This problem is particularly significant for resonant coiled electrically-small antennas, such as the normal-mode helical antenna (NMHA), in which the surface current distribution has a critical effect on the efficiency and Q-factor. A moment-method (MoM) formulation uses two orthogonal basis functions on the surface of the wire including its ends was thus developed to investigate this problem in details. The work detailed a more generalised theory and results of the work done by the present authors [47]. It should be noted that there are more advanced commercial codes now available which, in particular, implement patch modelling more effectively (e.g., FEKO [53], CST [54], HFSS [55] and IE3 [56]). In general MoM [57-63] has the ability to solve complex antenna pattern in frequency and time domain.

Basically, the original motivation for this work was to assess the degree of benefit that would be obtained if an antenna of this type were to be realised in high-temperature superconductor. Electrically-small antennas have a low radiation resistance that is easily swamped by ohmic loss resistance, resulting in a low efficiency. Superconductors have the potential to remove much of the loss and hence raise the efficiency significantly. There is then the possible disadvantage that the inherent Q-factor of the antenna may become very high: whether this is a real disadvantage depends on the nature of the system into which the antenna is proposed for deployment. To quantify the reduction in loss, and hence improvement in efficiency, which might accrue from the use of superconductor, it is necessary to quantify the surface loss P_s measured in W/m^2:

$$P_s = \frac{J_s^2}{\sigma_s} \tag{3.88}$$

where J_s is the surface current density (A/m) and σ_s is the surface conductivity (Ω). The self-resonant helix had already been identified as a convenient design of electrically-small antenna in which quite interesting results were reported for example, broad band V-helical antenna [64], circular NMHA [65], double pitch NMHA [66] and multiple pitches NMHA [67], however, for realisation in high-temperature superconductor, the superconducting element may be left electrically isolated. The detailed quantification of J_s in this particularly complex case was thus the main original objective of the work. The very detailed modelling procedure that has been developed has much wider uses, particularly in the accurate modelling of normally-conducting NMHAs, which see extensive use in mobile telecommunications.

Complete validation of the predictions of the procedure poses considerable difficulties, since it would require measurement of the surface current distribution on wire. This matter is an important topic for future work, but an adequate degree of validation can be claimed for the results that have been presented in this work from this type of modelling process.

3.8. Surface Current Formulation

Initially, the normal MoM approach is followed, but no attempt is made to approximate the surface current or the scattered field observation points to a single point on the wire cross-section. Instead, both are allowed to be completely general points on the wire surface and the surface current is allowed to have components both parallel to, and transverse to the wire axis. This leads to an equation of the form:

$$\sum_j I_j . L(\mathbf{J}_j) = (\mathbf{E}_i) \tag{3.89}$$

where I_j is a basis function for the surface current $\mathbf{J}_j$, $\mathbf{E}_i$ is the incident electric field strength and L is the integro-differential operator given by:

$$L(\mathbf{J}) = (j\omega\mathbf{A} + \nabla\phi)_{\tan} \tag{3.90}$$

where A and ϕ are the vector and scalar potentials. If a set of testing functions $\mathbf{W}_m$ is defined, equation (3.89) may be rewritten as:

$$\sum_j I_j \langle \mathbf{W}_m, L(\mathbf{J}_j) \rangle = \langle \mathbf{W}_m, \mathbf{E}_i \rangle \qquad for \quad j = 1,2,...N \tag{3.91}$$

where

$$\left\langle \mathbf{W}_m, L(\mathbf{J}_j) \right\rangle = \iint_{s} \iint_{s'} (\mathbf{W}_m . L(\mathbf{J}_j)) ds' ds = Z_{mj} \tag{3.92}$$

$$\left\langle \mathbf{W}_m, \mathbf{E}_i \right\rangle = \iint_{s'} (\mathbf{W}_m . \mathbf{E}_i) ds = V_m \tag{3.93}$$

where ds' and ds are the differential areas on the wire surface for the source and the observation points respectively, m = 1, 2, ... N is the index of the testing function and Z and V are the conventional abbreviations for the interaction matrix and excitation vector terms in the Method of Moments.

3.9. Evaluation of Impedance Matrix Elements

The impedance matrix elements Z_{mj} can be written using the closed surface integral identity [68] as follows:

$$Z_{mj} = j\omega\mu \iint_{s} \iint_{s'} \left(\mathbf{J}_j . \mathbf{W}_m - \frac{1}{k^2} (\nabla . \mathbf{J}_j)(\nabla . \mathbf{W}_m) \right) g(R) ds' ds \tag{3.94}$$

where $g(R)$ is the free-space Green function [69-71] and is given by the expression:

$$g(R) = \frac{e^{-jkR}}{4\pi R} \tag{3.95}$$

R is the distance between the observation and source points on the wire surface. Singular integral occurs when $R = 0$ (i.e., $g(R) \to \infty$). The coordinates for a point on the surface of the helix wire can be given by:

$$x(\phi, \alpha) = x' \cos(\phi) - y' \sin(\phi) \tag{3.96}$$

$$y(\phi,\alpha) = x'\sin(\phi) - y'\cos(\phi) \tag{3.97}$$

$$z(\phi,\alpha) = \frac{P}{2\pi}\phi + a\sin(\alpha)\cos(\delta) \tag{3.98}$$

where:

$$x' = b + a\cos(\alpha) \tag{3.99}$$

$$y' = -a\sin(\alpha)\sin(\delta) \tag{3.100}$$

$$\tan(\delta) = \frac{P}{2\pi b} \tag{3.101}$$

where a is the radius of the helix wire, b is the radius of the helix, P is the pitch distance between the turns, α is the azimuth angle of the circumferential cross-section wire and δ is the pitch angle. Equations (3.99), (3.100) and (3.101) are the exact coordinates of helix geometry. Defining two orthogonal directions on the surface of the helix wire as shown in Figure 3.22, the unit vectors of the curvilinear surface patches in both directions are:

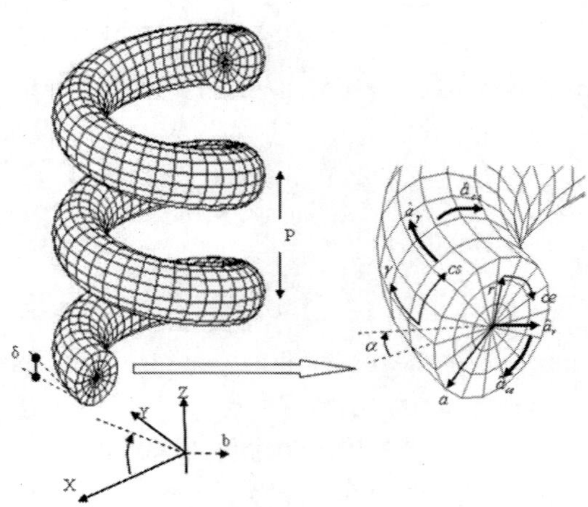

Figure 3.22. Basic geometry of the helical antenna driven by a voltage source at its centre. The directions of the orthogonal basis or test functions are shown on the right and represent the source or observations points on the wire surface and its ends.

$$\hat{a}_\gamma = -\sin(\varphi)\cos(\delta)\hat{a}_x + \cos(\varphi)\cos(\delta)\hat{a}_y + \sin(\delta)\hat{a}_z \tag{3.102}$$

$$\begin{aligned}\hat{a}_{cs} &= (-\sin(\alpha)\cos(\phi) + \cos(\alpha)\sin(\phi)\sin(\delta))\hat{a}_x \\ &\quad - (\sin(\alpha)\sin(\phi) + \cos(\alpha)\cos(\phi)\sin(\delta))\hat{a}_y + \cos(\alpha)\cos(\delta)\hat{a}_z\end{aligned} \tag{3.103}$$

where $\hat{a}_\gamma$ and $\hat{a}_{cs}$ are the unit vectors in the axial and circumferential surfaces of the wire respectively as shown in Figure 3.22. The differential lengths in both directions are:

$$d\gamma = b'' d\phi \tag{3.104}$$

$$dcs = a d\alpha \tag{3.105}$$

where $d\phi$ and $d\alpha$ are the differential lengths in ϕ and α respectively, and

$$b'' = \sqrt{\left(\frac{P}{2\pi}\right)^2 + b'^2} \tag{3.106}$$

$$b'^2 = (b + a\cos(\alpha))^2 \tag{3.107}$$

The coordinates of the start surface of the helix at $\phi = 0$ is given by:

$$x(r,\alpha) = b + r\cos(\alpha) \tag{3.108}$$

$$y(r,\alpha) = -r\sin(\alpha)\sin(\delta) \tag{3.109}$$

$$z(r,\alpha) = r\sin(\alpha)\cos(\delta) \tag{3.110}$$

where $0 \le r \le a$. Hence, the unit direction vectors of the basis function on the end surface can be expressed as:

$$\hat{a}_r = \cos(\alpha)\hat{a}_x - \sin(\alpha)\sin(\delta)\hat{a}_y + \sin(\alpha)\cos(\delta)\hat{a}_z \tag{3.111}$$

$$\hat{a}_{ce} = -\sin(\alpha)\hat{a}_x - \cos(\alpha)\sin(\delta)\hat{a}_y + \cos(\alpha)\cos(\delta)\hat{a}_z \tag{3.112}$$

where $\hat{a}_r$ and $\hat{a}_{ce}$ are the unit vectors in the radial and the circumferential directions on the end surface of the wire respectively as shown in Figure 3.22. The differential area on the end surface can be given by:

$$dA_{end} = rdrd\alpha \tag{3.113}$$

Similarly, the unit direction vectors, coordinators and differential area on the other end of the helix wire can be defined. Now, assume the surface current density over the wire surface can be expressed by two orthogonal current components in $\hat{a}_\gamma$ and $\hat{a}_{cs}$ (similarly at the surface end directions $\hat{a}_r$ and $\hat{a}_{ce}$). Then if the surface current is expanded over the wire surface using triangular basis functions in which the divergence of the current continuity is finite [72], then as example, these functions into the axial direction can be given by:

$$f(\gamma) = \begin{cases} f^+ = \dfrac{\gamma}{\gamma_o} \\ f^- = 1 - \dfrac{\gamma}{\gamma_o} \end{cases} \quad \textit{for} \quad 0 \le \gamma \le \gamma_o \quad \textit{and} \quad \alpha_1 \le \alpha \le \alpha_2 \tag{3.114}$$

$$f'(\gamma) = f'^{\pm}(\gamma) = \pm\frac{\gamma}{\gamma_o} \quad \textit{for} \quad 0 \le \gamma \le \gamma_o \quad \textit{and} \quad \alpha_1 \le \alpha \le \alpha_2 \tag{3.115}$$

where $f'(\gamma)$ is the differentiation of $f(\gamma)$ and γ_o is the axial length of the curvilinear patch presented in Figure 3.22 in the direction of γ for all angle values of α from α_1 to α_2.

A similar basis functions in the directions of $\hat{a}_{cs}$, $\hat{a}_r$ and $\hat{a}_{ce}$ can be given. The testing functions are chosen to be identical to the expansion basis functions (Galerkin's method) yielding a symmetric impedance matrix. Hence, by substituting equations (3.96)-(3.115) into equation (3.94), the impedance matrix elements can be found. As an example, the impedance element for basis and test function in the axial direction can be stated as follows:

$$Z_{\gamma\gamma'} = j\omega\mu \iint_{s} \iint_{s'} \left(\hat{\mathbf{a}}_{\gamma} . \hat{\mathbf{a}}_{\gamma'} f(\gamma) f(\gamma') - \frac{1}{k^2} f(\gamma) f(\gamma') \right) g(R) ds' ds \tag{3.116}$$

where:

$$ds = ab'' d\phi d\alpha \tag{3.117}$$

The other self and mutual impedance elements for all other basis directions can be obtained in a similar way. The magnitude of *R* is written as:

$$R = \sqrt{(x - x')^2 + (y - y')^2 + (z - z')^2} \tag{3.118}$$

Since *x*, *y* and *z* are the function of ϕ and α, then *R* is also a function of ϕ and α. By integrating equation (3.104), the following expression is achieved:

$$\gamma = b'' \phi \tag{3.119}$$

By substituting equations (3.106) and (3.107) into equation (3.119) gives:

$$\gamma = \left(\sqrt{\left(\frac{P}{2\pi} \right)^2 + (b + a\cos(\alpha))^2} \right) \phi \tag{3.120}$$

Let say there is another axial length γ_1 of the curvilinear patch, then:

$$\gamma_1 = \left(\sqrt{\left(\frac{P}{2\pi} \right)^2 + (b + a\cos(\alpha))^2} \right) \phi_1 \tag{3.121}$$

It can be clearly seen that α remains constant along the axial length. In other words:

$$f(\gamma) = \frac{\gamma}{\gamma_1} = \frac{\phi}{\phi_1} \tag{3.122}$$

3.10. SOLUTION FOR SINGULARITY

Equation (3.116) can be rewritten as:

$$Z_{\phi\alpha\phi'\alpha'} = j\omega\mu \int\limits_{\alpha}\int\limits_{\phi}\int\limits_{\alpha'}\int\limits_{\phi'} \frac{\mathbf{H}(\phi,\alpha,\phi',\alpha')e^{-jkR(\phi,\alpha,\phi',\alpha')}}{4\pi R(\phi,\alpha,\phi',\alpha')} d\phi' d\alpha' d\phi d\alpha \tag{3.123}$$

From this equation, singularity occurs when:

$$\phi = \phi' \quad and \quad \alpha = \alpha' \tag{3.124}$$

Moreover, when the surfaces of the patches are very close in proximity to each other still there will be pseudo-singularity. The condition occurs when:

$$\phi = \phi' + \delta\phi' \quad and \quad \alpha = \alpha' + \delta\alpha' \tag{3.125}$$

In this case, equation (3.123) cannot be integrated directly due to the problem of singularity. However, the obstacle can be alleviated by means of cancellation technique [73] with the introduction of additional function (see Appendix A). Consider a point *P* on a patch of helix wire as depicted in Figure 3.23. The point is considered to be anywhere on the patch which is located in proximity above the origin. It should be noted that point *S* is the starting point for integration process. *L* and *W* are the length and the width of the patch respectively. $\hat{a}_\gamma$ and $\hat{a}_{cs}$ are the unit vectors in the axial and circumferential surfaces of the wire respectively. $\hat{u}_p$ and $\hat{u}_t$ are the new unit vectors due to point *P* in the axial and circumferential surfaces of the wire respectively.

It can be shown that from Figure 3.23, the following expression can be obtained fundamentally:

$$\bar{r}_p = p\hat{u}_p + t\hat{u}_t \tag{3.126}$$

$$\bar{r}_{1s} = \bar{r}_1 - \bar{r}_p \tag{3.127}$$

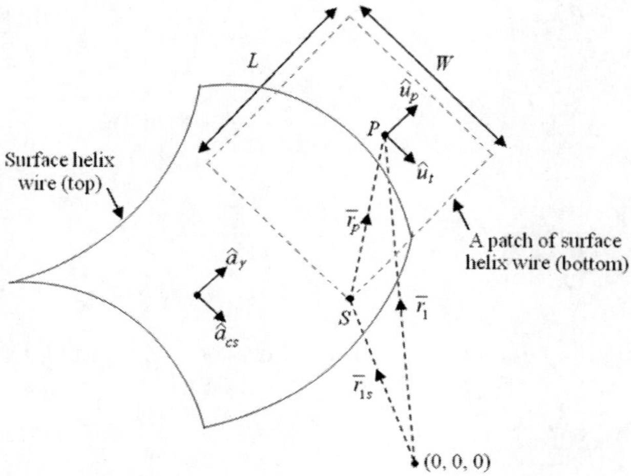

Figure 3.23. Point *P* on a patch of helix wire above the origin.

By substituting equation (3.126) into equation (3.127), the following expression can be achieved:

$$\bar{r}_{1s} = \bar{r}_1 - (p\hat{u}_p + t\hat{u}_t) \tag{3.128}$$

In general, equation (3.128) is used as the starting point for integration. Equation (3.123) can thus be expanded as:

$$\begin{aligned} Z_{\phi\alpha\phi'\alpha'} = j\omega\mu \Bigg[& \int_{\alpha}\int_{\phi}\int_{\alpha'}\int_{\phi'} \frac{\mathbf{H}(\phi,\alpha,\phi',\alpha')e^{-jkR(\phi,\alpha,\phi',\alpha')}}{4\pi R(\phi,\alpha,\phi',\alpha')} d\phi' d\alpha' d\phi d\alpha \\ & - \int_{\alpha}\int_{\phi}\int_{p'}\int_{t'} \frac{\mathbf{J}(\phi,\alpha,p',t')e^{-jkR(\phi,\alpha,p',t')}}{4\pi R(\phi,\alpha,p',t')} dp' dt' d\phi d\alpha \Bigg] \\ & + j\omega\mu \int_{\alpha}\int_{\phi}\int_{p'}\int_{t'} \frac{\mathbf{J}(\phi,\alpha,p',t')e^{-jkR(\phi,\alpha,p',t')}}{4\pi R(\phi,\alpha,p',t')} dp' dt' d\phi d\alpha \end{aligned} \tag{3.129}$$

The above equation can be solved using numerical solution:

$$Z_{\phi\alpha\phi'\alpha'} = j\omega\mu \sum_{i=1}^{n_\phi}\sum_{j=1}^{n_\alpha}\left[\sum_{k=1}^{n_{\phi'}}\sum_{l=1}^{n_{\alpha'}}\left(\frac{\mathbf{H}(\phi_i,\alpha_j,\phi'_k,\alpha'_l)e^{-jkR(\phi_i,\alpha_j,\phi'_k,\alpha'_l)}}{4\pi R(\phi_i,\alpha_j,\phi'_k,\alpha'_l)}\right)w_{\alpha'_l}w_{\phi'_k}\right.$$

$$\left.-\sum_{k=1}^{n_{p'}}\sum_{l=1}^{n_{t'}}\left(\frac{\mathbf{J}(\phi_i,\alpha_j,p'_k,t'_l)e^{-jkR(\phi_i,\alpha_j,p'_k,t'_l)}}{4\pi R(\phi_i,\alpha_j,p'_k,t'_l)}\right)w_{t'_l}w_{p'_k}\right]$$

$$+ j\omega\mu \sum_{i=1}^{n_\phi}\sum_{j=1}^{n_\alpha}\left(\iint_{p't'}\frac{\mathbf{J}(\phi_i,\alpha_j,p',t')e^{-jkR(\phi_i,\alpha_j,p',t')}}{4\pi R(\phi_i,\alpha_j,p',t')}dt'\,dp'\right)w_{\alpha_j}w_{\phi_i} \tag{3.130}$$

The last term of equation (3.130) can be solved using analytical solution:

$$I_{\phi\alpha p't'} = \iint_{p'\,t'}\frac{\mathbf{J}(\phi_i,\alpha_j,p',t')e^{-jkR(\phi_i,\alpha_j,p',t')}}{4\pi R(\phi_i,\alpha_j,p',t')}dt'\,dp' \tag{3.131}$$

For simplicity, the above equation can be reduced to the following format:

$$I = \iint_{p\,t}\frac{a_1 + a_2 t}{\sqrt{v^2 + (p+p_1)^2 + (t+t_1)^2}}dt\,dp \tag{3.132}$$

where a_1, a_2, v, p_1 and t_1 are constants. The analytical solution of the equation is given by the expression:

$$I = a_1 C + a_2 D + constant \tag{3.133}$$

The expression for C and D are given in the Appendix B.

3.11. Simulation and Results

Initially, simple antenna geometries such as dipole and loop antennas were investigated and discussed as special cases of more complex geometries such as the helix. The antenna geometries of the parallel dipoles and loops are shown in Figure 3.24. A similar procedure of placing orthogonal basis distribution over the wire surface and wire ends as for the dipoles are used. A computer program was written to implement the analysis given in the previous section. The surface patch subdivision was automatically generated by the program, subject to the number of the basis functions in both orthogonal directions. The impressed field E_i is modelled by a delta-function voltage source at the centre of the dipole and the helix whereas in loop it was placed at $\phi = 0°$.

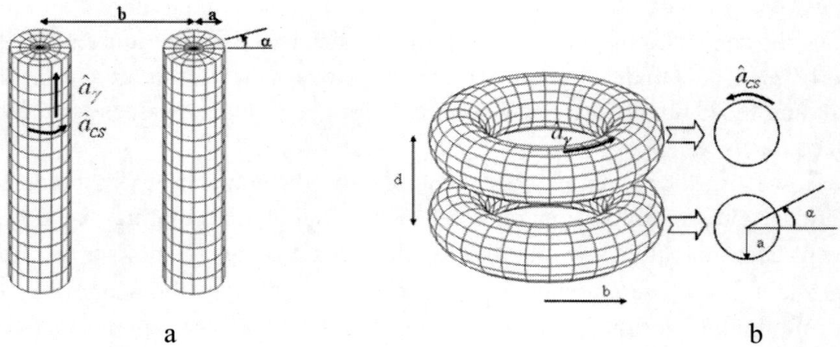

Figure 3.24. The geometry models of two parallel dipoles and loop antennas including the directions of the basis or test functions used; (a) dipoles, (b) loops. A simple axial excitation (in the γ-direction) was considered. Thus the impressed field can be given by:

$$E_i = \frac{1}{2\pi a}\delta(\gamma_c)\hat{a}_\gamma \tag{3.134}$$

where a is the radius of the wire antenna and γ_c is the half axial length of the dipole or the helix. The antenna wire for all geometries was assumed to be perfectly conducting and surrounded by free space. Several examples were used to investigate the surface current distribution of dipoles, loops and NMHA as predicted by the formulation, as follows.

The response of the input impedance of two parallel dipoles of 50 cm length and 5 mm wire radius separated by 15 mm is presented in Figure 3.25. In this example, both dipoles were centrally fed as presented in equation (3.47).

The axial and circumferential lengths were subdivided by 16 curvilinear patches of equal lengths in both directions. The attachment basis modes between the wire ends and the wire surface were placed subject to the corresponding number of axial and circumferential patches used. It is found that the results agree well with those calculated using NEC [51] (extended kernel solution), although it is to be expected that the latter will be less reliable, as less detail in the behaviour of the wire is taken into account. It is worth noting that Tulyathan and Newman [45], observed this behaviour on half wavelength dipoles when they ignored the circumferential surface current component.

For the same antenna geometry, the input impedance at 300 MHz (equivalent to half wavelength dipoles and 0.005 wavelength wire radius) versus the separated distance between the dipoles is shown in Figure 3.26. It is clearly seen that there is good agreement between the results of the present work and those obtained from NEC except for the reactance values for closely spaced distances.

However, the methods were completely different in their numerical solutions. The normalised magnitudes of the axial and circumferential surface current components of two parallel dipoles separated by 15 mm for the same wire radius as above are shown in Figure 3.27, versus α (the azimuth of the circumferential cross-section wire) at 6.25 cm and 4.6845 cm respectively, considered from bottom of the dipoles of their local axes (equivalent to γ length measured from the bottom of the dipole), for different operating frequencies.

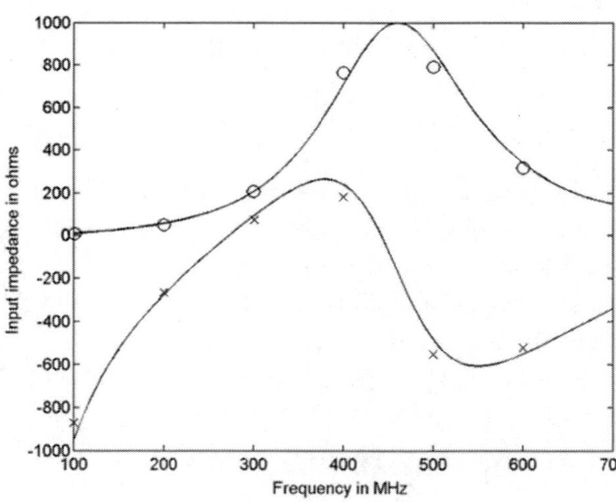

Figure 3.25. The input impedance versus frequency for two parallel dipoles separated by 15 mm; solid line: NEC [51], ('ooo' Resistance, 'xxx' Reactance present work).

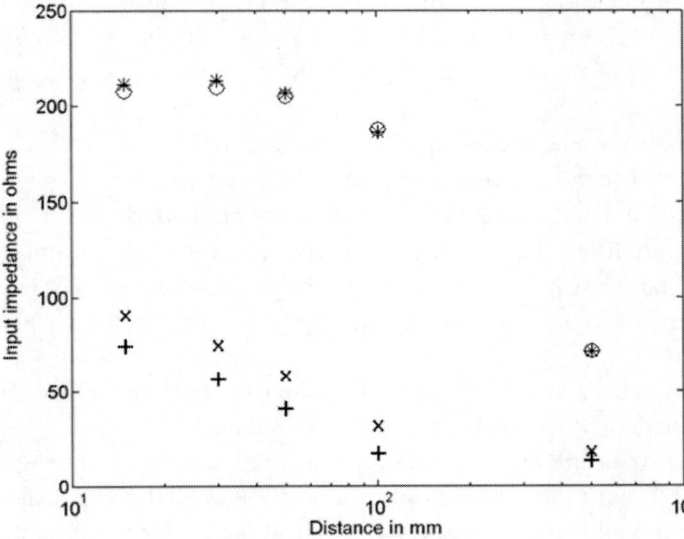

Figure 3.26. Input impedance at 300 MHz operating frequency of two parallel dipoles of 50 cm length and 5 mm wire radius versus the separated distance between them; ('***', '+++': Present work), ('ooo' and 'xxx': NEC).

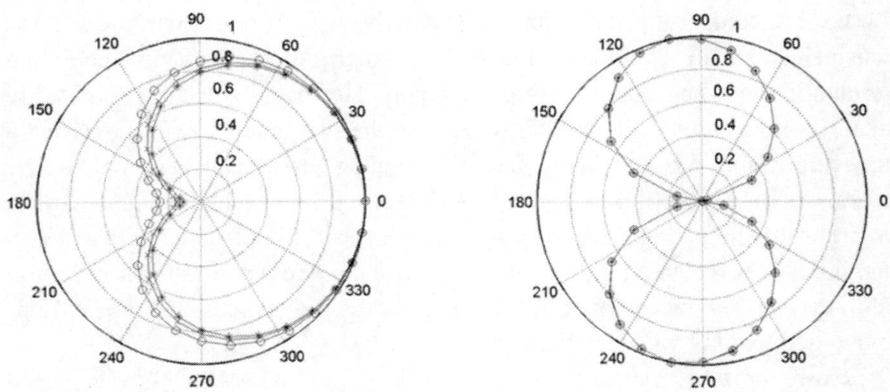

a b

Figure 3.27. The normalised magnitudes of the axial and circumferential surface current components of the antenna geometry given in Figure 3.24 separated by 15 mm versus α at γ= 6.25 cm for axial and γ= 4.6845 cm for circumferential from the bottom of the dipoles for different operating frequencies: 'xxx' 100 MHz, '***' 300 MHz, 'ooo' 500 MHz; (a) axial component, (b) circumferential component.

It is very interesting to note that the non-uniform variations of these currents over different frequencies have small marginal differences. The maximum ratio of the axial component to the circumferential component was around 34:1.

Similarly, these currents at 300 MHz (equivalent to half wavelength dipoles) for different separation distances are presented in Figure 3.28. It should be noted that the actual magnitudes of the circumferential component is inversely proportional to the distance between the dipoles in spite of their fixed variations shown in Figure 3.28(b). Also the axial component is still non-uniform even when the separation distance between the dipoles is 100 mm (0.1 wavelengths).

The normalised surface currents for a thicker wire of radius of 10 mm (0.01 wavelengths) for the same antenna geometry versus the separation distances between the dipoles are shown in Figure 3.29. Comparing Figures 3.28(a) and 3.29(a) the non-uniform effects on the axial components can be strongly seen on the thick wires, for example at the separation distance of 100 mm. It is clear from Figures 3.25 and 3.26 that the average current along the local axis of the dipoles will be similar to that computed using NEC, thus the expected field pattern will be similar, and is not reproduced here. The ratio of the power losses predicted from the non-uniform surface current distribution to those predicted from the average (or uniform) current distribution was considered as an equivalent measure of the improvement in modelling verisimilitude when using the new method. However, since the losses are small in most cases it is possible to assume that the antenna wire is perfectly conducting and surrounded by free space; then losses can be predicted by taking them as proportional to the surface current density squared. The variation of the power loss ratio against the separated distance between two parallel half wavelength dipoles, for two wire thicknesses, is shown in Figure 3.30. It can be noticed that for high separation distances, the power loss converged to unity value as is expected.

However, the variations show a significant power loss for closely spaced dipole antennas. For the following two antenna geometries we restrict our discussion on the presence of non-uniformity of the surface currents that clearly match the variations of the power loss ratios.

The normalised magnitudes of the axial and circumferential surface current components for single and two parallel loops versus α for different operating frequencies are shown in Figures 3.31 and 3.32 respectively. For both figures the loop radius and wire radius are 3 cm and 5 mm respectively.

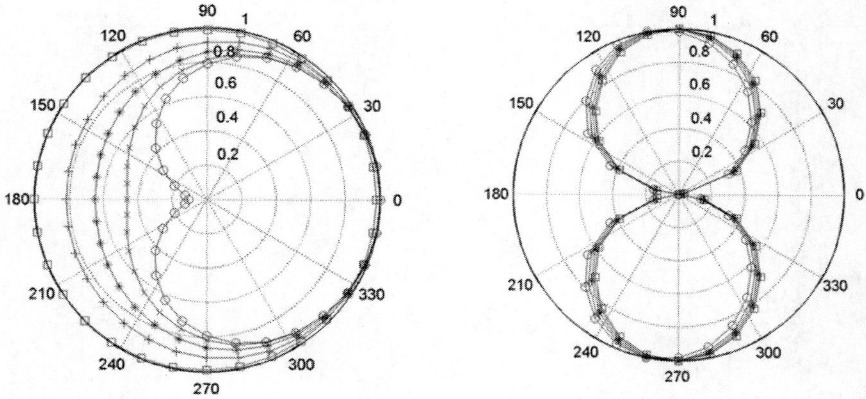

a b

Figure 3.28. The normalised magnitudes of the axial and circumferential surface current components versus α at similar location as Figure 3.27, for different separated distances between the dipoles at 300 MHz (equivalent half wavelength dipoles with 0.005λ wire radius): (a) axial component, (b) circumferential component. ('ooo': 15 mm, 'xxx': 30 mm, '***': 50 mm, '+++': 100 mm, '□□□':500 mm).

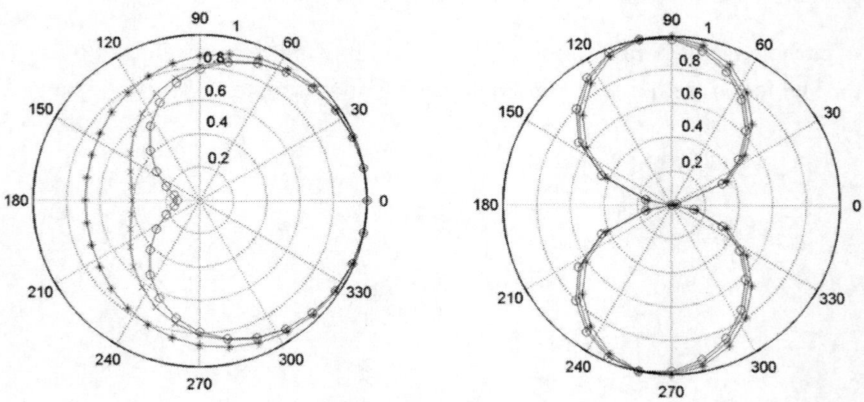

a b

Figure 3.29. The normalised magnitudes of the axial and circumferential surface current components versus α at similar locations as Figure 3.27, for different separated distances between the dipoles at 300 MHz (equivalent to half wavelength dipoles with 0.01λwire radius): (a) axial component, (b) circumferential component. ('ooo': 30 mm, 'xxx': 50 mm, '***': 100 mm).

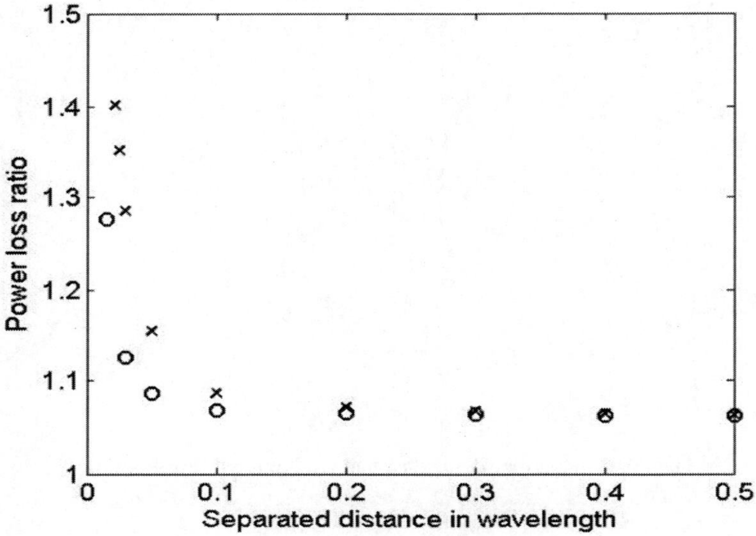

Figure 3.30. Power loss ratio of two parallel half wavelength dipoles for various separation distances; 'ooo': wire radius = 0.005λ, 'xxx': wire radius = 0.01λ.

In the case of parallel loops the separation distance was selected to be 15 mm. The loops are fed by a simple delta excitation source at $\phi = 0°$.

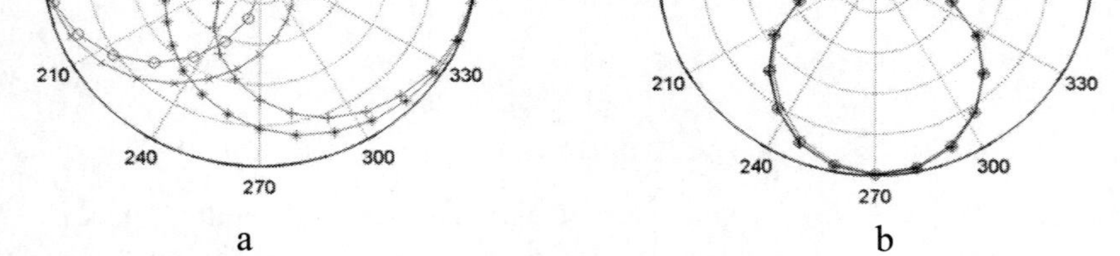

Figure 3.31. The normalised magnitudes of the axial and circumferential surface current components versus α at $\phi = 0°$: axial component and $\phi = 33.75°$: circumferential component, for a single loop antenna for different operating frequencies. The loop radius and wire radius are 3 cm and 5 mm respectively.(a) axial component, (b) circumferential component. ('xxx': 400 MHz, 'ooo': 600 MHz, '+++': 800 MHz, '***': 900 MHz).

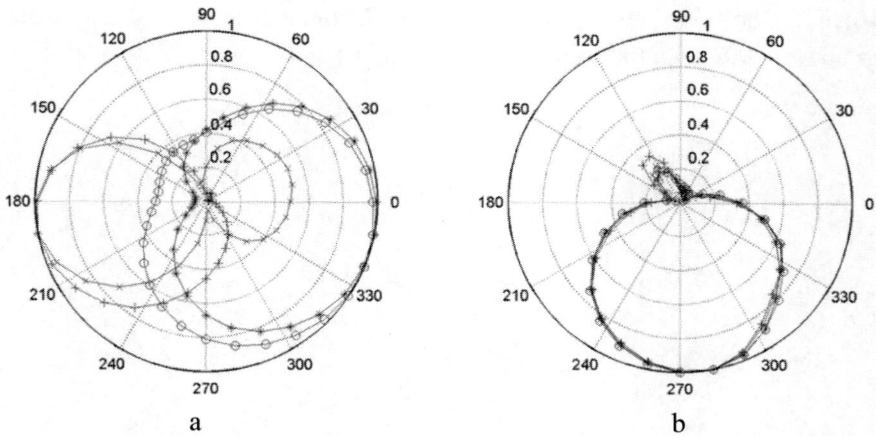

Figure 3.32. The normalised magnitudes of the axial and circumferential surface current components versus α at $\phi = 0°$: axial component and $\phi = 33.75°$: circumferential component, for two parallel loops separated by 15 mm and each has a radius of 3 cm and 5 mm wire radius versus different operating frequencies. (a) axial component, (b) circumferential component. ('+++': 400 MHz, 'xxx': 600 MHz, '***': 800 MHz, 'ooo': 900 MHz).

The axial component is taken at the source location whereas the circumferential is taken at $\phi = 33.75°$ (angles are simply used here to define the locations of the circumferential cross-section wire, and for this particular angle the γ length is 0.0177cm). It should be noted that the variations of the currents for two parallel loops are taken for the bottom loop as shown in Figure 3.24(b). It is clearly shown that the maximum variations of the axial currents for frequencies less that the expected parallel resonance frequency of the antenna structure are always pointed inside the loops geometries (i.e. $\alpha = 180°$). The circumferential component for the single loop antenna is similar to that computed on the two parallel dipoles, and its ratio compared to the axial component was found to be 41:1.

However, the same component for two parallel loops was reduced to minimum around $\alpha = 90°$, whereas its ratio to axial component was 28:1. Moreover, the normalised magnitudes of the axial and circumferential surface current components for two half wavelength parallel loops each of radius 0.0796 wavelength and 0.013 wavelength wire radius versus α for various separation distances d are shown in Figure 3.33. The locations of these currents are similar to those taken in Figures 3.31 and 3.32. The axial component reserved its variations for most of the distances considered in this example as in the case of the single loop antenna except when at very close distance. The variations of the circumferential component was also eliminated around $\alpha = 90°$ even the separation distance was $20a$.

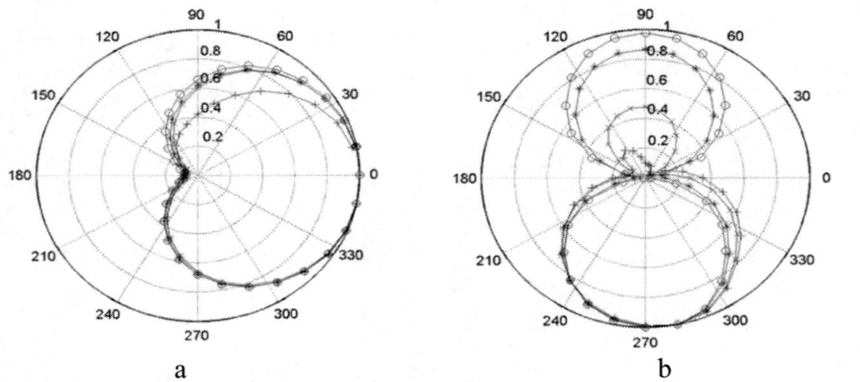

Figure 3.33. The normalised magnitudes of the axial and circumferential surface current components versus α at $\phi = 0°$: axial component and $\phi = 33.75°$: circumferential component, for two half wavelength parallel loops each of radius 0.0796 wavelength and 0.013 wavelength wire radius, versus the separation distance 'd'. (a) axial component, (b) circumferential component. ('+++': $d = 3a$, 'xxx': $d = 6a$, '***': $d = 20a$, 'ooo': $d = 100a$).

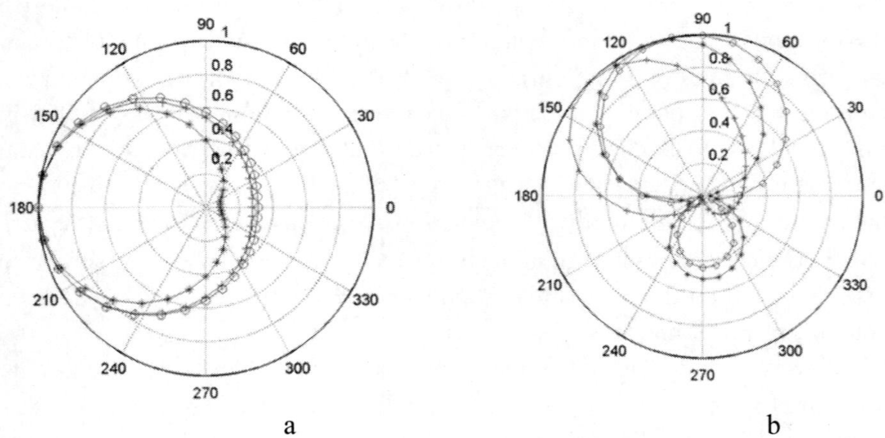

Figure 3.34. The normalised magnitudes of the axial and circumferential surface current components of a half wavelength single turn helix antenna versus α at different positions from the first end of the helix. The helix radius, wire radius and pitch distance are 0.0796λ, 0.013λ and $3a$ respectively. (a) axial component: '***':0.031λ, 'ooo':0.125λ, '+++':0.25λ (b) circumferential component: '***':0.0156λ, 'ooo':0.109λ, '+++':0.234λ.

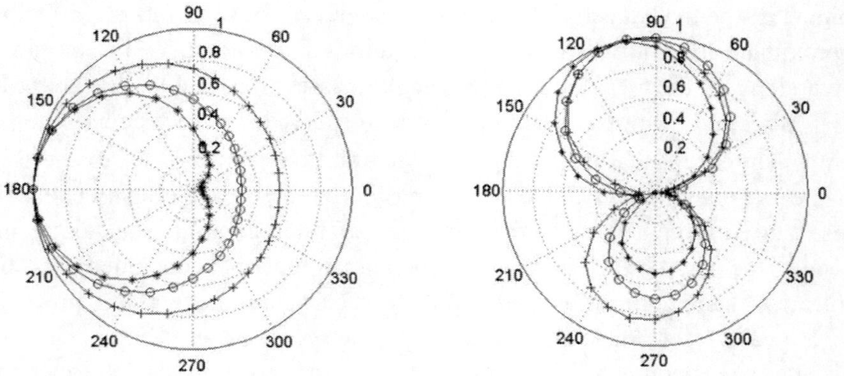

a b

Figure 3.35. The normalised magnitudes of the axial and circumferential surface current components of the same antenna geometry given in Figure 3.33, except that the operating frequency is half than that used in Figure 3.33 (i.e. operating wavelength 2λ). (a) axial component: '***':0.031λ, 'ooo':0.125λ, '+++':0.25λ (b) circumferential component: '***':0.0156λ, 'ooo':0.109λ, '+++':0.234λ.

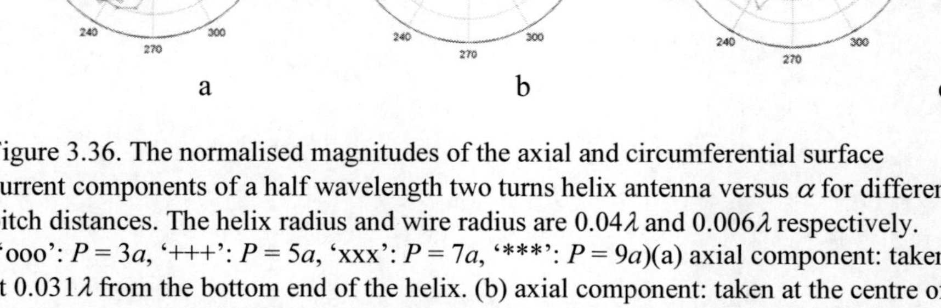

a b c

Figure 3.36. The normalised magnitudes of the axial and circumferential surface current components of a half wavelength two turns helix antenna versus α for different pitch distances. The helix radius and wire radius are 0.04λ and 0.006λ respectively. ('ooo': $P = 3a$, '+++': $P = 5a$, 'xxx': $P = 7a$, '***': $P = 9a$)(a) axial component: taken at 0.031λ from the bottom end of the helix. (b) axial component: taken at the centre of the helix. (c) circumferential component: taken at 0.023λ from the bottom end of the helix.

The normalised magnitudes of the axial and circumferential surface current components of a half wavelength single turn helix antenna versus α at different positions from the first end of the helix are shown in Figure 3.34. The helix radius, wire radius and pitch distance are 0.0796λ, 0.013λ and $3a$ respectively. It is shown that the strong effects of the axial component were pointed inside the helix for all locations presented. These results might help to approximate the equivalent of these variations into one curve that might be taken along all the local length of the helix to assess the total loss power in the axial direction. Similar variations were observed on the circumferential component except that the strong effect has shifted from $\alpha = 90°$ to $\alpha = 135°$ for the locations presented. However, considering the strong effect locations of these currents regardless of the locations of their maxima and taking into account the similarities in these variations it may be concluded that this indicates the approximate contribution of this current to the total loss power.

However, for the same antenna geometry given in Figure 3.34, the currents were computed at half the operating frequency (i.e. operating wavelength 2λ) as shown in Figure 3.35. It is clearly shown that the strong effects of these current variations are mostly similar to that presented in Figure 3.34. Moreover, a strong correlation can be found in the variations of the circumferential currents for different locations for this example. However, a one turn helix is not sufficient to permit comment on the currents variations of a multi-turn helix antenna, thus the following examples are considered. The normalised magnitudes of the axial and circumferential surface current components of a half wavelength of two, three, four and five turns helix antenna versus α for different pitch distances are shown in Figures 3.36, 3.37 and 3.38.

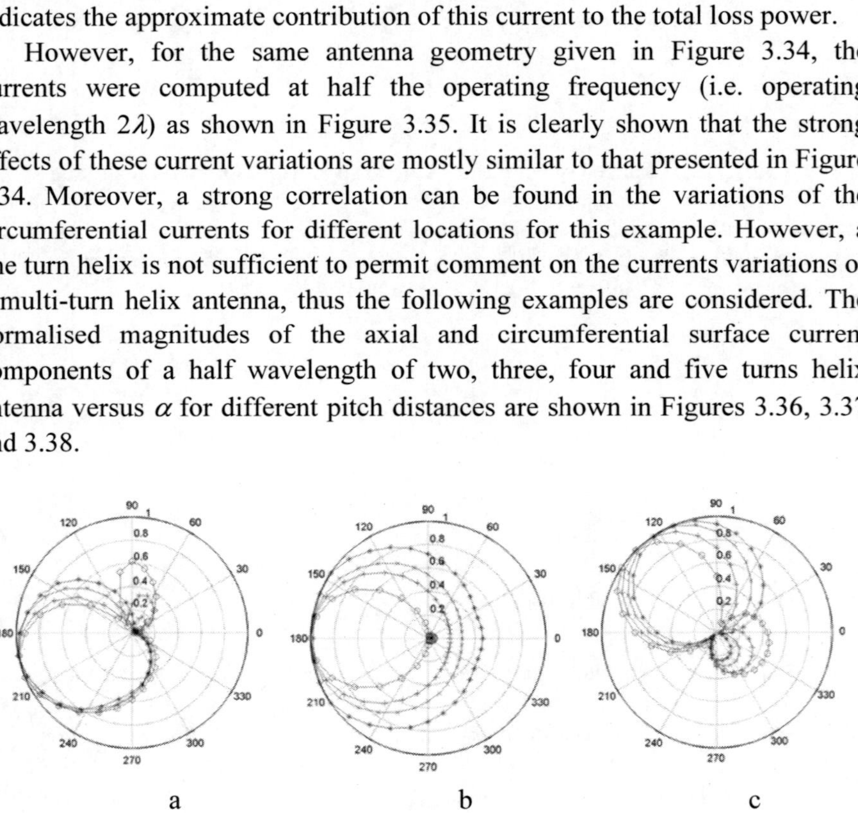

Figure 3.37. The normalised magnitudes of the axial and circumferential surface current components of a half wavelength three turns helix antenna versus α for different pitch distances. The helix radius and wire radius are 0.0265λ and 0.004λ respectively. ('ooo': $P = 3a$, '+++': $P = 5a$, 'xxx': $P = 7a$, '***': $P = 9a$); (a) axial component: taken at 0.0294λ from the bottom end of the helix. (b) axial component: taken at the centre of the helix. (c) circumferential component: taken at 0.022λ from the bottom end of the helix.

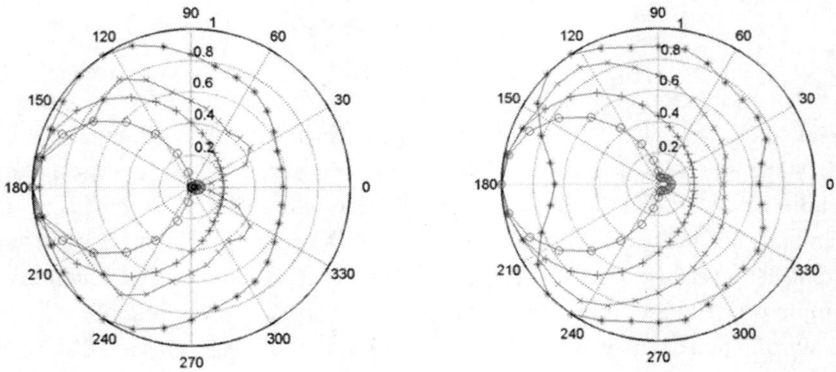

a b

Figure 3.38. The normalised magnitudes of the axial surface current components of a half wavelength four turns helix antenna versus α for different pitch distances. (a) four turns: the helix radius and wire radius are 0.02λ and 0.003λ respectively; (b) five turns: the helix radius and wire radius are 0.015λ and 0.002λ respectively; ('ooo': $P = 3a$, '+++': $P = 5a$, 'xxx': $P = 7a$, '***': $P = 9a$).

It is very clear that the axial and circumferential components were non-uniform even if the pitch distance between the helix turns was nine times the radius of the wire. Another interesting point is that the peak values of the axial component were pointed inside the helix (i.e. around $\alpha = 180°$) for all helices presented. This is clearly shown in the variations of these currents at the feed points on Figures 3.36(b), 3.37(b), 3.38(a) and 3.38(b) and the helix turns in Figures 3.36(a) (first turn) and 3.37(a) (first turn). The similarities of these variations permit the approximate calculation of the effective power loss in that particular direction. It was also observed that the maximum variations of the circumferential component on the first turn confined between $\alpha = 0°$ and 180° as shown in Figures 3.34(b), 3.35(b), 3.36(c) and 3.37(c). However, the maximum ratio of the axial component to the circumferential component for all pitch distances was found between 15:1 and 40:1 for all helices more than one turn.

General comments on the trend of these results are to predict the accurate or approximated equivalent power losses that are associated with non-uniform variations along the wire surfaces. Hence this will follow to affect the radiation efficiency of this kind of antennas.

CONCLUSION

The application of moment method in space domain has been presented in this chapter, for 1-D, 2-D and 3-D antennas. From the 1-D antenna, we can understand the very basic principles of the moment method since only one variable involved. The application of this case can be just for very thin wire antenna. It has been noticed that the current at the edges is vanishing due to the current has greater value at the edges. In the second case of 2-D antenna, an example of infinite cylinder has been studied. This can have an application for thin wire antenna but infinite, the most important concept presented here are the electric integral equation (EFIE) and the magnetic integral equation (MFIE) manipulation. For each type of polarization we have to choose the corresponding integral equation, also the concept of digitalizing the integral equation is another important task to understand. In the last section, the 3-D antennas have been treated, but only for body of revolution (BOR), where a 3-D problem is brought to 2-D problem.

An application has been presented concerning the reflector antenna fed by a dipole antenna, working at 30 GHz. The basis and testing functions are chosen as wavelets mainly to reduce the computation time and the size of the impedance matrix.

Finally, the radiation pattern of the reflector antenna was given for different size of the diameter, the co-polarization is maximum in front of the antenna and the cross-polarization is null in front of the antenna.

The surface current distributions on structures with closely spaced parallel wires, such as dipoles, loops and helical antennas, can be computed by using the method of moments with a general surface patch formulation. The current distribution varies substantially from the common assumption that it is uniform around the wire cross-section.

Transverse (circumferential) currents are shown to be present: they are relatively weak on thin wires (around 0.01λ wire radius) excited by axial component parallel to the local axis of the wire. The effect is still significant when the wire separation distance is relatively large.

In spite the strong variations of the axial and circumferential currents, it was found that the input impedance and the average value of the axial surface current are in reasonable good agreement with the results of thin wire codes such as NEC using an extended kernel solution. The power loss ratio resulting from use of non-uniform surface current, compared with the conventional uniform assumption of two parallel dipoles, shows a significant increase of power loss when they are closely separated. However, these current variations will dominate the radiation efficiency when predicting the accurate total power loss on these types of antennas and this can be important in some applications, e.g., highly resonant antennas and antennas realised in superconducting materials. As a matter of interest it was computed that the maximum ratio of the variations of the axial component to the circumferential component on a half wavelength helix of a few turns for different pitch distances was between 15:1 to 40:1. This behaviour is expected as the NMHA has a hybrid of dipole and loop behaviour. The modelling method employed a two-dimensional electric surface patch integral equation formulation solved by independent piecewise-linear basis function methods in the circumferential and axial directions of the wire. A similar orthogonal basis function was used on the end surface and appropriate attachments with the wire surface were employed to satisfy the requirements of current continuity. The results were stable and showed good agreement with less comprehensive earlier work by others.

REFERENCES

[1] M. Amir, S. Bedra, S. Benkouda, and T. Fortaki, "Bacterial foraging optimisation and method of moments for modelling and optimisation of microstrip antennas," *IET Microwaves, Antennas and Propagation*, vol. 8, no. 4, pp. 295-300, 2014.

[2] D. J. Ludick, R. Maaskant, D. B. Davidson, U. Jakobus, R. Mittra, and D. De Villiers, "Efficient analysis of large aperiodic antenna arrays using the domain Green's function method," *IEEE Transactions on Antennas and Propagation*, vol. 62, no. 4, part 1, pp. 1579 - 1588, 2014.

[3] S. R. Zang and J. R. Bergmann, "Analysis of omnidirectional dual-reflector antenna and feeding horn using method of moments," *IEEE Transactions on Antennas and Propagation*, vol. 62, no. 3, pp. 1534-1538, 2014.

[4] M. N. O. Sadiku, *Numerical techniques in electromagnetics,* New York: CRC Press, 2001.

[5] Q. Ye, Electromagnetic scattering by numerical methods for large structures, *PhD thesis*, Canada: Manitoba University, 2000.

[6] E. V. Jull, *Antenna handbook – Theory, applications and design*, Ch1, London, 1999.

[7] C. P. Davis, Understanding and improving moment method scattering solutions, *M. Sc. Thesis*, Birmingham: Birmingham University, 2004.

[8] M. A. AbdelMoneum, Z. Shen, J. L. Volakis, "Hybrid PO-MoManalysis of large axi-symmetric radomes", *IEEE Transactions on Antennas and Propagation*, vol. 49, no. 12, pp. 1657-1666, 2001.

[9] M. Djordjevic, "Higher order hybrid method of moments-physical optics modeling technique for radiation and scattering from large perfectly conducting surface," *IEEE Transactions on Antennas and Propagation*, vol. 53, no. 2, pp. 800-813, 2005.

[10] R. F. Harrington, *Field computation by moment method*, New York: IEEE Press, 1993.

[11] E. D. Caswell, Analysis of helix antenna using a moment method approach with curved basis and testing functions, *M. Sc. Thesis*, Blacksburg: Virginia Polytechnic, 1998.

[12] J. Yang and P.-S. Kildal, Implementation of formulations for dielectric material and perfect conductors in G2DMULT, *IEEE Antennas and Propagation Society International Symposium*, vol. 4, pp. 830-833, 2001.

[13] U. C. Resende and F. J. S. Moreira, "Optimal number of basis functions in the MoM solutions for bodies of revolution," *Journal of Microwaves and Optoelectronics*, vol. 6, no. 1, pp. 220-235, 2007.

[14] R. V. Sabariego and L. Landesa, "Far field decoupled basis for the moment method 2D case," *Progress in Electromagnetic Research, PIER 24*, pp. 163-183, 1999.

[15] F. L. Teixeira and R. J. Bergmann, "Moment method analysis of circularly symmetric reflectors using band limited basis functions," *IEE Proceedings on Microwave, Antennas and Propagation*, vol. 144, no. 3, pp. 620-625, 1997.

[16] W.-B. Ewe and L.-W. Li, "Analysis of reflector and horn antennas using adaptive integral method," *IEICE Transaction on Communication*, vol. E88-B, no. 6, pp. 2327-2333, 2005.

[17] G. W. Pan, *Wavelet in electromagnetics and device modeling*, John Wiley and Sons Publication, 2001.

[18] J. C. Goswamiand A. K. Chan, *"Fundamentals of wavelets, theory algorithms and applications"*, Wiley Inter Science Publications, 2000.

[19] A. R. Baghai-Wadji and G. G. Walter, "Green's function-based wavelets," *IEEE Ultrasonic Symposium*, pp. 537-546, 2000.

[20] Z. Baharav and Y. Leviatan, "Impedance matrix compression (IMC) using iteratively selected wavelet basis," *IEEE Transactions on Antennas and Propagation*, vol. 46, no. 2, pp. 226-233, 1998.

[21] M. Lashab, C. Zebiri and F. Benabdelaziz, "Wavelet-based moment method and physical optics use on large reflector antennas," *Progress in Electromagnetic Research M, PIER M*, vol. 2, pp. 189-200, 2008.

[22] R. Bancroft, *Understanding electromagnetic scattering using the moment method*, London: Artech House Boston, 1999.

[23] C. A. Balanis, *Advanced engineering electromagnetics*, Arizona State University: John Wiley and Sons Publication, 1989.

[24] S. Sefi, Computational electromagnetics: Software development and high frequency modeling of surface current on perfect conductors, *PhD thesis*, Sweden: Stockholm University, 2005.

[25] S. Hatamzadeh-Varmazyar and M. Naser-Moghadasi, "An integral equation modeling of electromagnetic scattering from the surfaces of arbitrary resistance distribution," *Progress In Electromagnetics Research B*, vol. 3, pp. 157-172, 2008.

[26] A. G. Papagiannakis, "Application of point-matching MoMreduced scheme to scattering from finite cylinders," *IEEE Transactions on Microwave Theory and Techniques*, vol. 45, no. 9, pp. 1545-1553, 1997.

[27] A. El-Hajj, K. Y. Kabalan and G. Nehmetallah, "Moment method solution of partially filled parallel plate fed through infinite conducting plane (TE case)," *The Arabian Journal for Science and engineering*, vol. 28, no. 2B, pp. 221-232, 2003.

[28] L. Gurel and O. Ergul, "Singularity of the magnetic-field integral equation and its extraction," *IEEE Antennas and Wireless Propagation Letters,* vol. 4, pp. 229-232, 2005.

[29] U. C. Resende and F. J. S. Moreira, "Efficient evaluation of singular integral equations in moment method analysis of bodies of revolutions," *Journal of Microwaves and Optoelectronics*, vol. 6, no. 2, 2007.

[30] H. Esteban and J. M. Lopez, "A generalized method for characterizing two dimensional scattering problems with spectral techniques," *Microwave and Optical Technology Letters*, vol. 14, no. 1, 1997.

[31] C. Su and T. K. Sarkar, "Electromagnetic scattering from coated strips utilizing the adaptive multi-scale moment method," *Progress In Electromagnetic Research, PIER 18*, pp. 173-208, 1998.

[32] M. Lashab, F. Benabdelaziz and C. Zebiri, "Analysis of electromagnetic scattering from reflector and cylindrical antennas using wavelet-based moment method," *Progress in Electromagnetic Research, PIER 76*, pp. 357-368, 2007.

[33] W. Quan, Fast wavelet solution for electromagnetic integral equation, *PhD Thesis,* Canada: University of Manitoba, 2003.

[34] G. Wang, "Application of wavelets on the interval to numerical analysis of integral equations in electromagnetic scattering problems," *International Journal for Numerical Methods in Engineering,* vol. 40, pp. 1-13, 1997.

[35] G. Wang, "Analysis of electromagnetic scattering from conducting bodies of revolution using orthogonal wavelet expansions," *IEEE Transactions on Electromagnetic Compatibility*, vol. 40, no. 1, pp. 1-11, 1998.

[36] K. F. Sabet, J. C. Cheng and L. P. B. Katehi, "Efficient wavelet-based modeling of printed circuit antenna arrays," *IEE Proceedings on Microwave, Antennas and Propagation*, vol. 146, no. 4, pp. 298-304, 1999.

[37] M. R. Zunoubiand A. A. Kishk, "A combined BI-CGSTAB (*l*) and wavelet transform method for EM problems using method of moment," *Progress In Electromagnetic Research, PIER* 52, pp. 205-224, 2005.

[38] G. Wang, "Application of wavelet to the analysis of arbitrary thin wire loop antennas and scatterers," *International Journal for Numerical Modeling*, vol. 10, pp. 193-204, 1997.

[39] Y. Tretiakov and G. Pan, "Malvarwavelet based Pocklingtonequation solution to thin wire antennas and scatterers," *Progress In Electromagnetics Research, PIER* 47, pp. 123-133, 2004.

[40] A. P. Anyutin and A. G. Kyurkchan, "Application of the wavelet in 2D and 3D scattering or radiation problems," *IEEE Transactions on Antennas and Propagation*, vol. 15, no. 5, pp. 452-455, 2003.

[41] A. Capozzoli and G. D'Elia, "Global optimization and antennas synthesis application to reflector antennas," *Progress In Electromagnetics Research, PIER 56*, pp. 233-261, 2006.

[42] G. Smith, "The proximity effect in systems of parallel conductors and electrically small multiturn loop antennas," *Tech. Rep. 624, Division of Engineering and Applied Physics,* Harvard University, US, 1971.

[43] G. Smith, "The proximity effect in systems of parallel conductors," *Journal of Applied Physics*, vol. 43, pp. 2196-2203, 1972.

[44] G. O. Olaofe, "Scattering by two cylinders," *Radio Science*, vol. 5, pp. 1351-1360, 1970.

[45] P. Tulyathan and E. H. Newman, "The circumferential variation of the axial component of current in closely spaced thin-wire antennas," *IEEE Transactions on Antennas and Propagation*, vol. AP-27, pp. 46-50, 1979.

[46] R. A. Abd-Alhameed and P. S. Excell, "The complete surface current for NMHA using sinusoidal basis functions and Galerkins solution," *IEE Proceedings of Science, Measurement and Technology on Computational Electromagnetics*, vol. 149, pp. 272-276, 2002.

[47] R. A. Abd-Alhameed and P. S. Excell, "Surface current distribution on closely parallel wires within antennas," *The Second European Conference on Antennas and Propagation, (EuCAP 2007)*, pp. 1-6, 2007.

[48] R. A. Abd-Alhameed and P. S. Excell, "Accurate power loss computation of closely spaced radiating wire elements for mobile phone MIMO application," *IEEE International Conference on Signal Processing and Communication,* Dubai, United Arab Emirates, pp. 1-4, 2007.

[49] R. A. Abd-Alhameed and P. S. Excell, "Non-uniform surface current distribution on parallel wire loop antennas using curved patches in the method of moments," *IET Science, Measurement and Technology*, vol. 2, pp. 493-498, 2008.

[50] A. R. Djordjevic, M. B. Bazdar, V. V. Petrovic, D. I. Olcan, T. K. Sarkar, and R. F. Harrington, *Analysis of wire antennas and scatterers,* Boston, US: Artech House, 1990.

[51] G. J. Burke and A. J. Poggio, "Numerical electromagnetics code (NEC): method of moments," *US Naval Ocean Systems Centre, Rep. No. TD116*, 1981.

[52] J. H. Richmond, "Radiation and scattering by thin-wire structures in the complex frequency domain," *NASA Rept. No. CR-2396*, 1974.

[53] FEKO: EM Software and Systems S. A., (Pty) Ltd, Stellenbosch, South Africa.

[54] CST: Computer Simulation Technology Corporation, Microwave Studio, Version 5.0, German.

[55] HFSS: v. 10, Ansoft [online], www.ansoft.com.

[56] IE3D: Release 12, Zeland Software Inc., Fremont CA, US, 2007.

[57] K. Tap, P. Pathak and R. Burkholder, "Complex source beam - moment method procedure for accelerating numerical integral equation solutions of radiation and scattering problems," *IEEE Transactions on Antennas and Propagation*, vol. 62, no. 4, pp. 2052 - 2062, 2014.

[58] X.-M. Pan and X.-Q. Sheng, "Improved algebraic preconditioning for MoM solutions of large-scale electromagnetic problems," *IEEE Antennas and Wireless Propagation Letters*, vol. 13, pp. 106-109, 2014.

[59] M. A. Echeverri Bautista, M. A. Francavilla, F. Vipiana and G. Vecchi, "A hierarchical fast solver for EFIE-MoM analysis of multiscale structures at very low frequencies," *IEEE Transactions on Antennas and Propagation*, vol. 62, no. 3, pp. 1523-1528, 2014.

[60] M. Shafieipour, I. Jeffrey, J. Aronsson and V. I. Okhmatovski, "On the equivalence of RWG method of moments and the locally corrected nyström method for solving the electric field integral equation," *IEEE Transactions on Antennas and Propagation*, vol. 62, no. 2, pp. 772-782, 2014.

[61] A. Schroder, G. A. Rasek, H.-D. Bruns, Z. Reznicek, J. Kucera, S. E. Loos, and C. Schuster, "Analysis of high intensity radiated field coupling into aircraft using the method of moments," *IEEE Transactions on Electromagnetic Compatibility*, vol. 56, no. 1, pp. 113-122, 2014.

[62] N. Hendijani, J. Cheng and R. J. Adams, "Combined field integral equation using a constraint-based Helmholtz decomposition," *IEEE Transactions on Antennas and Propagation*, vol. 62, no. 3, pp. 1500-1503, 2014.

[63] J. Alvarez, L. D. Angulo, M. R. Cabello, A. Rubio Bretones and S. G. Garcia, "An analysis of the leap-frog discontinuous Galerkin method for Maxwell's equations," *IEEE Transactions on Microwave Theory and Techniques*, vol. 62, no. 2, pp. 197-207, 2014.

[64] D. A. E. Mohamed, "Comprehensive analysis of broad band wired antennas," *National Radio Science Conference (NRSC)*, pp. 1-10, 2009.

[65] W. G. Hong, Y. Yamada and N. Michishita, "Low profile small normal mode helical antenna achieving long communication distance," *International Workshop on Antenna Technology: Small Antennas and Novel Metamaterials (iWAT),* Chiba, Japan, pp. 167-170, 2008.

[66] H. Mimaki and H. Nakano, "Double pitch helical antenna," *IEEE Antennas and Propagation Society International Symposium*, vol. 4, pp. 2320-2323, 1998.

[67] S. Ooi, "Normal mode helical antenna broad-banding using multiple pitches," *IEEE Antennas and Propagation Society International Symposium*, vol. 1, pp. 860-863 2003.

[68] W. L. Stutzman and G. A. Thiele, *Antenna theory and design*, 2nd ed. New York: John Wiley and Sons, 1998.

[69] K. Sawaya, "Antenna design by using method of moments," *IEICE Transaction Communication*, vol. E88-B, pp. 1766-1773, 2005.

[70] R. E. Collin, *Field theory of guided waves*, 2nd ed. New York: The Institute of Electrical and Electronics Engineers Inc., 1995.

[71] A. Ishimaru, *Electromagnetic wave propagation, radiation and scattering*. Englewood Cliffs, New Jersey: Prentice Hall, 1991.

[72] M. I. Aksun and R. Mittra, "Choices of expansion and testing functions for the method of moments applied to a class of electromagnetic problems," *IEEE Transactions on Microwave Theory and Techniques*, vol. 41, pp. 503-509, 1993.

[73] M. A. Khayat and D. R. Wilton, "Numerical evaluation of singular and near-singular potential integrals," *IEEE Transactions on Antennas and Propagation*, vol. 53, pp. 3180-3190, 2005.

APPENDIX A

To find the tangent surface equation at a point on the curved surface can be stated in the following. For simplicity, assume each point on the curved surface S for all 3-D axes can be represented by two variables as u and v given by:

$$x = f(u,v), \qquad x_o = f(u_o, v_o) \tag{A.1}$$

$$y = g(u,v), \qquad y_o = g(u_o, v_o) \tag{A.2}$$

$$z = h(u,v), \qquad z_o = h(u_o, v_o) \tag{A.3}$$

It should be noted that the above equations must be differentiable at any given point such as (x_o, y_o, z_o). Hereafter, these should be differentiable at (u_o, v_o). Thus, the partial differentiation with respect to u and v evaluated at (u_o, v_o) is given by:

$$\underline{a} = [f_u \quad g_u \quad h_u] \tag{A.4}$$

$$\underline{b} = [f_v \quad g_v \quad h_v] \tag{A.5}$$

where $R_p = \partial R/\partial p$, R could be f or g or h and p might be u or v. The normal vector to the surface (i.e. orthogonal to the surface), can be expressed as:

$$\begin{bmatrix} L \\ M \\ N \end{bmatrix} = \underline{a} \times \underline{b} \qquad \text{(A.6)}$$

The above notation can be extended using the cross product as in the following matrix:

$$\begin{bmatrix} L \\ M \\ N \end{bmatrix} = \begin{bmatrix} (g_u \times h_v) - (g_v \times h_u) \\ (f_v \times h_u) - (f_u \times h_v) \\ (f_u \times g_v) - (f_v \times h_u) \end{bmatrix} \qquad \text{(A.7)}$$

The total length of the normal vector can be given by:

$$D = \sqrt{L^2 + M^2 + N^2} \qquad \text{(A.8)}$$

The tangent plane should satisfy the following formula:

$$\frac{L}{D}x + \frac{M}{D}y + \frac{N}{D}z = Lx_o + My_o + Nz_o \qquad \text{(A.9)}$$

Thus, the tangent surface for a point at (x_o, y_o, z_o) can simply be expressed by the following:

$$L(x - x_o) + M(y - y_o) + N(z - z_o) = 0 \qquad \text{(A.10)}$$

Of course the result is equivalent to the following necessary condition:

$$F(x, y, z) = 0 \qquad \text{(A.11)}$$

The above equations are also applicable to direct surface equation such as that given by the unit sphere at the origin:

$$x^2 + y^2 + z^2 - 1 = 0 \qquad \text{(A.12)}$$

The equation of the tangent plane at (x_o, y_o, z_o):

$$\left.\frac{\partial F}{\partial x}\right|_{x_o,y_o,z_o}(x-x_o)+\left.\frac{\partial F}{\partial y}\right|_{x_o,y_o,z_o}(y-y_o)+\left.\frac{\partial F}{\partial z}\right|_{x_o,y_o,z_o}(z-z_o)=0 \quad \text{(A.13)}$$

Keep in mind the required transformation can also be implemented for other coordinates such as between the spherical and the Cartesian coordinates, given by:

$$\begin{matrix} r_o & & x_o \\ \theta_o & \rightarrow & y_o \\ \phi_o & & z_o \end{matrix} \quad \text{(A.14)}$$

Again the above derivation can be applied to simple surfaces such as the one given by the following example. Surface equation is $z = 4x^3y^2 + 2y$ and the tangent at point (1, -2, 12) can be derived as:

$$f(x, y) = 4x^3y^2 + 2y \quad \text{(A.15a)}$$

$$\frac{\partial f}{\partial x} = 12x^2y^2 \quad \text{(A.15b)}$$

$$\frac{\partial f}{\partial y} = 8x^3y + 2 \quad \text{(A.15c)}$$

Substituting the value of x_o = 1 and y_o = -2 to equation (A.15b) and (A.15c) gives:

$$\left.\frac{\partial f}{\partial x}\right|_{x_o,y_o} = 48 \quad \text{(A.15d)}$$

$$\left.\frac{\partial f}{\partial y}\right|_{x_o, y_o} = -14 \tag{A.15e}$$

Hence, the tangent plane is given by the expression:

$$48(x-1)-14(y+2)-(z-12)=0 \tag{A.15f}$$

$$48x-14y-z=64 \tag{A.15g}$$

Finally, the tangent surface of helix wire can be derived as follows. Given the coordinates of the helix by the following:

$$x=(b+a\cos\theta)\cos\phi \tag{A.16}$$

$$y=(b+a\cos\theta)\sin\phi \tag{A.17}$$

$$z=\frac{P}{2\pi}\phi+a\sin\theta \tag{A.18}$$

$$\begin{bmatrix} L \\ M \\ N \end{bmatrix} = \begin{bmatrix} \left.\frac{\partial x}{\partial \theta}\right|_{\theta_o, \phi_o} \\ \left.\frac{\partial y}{\partial \theta}\right|_{\theta_o, \phi_o} \\ \left.\frac{\partial z}{\partial \theta}\right|_{\theta_o, \phi_o} \end{bmatrix} \times \begin{bmatrix} \left.\frac{\partial x}{\partial \phi}\right|_{\theta_o, \phi_o} \\ \left.\frac{\partial x}{\partial \phi}\right|_{\theta_o, \phi_o} \\ \left.\frac{\partial x}{\partial \phi}\right|_{\theta_o, \phi_o} \end{bmatrix} \tag{A.19}$$

APPENDIX B

The expression for C in equation (3.133) in Chapter 3 is given by:

$$
\begin{aligned}
C = {} & \log(t_1+t+(v^2+p^2+2pp_1+p_1^2+t^2+2tt_1+t_1^2)^{\frac{1}{2}})p-p+\frac{1}{2}p_1 \\
& \times\log(v^2+p^2+2pp_1+p_1^2)+v\times\operatorname{atan}(\frac{1}{2}(2p+2p_1)/v)+\frac{1}{2}t/((t+t_1)^2)^{\frac{1}{2}} \\
& \times\log((2t^2+4tt_1+2t_1^2+2(-v^2)^{\frac{1}{2}}(p+p_1-(-v^2)^{\frac{1}{2}})+2((t+t_1)^2)^{\frac{1}{2}}((p+p_1-(-v^2)^{\frac{1}{2}})^2 \\
& +2(-v^2)^{\frac{1}{2}}(p+p_1-(-v^2)^{\frac{1}{2}})+t^2+2tt_1+t_1^2)^{\frac{1}{2}})/(p+p_1-(-v^2)^{\frac{1}{2}}))p_1-\frac{1}{2}t(-v^2)^{\frac{1}{2}}/((t+t_1)^2)^{\frac{1}{2}} \\
& \times\log((2t^2+4tt_1+2t_1^2+2(-v^2)^{\frac{1}{2}}(p+p_1-(-v^2)^{\frac{1}{2}})+2((t+t_1)^2)^{\frac{1}{2}}((p+p_1-(-v^2)^{\frac{1}{2}})^2 \\
& +2(-v^2)^{\frac{1}{2}}(p+p_1-(-v^2)^{\frac{1}{2}})+t^2+2tt_1+t_1^2)^{\frac{1}{2}})/(p+p_1-(-v^2)^{\frac{1}{2}}))+\frac{1}{2}t(-v^2)^{\frac{1}{2}}/((t+t_1)^2)^{\frac{1}{2}} \\
& \times\log((2t^2+4tt_1+2t_1^2-2(-v^2)^{\frac{1}{2}}(p+p_1+(-v^2)^{\frac{1}{2}})+2((t+t_1)^2)^{\frac{1}{2}}((p+p_1+(-v^2)^{\frac{1}{2}})^2 \\
& -2(-v^2)^{\frac{1}{2}}(p+p_1+(-v^2)^{\frac{1}{2}})+t^2+2tt_1+t_1^2)^{\frac{1}{2}})/(p+p_1+(-v^2)^{\frac{1}{2}}))+\frac{1}{2}t/((t+t_1)^2)^{\frac{1}{2}} \\
& \times\log((2t^2+4tt_1+2t_1^2-2(-v^2)^{\frac{1}{2}}(p+p_1+(-v^2)^{\frac{1}{2}})+2((t+t_1)^2)^{\frac{1}{2}}((p+p_1+(-v^2)^{\frac{1}{2}})^2 \\
& -2(-v^2)^{\frac{1}{2}}(p+p_1+(-v^2)^{\frac{1}{2}})+t^2+2tt_1+t_1^2)^{\frac{1}{2}})/(p+p_1+(-v^2)^{\frac{1}{2}}))p_1+\frac{1}{2}t_1/((t+t_1)^2)^{\frac{1}{2}} \\
& \times\log((2t^2+4tt_1+2t_1^2+2(-v^2)^{\frac{1}{2}}(p+p_1-(-v^2)^{\frac{1}{2}})+2((t+t_1)^2)^{\frac{1}{2}}((p+p_1-(-v^2)^{\frac{1}{2}})^2 \\
& +2(-v^2)^{\frac{1}{2}}(p+p_1-(-v^2)^{\frac{1}{2}})+t^2+2tt_1+t_1^2)^{\frac{1}{2}})/(p+p_1-(-v^2)^{\frac{1}{2}}))p_1-\frac{1}{2}t_1(-v^2)^{\frac{1}{2}}/((t+t_1)^2)^{\frac{1}{2}} \\
& \times\log((2t^2+4tt_1+2t_1^2+2(-v^2)^{\frac{1}{2}}(p+p_1-(-v^2)^{\frac{1}{2}})+2((t+t_1)^2)^{\frac{1}{2}}((p+p_1-(-v^2)^{\frac{1}{2}})^2 \\
& +2(-v^2)^{\frac{1}{2}}(p+p_1-(-v^2)^{\frac{1}{2}})+t^2+2tt_1+t_1^2)^{\frac{1}{2}})/(p+p_1-(-v^2)^{\frac{1}{2}}))t \\
& \times\log(p+p_1+(v^2+p^2+2pp_1+p_1^2+t^2+2tt_1+t_1^2)^{\frac{1}{2}})+t_1 \\
& \times\log(p+p_1+(v^2+p^2+2pp_1+p_1^2+t^2+2tt_1+t_1^2)^{\frac{1}{2}})+\frac{1}{2}t_1(-v^2)^{\frac{1}{2}}/((t+t_1)^2)^{\frac{1}{2}}\ldots
\end{aligned}
$$

$$\ldots\times\log((2t^2+4tt_1+2t_1^2-2(-v^2)^{\frac{1}{2}}(p+p_1+(-v^2)^{\frac{1}{2}})+2((t+t_1)^2)^{\frac{1}{2}}((p+p_1+(-v^2)^{\frac{1}{2}})^2$$
$$-2(-v^2)^{\frac{1}{2}}(p+p_1+(-v^2)^{\frac{1}{2}})+t^2+2tt_1+t_1^2)^{\frac{1}{2}})/(p+p_1+(-v^2)^{\frac{1}{2}}))+\frac{1}{2}t_1/((t+t_1)^2)^{\frac{1}{2}}$$
$$\times\log((2t^2+4tt_1+2t_1^2-2(-v^2)^{\frac{1}{2}}(p+p_1+(-v^2)^{\frac{1}{2}})+2((t+t_1)^2)^{\frac{1}{2}}((p+p_1+(-v^2)^{\frac{1}{2}})^2$$
$$-2(-v^2)^{\frac{1}{2}}(p+p_1+(-v^2)^{\frac{1}{2}})+t^2+2tt_1+t_1^2)^{\frac{1}{2}})/(p+p_1+(-v^2)^{\frac{1}{2}}))p_1$$

The expression for D in equation (3.133) in Chapter 3 is given by:

$$D=\frac{1}{4}(2p+2p_1)(v^2+p^2+2pp_1+p_1^2+t^2+2tt_1+t_1^2)^{\frac{1}{2}}+\frac{1}{2}$$
$$\times\log(p+p_1+(v^2+p^2+2pp_1+p_1^2+t^2+2tt_1+t_1^2)^{\frac{1}{2}})v^2-\frac{1}{2}$$
$$\times\log(p+p_1+(v^2+p^2+2pp_1+p_1^2+t^2+2tt_1+t_1^2)^{\frac{1}{2}})t_1^2+\frac{1}{2}$$
$$\times\log(p+p_1+(v^2+p^2+2pp_1+p_1^2+t^2+2tt_1+t_1^2)^{\frac{1}{2}})t^2-t_1p$$
$$\times\log(t_1+t+(v^2+p^2+2pp_1+p_1^2+t^2+2tt_1+t_1^2)^{\frac{1}{2}})+t_1p-\frac{1}{2}t_1tp_1/((t+t_1)^2)^{\frac{1}{2}}$$
$$\times\log((2t^2+4tt_1+2t_1^2+2(-v^2)^{\frac{1}{2}}(p+p_1-(-v^2)^{\frac{1}{2}})+2((t+t_1)^2)^{\frac{1}{2}}((p+p_1-(-v^2)^{\frac{1}{2}})^2+2(-v^2)^{\frac{1}{2}}$$
$$\times(p+p_1-(-v^2)^{\frac{1}{2}})+t^2+2tt_1+t_1^2)^{\frac{1}{2}})/(p+p_1-(-v^2)^{\frac{1}{2}})+t^2+2tt_1+t_1^2)^{\frac{1}{2}}))-\frac{1}{2}t_1tp_1/((t+t_1)^2)^{\frac{1}{2}}$$
$$\times\log((2t^2+4tt_1+2t_1^2-2(-v^2)^{\frac{1}{2}}(p+p_1+(-v^2)^{\frac{1}{2}})+2((t+t_1)^2)^{\frac{1}{2}}((p+p_1+(-v^2)^{\frac{1}{2}})^2-2(-v^2)^{\frac{1}{2}}$$
$$\times(p+p_1+(-v^2)^{\frac{1}{2}})+t^2+2tt_1+t_1^2)^{\frac{1}{2}})/(p+p_1+(-v^2)^{\frac{1}{2}}))-\frac{1}{2}t_1^2p_1/((t+t_1)^2)^{\frac{1}{2}}$$
$$\times\log((2t^2+4tt_1+2t_1^2+2(-v^2)^{\frac{1}{2}}(p+p_1-(-v^2)^{\frac{1}{2}})+2((t+t_1)^2)^{\frac{1}{2}}((p+p_1-(-v^2)^{\frac{1}{2}})^2+2(-v^2)^{\frac{1}{2}}$$
$$\times(p+p_1-(-v^2)^{\frac{1}{2}})+t^2+2tt_1+t_1^2)^{\frac{1}{2}})/(p+p_1-(-v^2)^{\frac{1}{2}}))-\frac{1}{2}t_1^2p_1/((t+t_1)^2)^{\frac{1}{2}}$$
$$\times\log((2t^2+4tt_1+2t_1^2-2(-v^2)^{\frac{1}{2}}(p+p_1+(-v^2)^{\frac{1}{2}})+2((t+t_1)^2)^{\frac{1}{2}}((p+p_1+(-v^2)^{\frac{1}{2}})^2+2(-v^2)^{\frac{1}{2}})^2$$
$$-2(-v^2)^{\frac{1}{2}}(p+p_1+(-v^2)^{\frac{1}{2}})+t^2+2tt_1+t_1^2)^{\frac{1}{2}})/(p+p_1+(-v^2)^{\frac{1}{2}}))-\frac{1}{2}t_1p_1$$
$$\times\log((v^2+p^2+2pp_1+p_1^2+\frac{1}{2}t_1t(-v^2)^{\frac{1}{2}}/((t+t_1)^2)^{\frac{1}{2}}\ldots$$

$$\ldots\times\log((2t^2+4tt_1+2t_1^2+2(-v^2)^{\frac{1}{2}}(p+p_1-(-v^2)^{\frac{1}{2}})+2((t+t_1)^2)^{\frac{1}{2}}((p+p_1-(-v^2)^{\frac{1}{2}})^2+2(-v^2)^{\frac{1}{2}}$$

$$\times(p+p_1-(-v^2)^{\frac{1}{2}})+t^2+2tt_1+t_1^2)^{\frac{1}{2}})/(p+p_1-(-v^2)^{\frac{1}{2}}))-\frac{1}{2}t_1t(-v^2)^{\frac{1}{2}}/((t+t_1)^2)^{\frac{1}{2}}$$

$$\times\log((2t^2+4tt_1+2t_1^2-2(-v^2)^{\frac{1}{2}}(p+p_1+(-v^2)^{\frac{1}{2}})+2((t+t_1)^2)^{\frac{1}{2}}((p+p_1+(-v^2)^{\frac{1}{2}})^2-2(-v^2)^{\frac{1}{2}}$$

$$\times(p+p_1+(-v^2)^{\frac{1}{2}})+t^2+2tt_1+t_1^2)^{\frac{1}{2}})/(p+p_1+(-v^2)^{\frac{1}{2}}))+\frac{1}{2}t_1^2(-v^2)^{\frac{1}{2}}/((t+t_1)^2)^{\frac{1}{2}}$$

$$\times\log((2t^2+4tt_1+2t_1^2+2(-v^2)^{\frac{1}{2}}(p+p_1-(-v^2)^{\frac{1}{2}})+2((t+t_1)^2)^{\frac{1}{2}}((p+p_1-(-v^2)^{\frac{1}{2}})^2+2(-v^2)^{\frac{1}{2}}$$

$$\times(p+p_1-(-v^2)^{\frac{1}{2}})+t^2+2tt_1+t_1^2)^{\frac{1}{2}})/(p+p_1-(-v^2)^{\frac{1}{2}}))-\frac{1}{2}t_1^2(-v^2)^{\frac{1}{2}}/((t+t_1)^2)^{\frac{1}{2}}$$

$$\times\log((2t^2+4tt_1+2t_1^2-2(-v^2)^{\frac{1}{2}}(p+p_1+(-v^2)^{\frac{1}{2}})+2((t+t_1)^2)^{\frac{1}{2}}((p+p_1+(-v^2)^{\frac{1}{2}})^2-2(-v^2)^{\frac{1}{2}}$$

$$\times(p+p_1+(-v^2)^{\frac{1}{2}})+t^2+2tt_1+t_1^2)^{\frac{1}{2}})/(p+p_1+(-v^2)^{\frac{1}{2}}))-t_1v\times\operatorname{atan}(\frac{1}{2}(2p+2p_1)/v)$$

In: Development of Complex Electromagnetic … ISBN: 978-1-61122-013-1
Editors: K. N. Ramli et al.

Chapter 4

QUASI-STATIC FINITE-DIFFERENCE TIME-DOMAIN SUBGRIDDING TECHNIQUE

***K. N. Ramli*[1]*, R. A. Abd-Alhameed*[2] *and P. S. Excell*[3]**

[1]Faculty of Electrical and Electronics Engineering,
Universiti Tun Hussein Onn Malaysia,
Parit Raja, Batu Pahat, Johor, Malaysia
[2]Mobile and Satellite Communications Research Centre,
Bradford University, Bradford, UK
[3]Institute for Arts, Science and Technology,
Glyndwr University, Wrexham, UK

ABSTRACT

A new approach to the modeling of electromagnetic wave propagation and penetration in and around electrically small objects is presented. The traveling electromagnetic wave from a source is simulated by the finite-difference time-domain solution of Maxwell's equations, and a sub-gridding technique is imposed at points of interest in order to observe the electromagnetic field at high resolution.

The computational burden caused by the requirement for a large number of time steps has been ameliorated by implementing the state-of-the-art quasi-static approach. The method is demonstrated by finding the induced electromagnetic fields near a buried pipeline that runs parallel to 400-kV power transmission lines; results are presented and discussed.

4.1. INTRODUCTION

The sharing of route by overhead high voltage power transmission lines (OHTL) and buried utility pipelines has become quite common nowadays [1].

Most of the countries in the world typically follow the same procedure to distribute the energy to rural and urban territories. The primary reason of this tendency is the restriction imposed by government and private organisations concerning the environmental effect due to the development of the facilities needed on a given location of interest.

The numerous constraints become more alarming as new construction reaches high density urban areas. This sharing of way prompts the question to the public concern as to the effect of OHTL has on underground pipelines. In some situations, the underground utility pipelines may be very close in proximity to OHTL. It is thus necessary to take into account the electromagnetic fields induced above the pipelines [1]. Electromagnetic interference can be generated in the pipelines due to the electromagnetic induction between the underground pipelines and OHTL when they are close to each other in the vicinity. In general, the interference can be composed into two; namely conductive interference and inductive interference. Conductive interference is the potential increase of the ground in the vicinity of the pipelines. It happens from a large current injected into the ground from the transmission line particularly due to the lightning strikes between the tower and the overhead transmission line. In contrast, inductive interference is the voltage generated in the pipelines due to the induction of the electromagnetic field of the OHTL.

This work presents the development of a new approach of modelling the source excitation and the penetration of structures by continuous propagating electromagnetic plane waves. The technique incorporates the solution of time-dependent Maxwell's equations and the initial value problem as the structures are illuminated by the plane waves. The propagation of waves from source excitation is simulated by solving a finite-difference Maxwell's equation in the time domain. The ultimate objective of research in this area is to access the appropriateness of the method in determining the amount of electromagnetic penetrating fields between OHTL and underground utility pipeline. The three-phase OHTL are modelled as the AC sources and the pipeline as the dielectric material. In this case, the pipeline is defined as the fine grid and the residual spaces as the coarse grid in the computational spaces. The fields between these two grids are unknown in nature and have to be calculated. Interpolation algorithm is thus required between the grids. The aim of the present work is to develop the general code for solving the electric and magnetic fields within arbitrary metal or dielectric structures, while maintaining a boundary of uncertainty low reflection level in two-dimensional approach.

In this work, finite-difference time-domain (FDTD) method has been used due to its easiness and potentiality to treat complex geometry structures in the huge calculation region [2-12]. This method of solving Maxwell's differential equations was first proposed in two-dimensional problems [13] and then utilized in three-dimensional applications [14]. However, the standard FDTD method is incompetent if the details of the geometry need to be modelled due to a global fine mesh. As a result, the total number of cells increases dramatically. The time step must be reduced to fulfill the Courant stability condition causing the computational time to increase significantly. The discretisation of time step is crucial for accurate determination of the scheme and has to be small enough to resolve different dielectric or metal structures.

A frequency domain integral equation such as method of moments is well suited for modelling complex of antennas in free space [15]. In other words, its strength is solving PEC structures effectively. Generally, it employs a method of weighted residuals.

All such techniques begin by establishing trial functions with one or more adjustable parameters, and the residuals are obtained from the differences between trial and true solutions. The parameters are found using minimisation to give the best fit of the trial function. In contrast, the time domain FDTD technique is best suited for modelling electromagnetic fields inside and outside inhomogeneous media, in particular the ground.

The presence of arbitrary inhomogeneous objects inside the computational domain does not seriously impact the number of unknowns to be determined.

Many researchers in the past have been prompted to investigate subgridding technique as an analysis approach in the interaction between source and scatterer [16-21]. In general, this technique is used to condense the lattice at the point of interest locally and does not require any analytical formula to be taken into account, and hence it is appropriate for objects of any shape. The residual of the space is filled with coarse grids. The fields on the boundary between coarse and fine grids are coupled using spatial and temporal interpolations. The regions of the coarse and fine grids are computed by the FDTD method and are kept in time step. A stable subgridding algorithm can refine the mesh locally and improve the accuracy of the result without increasing the computational efforts significantly. It is hence very useful for FDTD code.

Chilton and Lee [22] developed a subgridding method based on multigrid finite element principles in 2-D. The work was then extended to 3-D. The method was applied to the parallel plate waveguide with different dielectric material and shape for validation purposes. Later, they developed a higher order FDTD with appropriate media boundary conditions. The method which was called conservative subgridding was derived for Lobatto Cell [23]. Lin and Kuo [24] used Crank-Nicolson algorithm in the subgridding scheme to handle problems involving fine structural features. Numerical results were presented for unilateral fin-line and dielectric cylinder scattering. Some attempts have been taken in the analysis of cavity with one and two perfect electric conductor (PEC) fins [25, 26], metamaterial slab [27], electrostatic plate condensator, circular coaxial waveguide and circular waveguide [28], PEC cylinder with a current filament in free space [29, 30], lossy dielectric block located near a short linear antenna [31] and passive equivalent circuit [32].

FDTD technique has been applied to the high voltage power transmission line analysis in the published literature. The computation of transient electromagnetic fields due to switching within a typical high-voltage air-insulated substation (AIS) was done by using FDTD method [33]. Dedkova and Kriz [34] proposed a new effective approach to evaluate the distribution of voltage and current along the nonlinear transmission line by using FDTD method. An improved technique was proposed by Tang et al. [35] to calculate the transient inductive interference in underground metallic pipelines due to a fault in nearby power lines. The frequency-dependent problem in the analysis of transient interference was solved in phase domain based on FDTD method. Lu and Cui [36] used FDTD method to calculate the wave processes of voltage and current distributed along the three-phase 500 kV busbars and the power lines without load in the substation of multi-conductor transmission lines (MTL). The iterative formulas were presented to determine the boundary conditions at the node of the branches. The work was extended to transmission line network and non-uniform line [37]. Vector fitting method was adopted in FDTD to treat with the frequency dependent parameters [38]. In this case, the corresponding voltage and current recursion formulations in FDTD technique were presented based on the recursive algorithm for time domain convolution. The comparison of transient analysis method using Bergeron's method, FDTD method and time-domain finite-element (TDFE) was discussed in [39]. Numerical results of MTL simulations based on Laplace transform and FDTD method was presented and compared in [40].

The method mentioned above has two main advantages relative to the other modelling approaches. First, it is simple to implement for complicated dielectric or metal structures due to arbitrary electrical parameters can be assigned to each cell on the grid.

Second, the entire computational spaces need not to be discretised with a fine grid as it put unreasonable burden on the computer processing time. The ultimate objective of research in this area is to access the appropriateness of the method in determining the amount of electromagnetic penetrating fields between transmission lines and underground utility pipeline. The three-phase transmission lines are modelled as the AC sources and the pipeline as the dielectric material. In this case, the pipeline is defined as the fine grid and the residual spaces as the coarse grid in the computational spaces. The fields between these two grids are unknown in nature and have to be calculated. Interpolation algorithm is thus required between the grids. The aim of the present work is to develop the general code for solving the electric and magnetic fields within arbitrary metal or dielectric structures, while maintaining a boundary of uncertainty low reflection level in two-dimensional approach.

4.2. Quasi-Static Theorem

FDTD technique is not a practical scheme when low frequency is applied due to lengthy simulation time. The computer burden cannot be solved even for reasonable spatial resolution. For example, the spatial resolution of $\Delta x = \Delta y = \Delta z = 1$ cm is given at 50 Hz power line frequency. The duration of time step required is given from Courant stability criterion, $\Delta t = \Delta x / c\sqrt{3} = 1.92$ ns. In order to cover one complete cycle, the number of time steps needed is $N = 1/(f\Delta t) = 1.0 \times 10^7$. It needs many years to complete the simulation even when run on fast machine. However, a method known as quasi-static approximation proposed by Moerloose et al. [41] solve the difficulties. The formulation takes into account the wavelength which is much greater than the object of study. The primary advantage of quasi-static formulation is thus reducing the long simulation time constraint. There have been some early efforts of using quasi-static idea to study the interaction between living tissues exposed to extremely low frequency (ELF) electric fields in the published literature such as [42-44]. The basis of their research dominantly verifies the effectiveness of quasi-static scheme at very low frequency. FDTD comes into play when researches in [45-

48] applied the same knowledge. In this case, the dimension of the object of study is at a fraction of the wavelength. An attempt to gain the scheme of using the same technique but at much higher frequencies of 900 MHz and 1800 MHz has been made by See et al. in 2003 [49-50].

The theoretical method discussed by the authors in [42-44] has been realised in the present work to approximate the quasi-static FDTD subgridding. In general, two conditions must be satisfied before applying the quasi-static formulation:

i The size of the object is a factor of 10 or more, smaller than the wavelength.

ii $|\sigma + j\omega\varepsilon| >> \omega\varepsilon_o$

where:

σ is the conductivity of the object (S/m)
ε is permittivity of the object (F/m)
ω is the angular frequency (measured in radians per second, with units s^{-1})
ε_o is the free space permittivity (8.85×10^{-12} F/m)

From the conditions stated above, the electric field components tangent to the surface of the structure and the internal fields are roughly zero compared to the applied field. The external electric field components can be viewed as orthogonal to the structure. From Maxwell's equation with div $D = \rho$, the boundary condition for the normal electric field components at the surface of the region of interest is given by the expression [43-45]:

$$j\omega\varepsilon_o \hat{n} \bullet \vec{E}_{air} = (\sigma_{tissue} + j\omega\varepsilon_{tissue})\hat{n} \bullet \vec{E}_{tissue} \quad (4.1)$$

The '•' symbol in equation (4.1) basically refers to vector dot product. From this equation with the two stated conditions satisfied, the scaling relationship can be deduced [43-45]:

$$\vec{E}_{tissue}(f) = \left(\frac{\omega}{\omega'}\right)\left[\frac{\sigma'(f') + j\omega'\varepsilon(f')}{\sigma(f) + j\omega\varepsilon(f)}\right]\vec{E}'_{tissue}(f') \quad (4.2)$$

where:

$\vec{E}_{tissue}(f)$ is the resultant internal electric field (V/m)
$\vec{E}'_{tissue}(f')$ is the scaling internal electric field (V/m)
f is the frequency of interest (Hz)
f' is the scaling frequency (Hz)
ω is the angular frequency of interest (s^{-1})
ω' is the scaling angular frequency (s^{-1})
σ is the conductivity of the object (S/m)
σ' is the scaling conductivity of the object (S/m)

Assuming that $\omega\varepsilon(f) << \sigma(f)$ and $\omega'\varepsilon'(f') << \sigma'(f')$, then equation (4.2) can be approximated as [43-45]:

$$\vec{E}_{tissue}(f) \cong \left[\frac{f\sigma'(f')}{f'\sigma(f)}\right]\vec{E}'_{tissue}(f') \tag{4.3}$$

It can be concluded that from this equation, a higher working scaling frequency f' which falls within the quasi-static region can be chosen to excite the model in order to reduce the computational burden. Hence, the scaling internal electric field which is calculated at much higher frequency of 460 kHz can be shifted back to the actual 50 Hz power line frequency.

4.3. Validation of Subgridding Method

The main grid of the computational domain was basically divided into subgrids, and the missing fields on the boundary between them are predicted using temporal and spatial interpolations. The average electrical characteristics were considered between the main grid cells and fine grid (subgridding cells) when dielectric material (in this case the pipeline) was present on the interface surface. The field components will be updated on both the main grid and subgridding cells as shown in Figure 4.1.

Subgrid technique was validated by illustrating example in two cases. Case 1 was considered when the observed field was located inside subgrid area with two conditions: (a) Without subgrid, and (b) With subgrid, as shown in Figure 2 (refers to the x symbol inside the grid region). Case 2 was considered when the observed field was located outside subgrid area with two conditions:

(a) Without subgrid, and (b) With subgrid, as depicted in Figure 4.2 (refers to the x symbol outside the grid region).

The problem space was excited by sinusoidal wave and Gaussian pulse at 1800 MHz. The electric fields at the same point for case 1 and case 2 were observed and compared as illustrated in Figures 4.3(a).

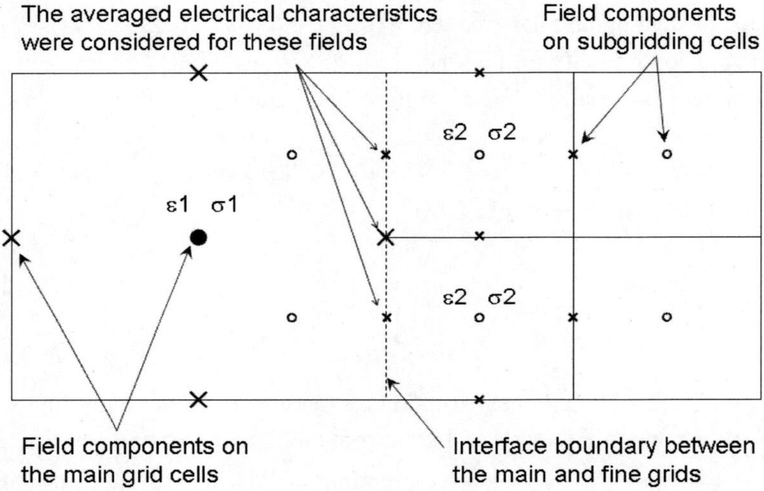

Figure 4.1. Subgridding model with field components at the main and fine grid.

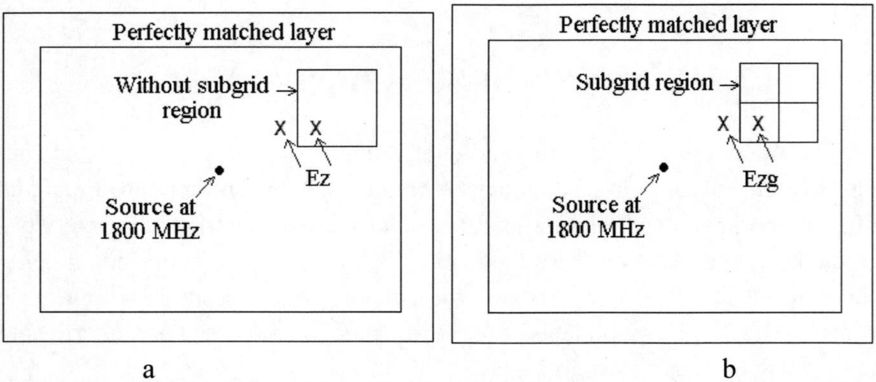

Figure 4.2. The observed field was located inside/outside subgrid area: (a) Without subgrid, (b) With subgrid.

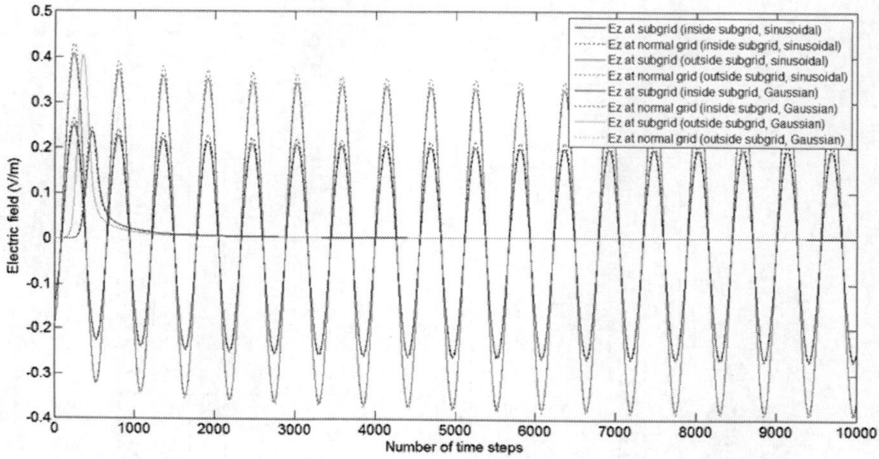

a

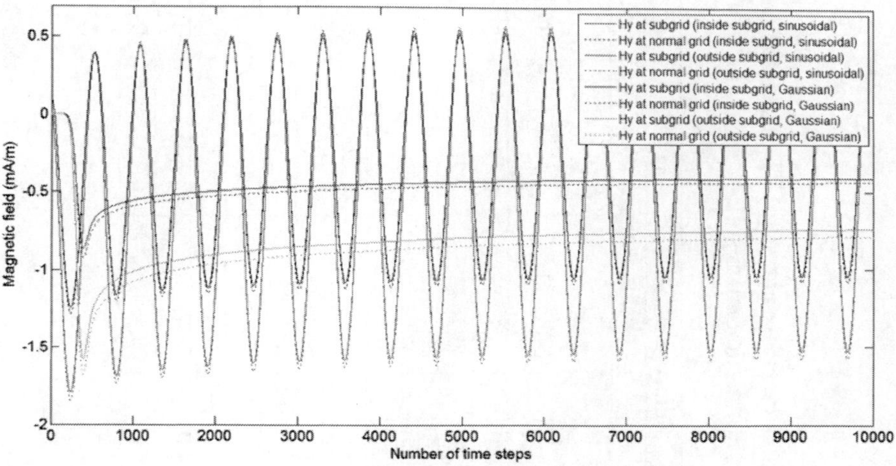

b

Figure 4.3. The observed fields: (a) The electric field inside subgrid and at normal grid, (b) The magnetic field inside subgrid and at normal grid.

The corresponding electric field values in numerical is shown in Table 4.1(a). The magnetic fields at the same point for case 1 and case 2 were also observed and compared as illustrated in Figures 4.3(b). The corresponding magnetic field values in numerical is shown in Table 4.1(b).

Table 4.1. The observed fields in numerical values: (a) The electric field inside subgrid and at normal grid, (b) The magnetic field inside subgrid and at normal grid

Number of Time Steps	Sinusoidal				Gaussian			
	Inside Subgrid		Outside Subgrid		Inside Subgrid		Outside Subgrid	
	E_z at Subgrid	E_z at Normal Grid	E_z at Subgrid	E_z at Normal Grid	E_z at Subgrid	E_z at Normal Grid	E_z at Subgrid	E_z at Normal Grid
1	-0.1400	-0.1400	-0.1900	-0.1900	-0.1900	0.0087	-0.1900	0.0087
1000	-0.1523	-0.1527	-0.2088	-0.2095	-0.2088	0.0191	-0.2088	0.0135
2000	0.1067	0.1154	0.1881	0.2014	0.1881	0.0061	0.1881	0.0043
3000	0.1994	0.2113	0.3301	0.3483	0.3301	0.0035	0.3301	0.0025
4000	-0.0135	-0.0091	0.0038	0.0106	0.0038	0.0025	0.0038	0.0018
5000	-0.2440	-0.2477	-0.3494	-0.3550	-0.3494	0.0019	-0.3494	0.0013
6000	-0.1781	-0.1794	-0.2483	-0.2504	-0.2483	0.0014	-0.2483	0.0010
7000	0.0899	0.0980	0.1624	0.1747	0.1624	0.0012	0.1624	0.0008
8000	0.1869	0.1983	0.3110	0.3285	0.3110	0.0010	0.3110	0.0007
9000	-0.0233	-0.0192	-0.0111	-0.0049	-0.0111	0.0008	-0.0111	0.0006
10000	-0.2520	-0.2559	-0.3616	-0.3676	-0.3616	0.0007	-0.3616	0.0005

a

Number of Time Steps	Sinusoidal				Gaussian			
	Inside Subgrid		Outside Subgrid		Inside Subgrid		Outside Subgrid	
	H_y at Subgrid	H_y at Normal Grid	H_y at Subgrid	H_y at Normal Grid	H_y at Subgrid	H_y at Normal Grid	H_y at Subgrid	H_y at Normal Grid
1	0.1000	0.1000	0.0067	0.0067	0.0039	0.0039	0.0055	0.0055
1000	0.1428	0.1441	0.0571	0.0588	0.0571	0.0588	0.0571	0.0588
2000	-0.7502	-0.7757	-1.1335	-1.1675	-1.1335	-1.1675	-1.1335	-1.1675
3000	-1.0701	-1.1052	-1.5602	-1.6070	-1.5602	-1.6070	-1.5602	-1.6070
4000	-0.3365	-0.3496	-0.5820	-0.5995	-0.5820	-0.5995	-0.5820	-0.5995
5000	0.4586	0.4693	0.4781	0.4925	0.4781	0.4925	0.4781	0.4925
6000	0.2316	0.2356	0.1755	0.1808	0.1755	0.1808	0.1755	0.1808
7000	-0.6922	-0.7160	-1.0563	-1.0879	-1.0563	-1.0879	-1.0563	-1.0879
8000	-1.0271	-1.0609	-1.5028	-1.5479	-1.5028	-1.5479	-1.5028	-1.5479
9000	-0.3028	-0.3149	-0.5371	-0.5532	-0.5371	-0.5532	-0.5371	-0.5532
10000	0.4859	0.4975	0.5145	0.5300	0.5145	0.5300	0.5145	0.5300

b

The electric fields in subgrid region (E_{zg}) and at normal grid (E_z) for case 1 were found to be identical to each other to confirm the proof of concept. The electric fields E_z with and without subgrid for case 2 were also found to be identical to each other. A similar explanation also applies for the magnetic fields for both case 1 and 2.

The results in Figure 4.4 illustrate the stability of the simulation inside the problem space. The electric field remained at 0.23 V/m when using different values of subgrid cells to justify that the results converge with the mesh size in the computational domain.

4.4. INTERACTION BETWEEN OHTL AND BURIED PIPELINE

A source code was written to implement the design and analysis of the interaction between overhead high voltage power transmission lines and buried utility pipeline. Fortran 90 was used as a programming language platform. The work was devoted to 2-D TM case.

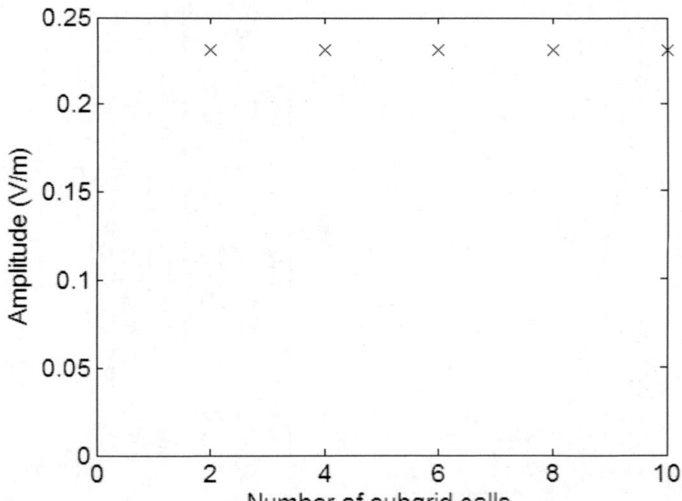

Figure 4.4. Electric field distribution for different numbers of subgrid cells in one main FDTD cell.

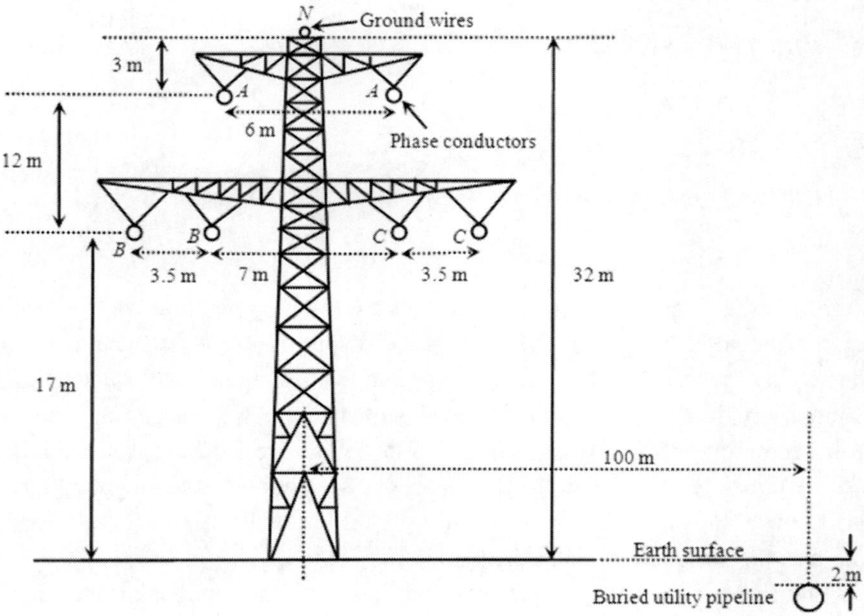

Figure 4.5. Outline of standard circuit 400 kV steel lattice transmission high voltage suspension towers with normal span of 300 m (low height construction design, not to scale) [51, 52].

Figure 4.5 illustrates the cross section and the dimension of a common corridor in which a buried utility pipeline runs parallel to a 400 kV overhead power transmission line. It is designed with low height construction. The height from the ground to the bottom conductors is 17 m. The distance from the overhead ground wire to the earth surface is 32 m. Phase *A* conductors at the top were collocated horizontally with a separation of 3.0 m. The bottom conductors of phase *B*, phase *B* to *C* and phase *C* were collocated horizontally with a separation of 1.75 m, 3.5 m and 1.75 m between two adjacent conductors respectively. The three phase steel lattice transmission high voltage suspension tower was designed with six cables. These cables were used as the source signal which propagates inside the problem space. Each of the two cables carries the same phase of the AC current.

The general equations of phase *A*, phase *B* and phase *C* cables were given respectively by the expressions:

$$Phase\ A = \sin(2\pi ft) \tag{4.4}$$

$$Phase\ B = \sin\left(2\pi ft + \frac{2}{3}\pi\right) \tag{4.5}$$

$$Phase\ C = \sin\left(2\pi ft + \frac{4}{3}\pi\right) \tag{4.6}$$

where f is the frequency (Hz) and t is the time (s). The pipeline was separated at a distance of 100 m from the steel lattice suspension towers and buried 2 m beneath the surface of the earth. It was made from metal with a very high conductivity of 4.75×10^6 S/m. The radius of the pipeline was 25 cm. The soil in the common corridor was designed to be inhomogeneous. It was modelled with different relative permittivity by means of random number generator. It was known that the relative permittivity of soil varies from 1 to 5 at 460 kHz, where as the conductivity is kept at 2.0×10^{-3} S/m [53].

Figure 4.6 represents a histogram that indicates the number of occurrance that soil with their respective value of random relative permittivity from 1 to 5. Figure 4.7 depicts the cumulative distribution function of soil relative permittivity.

The plot is basically based on equation (7). The representation of Figures 4.6 and 4.7 clearly indicates that the soil was designed as arbitrarily inhomogeneous.

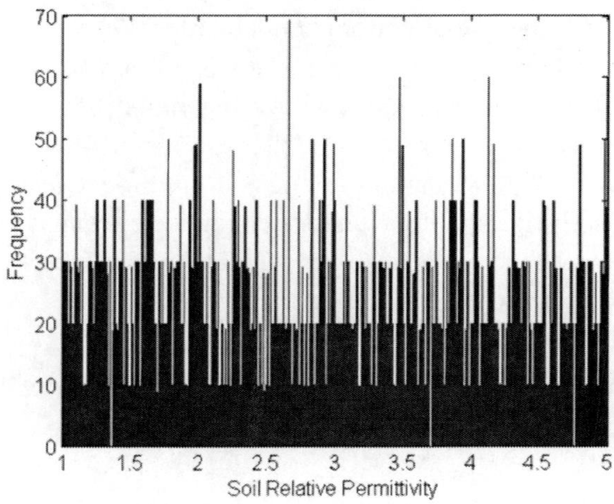

Figure 4.6. Histogram of soil relative permittivity.

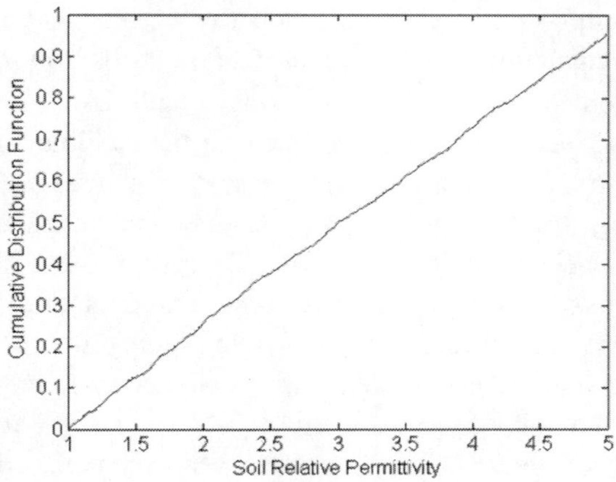

Figure 4.7. Cumulative distribution function of soil relative permittivity.

The cumulative distribution function (CDF) is given by:

$$CDF(x) = \int_{1}^{x_1} \frac{1}{4}\, dx \tag{4.7}$$

where x_1 is the variable of which is unknown in nature and must be determined by using random number generator. Basically, random number generator is given by the expression:

$$R_g(0 \rightarrow 1) = \frac{1}{4}(x_1 - 1) \tag{4.8}$$

where $R_g(0 \rightarrow 1)$ is the random number generator that generates random number from zero to one. Rearranging equation (4.8), the x_1 term can be deduced as:

$$x_1 = 4R_g(0 \rightarrow 1) + 1 \tag{4.9}$$

In addition, the conductivity of soil mainly depends on the water content in it and slightly on the granularity. In general, its value was very small typically in the order of 2.0×10^{-3} S/m or less [53]. Homogeneous soil also was used in order to compare the correlation between the neighbouring cells. It was modelled with $\varepsilon_r = 3.0$ while its conductivity remained the same value as before. The computational region at the coarse grids was discretised at a spatial resolution of 2,609 cells per wavelength ($\Delta y = \Delta z = 25$ cm). Subgridding involves local mesh refinement in the pipeline and some part of the ground in order to determine the propagation of the waves inside that area while observing the change in the electric and magnetic fields. The computational space for main region was 521 cells × 185 cells (130.25 m × 46.25 m) as shown in Figure 4.8. The subgrid computational space was 40 subgrid cells × 40 subgrid cells as illustrated in Figure 4.9.

In other words, there were many colours that represent random distribution of the soil. In contrast, the non-colour (white area) depicts the pipeline. The values for the other parameters were summarised in Table 4.2.

The distribution of ground surrounding the pipeline was generated by using random number to simulate the inhomogeneity of the media. The fine grids was discretised at a spatial resolution of 10,435 cells per wavelength ($\Delta y = \Delta z = 6.25$ cm). In other words, the ratio of the coarse to the fine grids was 4:1. The length of the coarse grids remained at 3.83×10^{-4} of the wavelength. The length of the fine grids remained at 9.58×10^{-5} of the wavelength. The induced EM fields section above the pipeline were observed for 30 cells × 20 cells (7.5 m × 5 m). The Courant stability condition for 2-D case is given by:

$$\Delta t \leq \frac{h}{c\sqrt{2}} \tag{4.10}$$

where h is the spatial homogeneous FDTD grid ($h = \Delta y = \Delta z$) and c is the speed of light in free space.

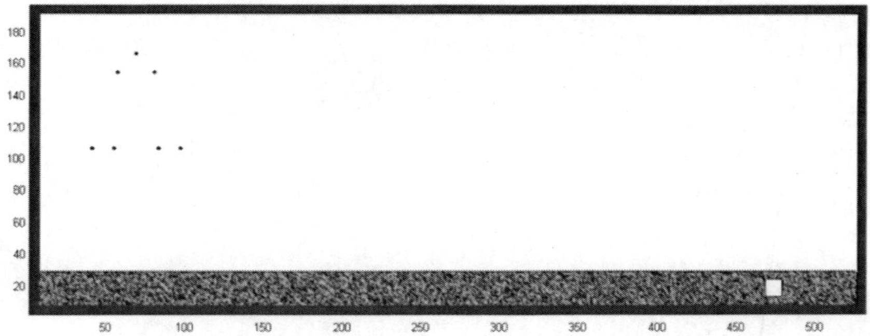

Figure 4.8. The main region in the computational domain for 400 kV steel lattice transmission high voltage suspension towers model.

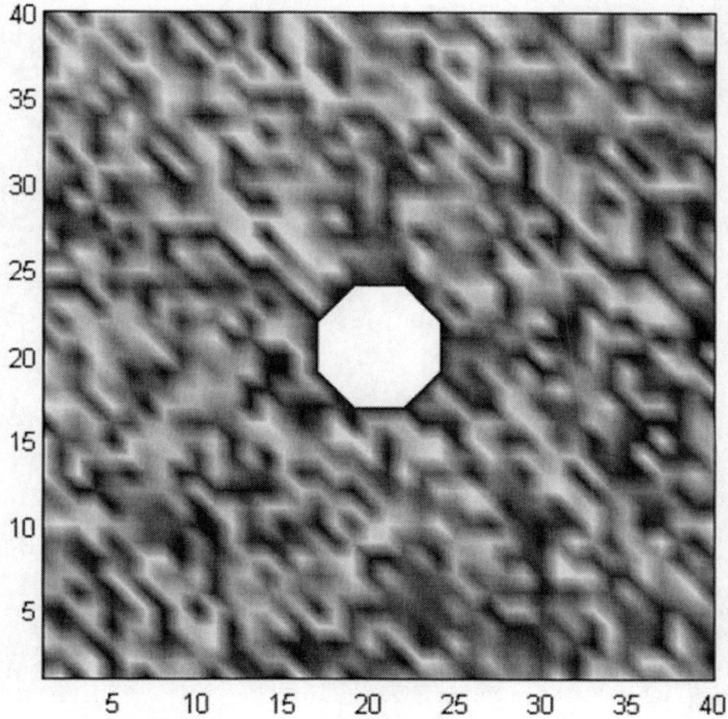

Figure 4.9. The subgrid region in the computational domain for 400 kV steel lattice transmission high voltage suspension towers model.

Table 4.2. FDTD simulation parameters

Parameter	Measurement
Source frequency	460 kHz
Coarse grids	25 cm
Fine grids	6.25 cm
Refinement factor	4
Time step	0.4 ns
Number of time steps	21,160
Number of cycles	4
Subgrid spatial resolution	40 subgrid cells × 40 subgrid cells
Induced EM fields spatial resolution	30 cells × 20 cells

According to this equation, the time step was set at 0.4 ns to satisfy the Courant stability condition. The simulation was run for 21,160 time steps to allow for the wave to fully traverse the spatial domain for 4 cycles.

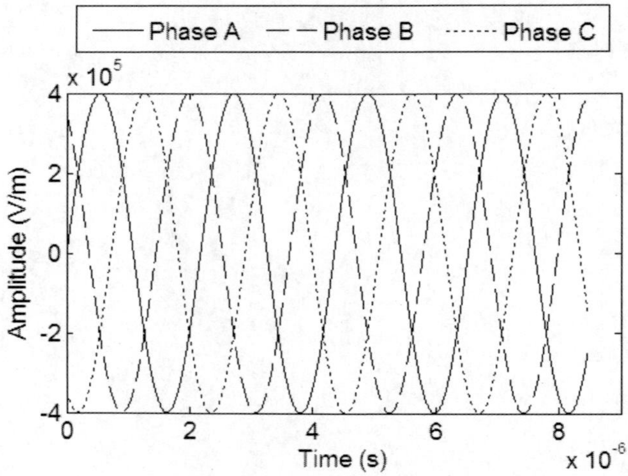

Figure 4.10. Three-phase sinusoidal sources driven from 400 kV steel lattice transmission high voltage suspension towers.

4.5. Near Field Simulation

The interaction results between 400 kV steel suspension towers and metallic cylinder are presented in this section. The frequency of 460 kHz was employed to get the fields before they were altered back to 50 Hz. Figure 4.10 shows the three-phase 400 kV sinusoidal sources separated by 120° phase shift.

The fields E_{zg}, H_{yg} and H_{xg} were observed at point (31.25 cm, 31.25 cm) within subgrid section. These fields distribution were plotted in Figure 4.11 for homogeneous and random distribution soil.

The EM wave which travels from the suspension tower to the pipeline varies from 2.24×10^{-7} V/m (-133 dBV/m) to 6.3×10^{3} V/m (76 dBV/m).

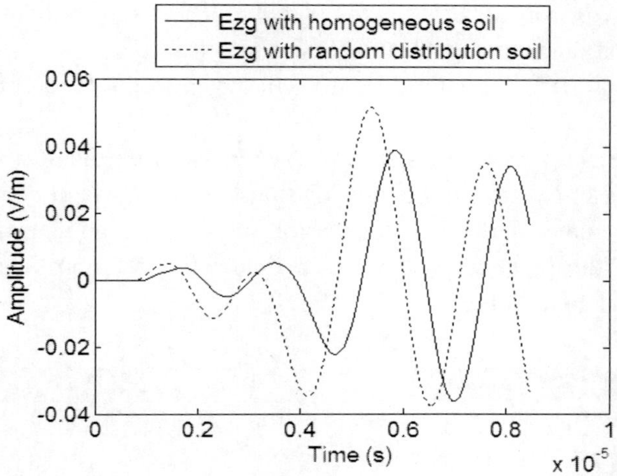

a

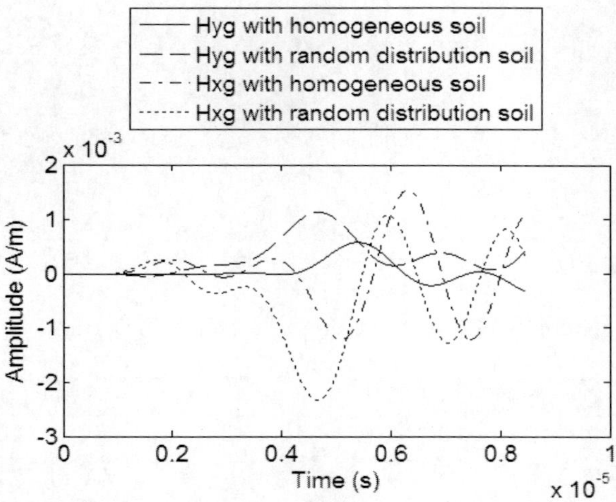

b

Figure 4.11. The amplitude of EM fields plotted against time inside subgrid region: (a) Electric field E_{zg}, (b) Magnetic field H_{yg} and H_{xg}.

The distribution of the electric field E_z, magnetic fields H_y and H_x through the simulated FDTD computational space were given in Figures 4.12, 4.13 and 4.14 respectively.

In subgrid region, the distribution of E_{zg}, H_{yg} and H_{xg} were demonstrated in Figures 4.15, 4.16 and 4.17 respectively. Here, the fields inside the metallic pipeline were also found to be zero. The reason for this phenomenon was due to the excess electrons at the surface of the metal preventing any incoming propagating waves from penetrating the pipeline. It was shown that the electric field distribution surrounding the pipeline alters from 7.9 × 10^{-7} V/m (-122 dBV/m) to 7.9 × 10^{-4} V/m (-62 dBV/m) which shows good conformity with [54].

The difference in relative distance of each phase from the nearby pipeline can create phase imbalance in the transmission line. Under fault condition, the currents on the faulty phases of transmission lines were high causing induced AC voltage on the pipeline. The induced field will not contribute to shock hazard in normal condition.

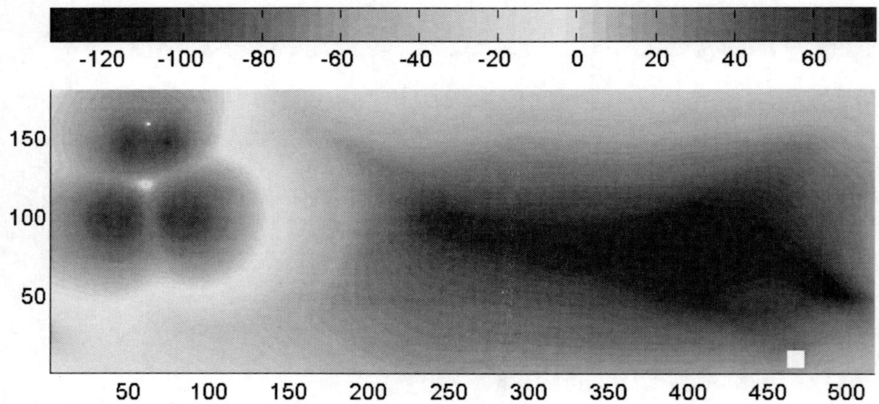

Figure 4.12. The electric field E_z distribution in the main FDTD grid.

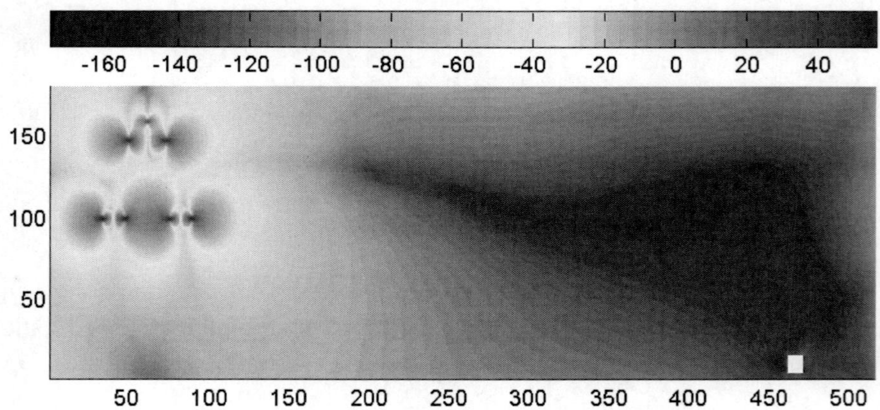

Figure 4.13. The magnetic field H_y distribution in the main FDTD grid.

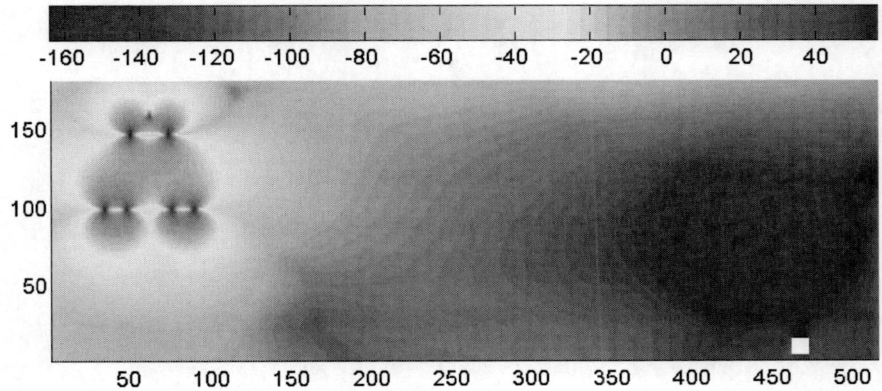

Figure 4.14. The magnetic field H_x distribution in the main FDTD grid.

Figures 4.18, 4.19 and 4.20 illustrate the induced EM fields for E_z, H_y and H_x respectively. As can be seen, the current amplitude induced on the pipe varies (H_y magnetic fields in figure 15(b)) from 1.8×10^{-8} A/m (-155 dBA/m) to 1.3×10^{-7} A/m (-138 dBA/m), and (H_x magnetic fields in figure 15(c)) from 5.6×10^{-8} A/m (-145 dBA/m) to 4.0×10^{-7} A/m (-128 dBA/m). It is noteworthy that these currents do not produce any sudden risk to a person nearby.

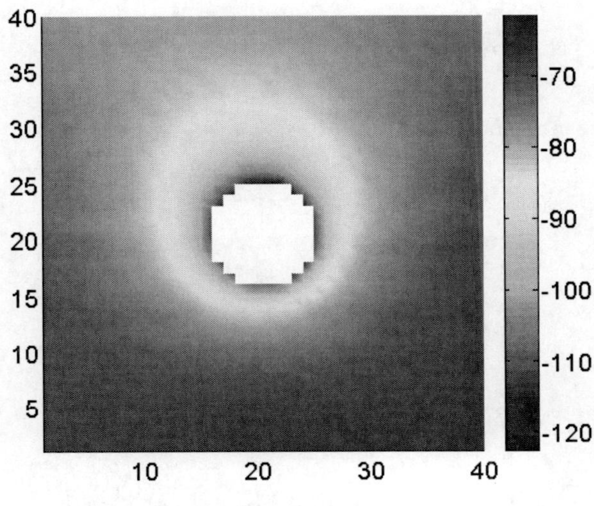

Figure 4.15. The electric field E_{zg} distribution in the subgrid section.

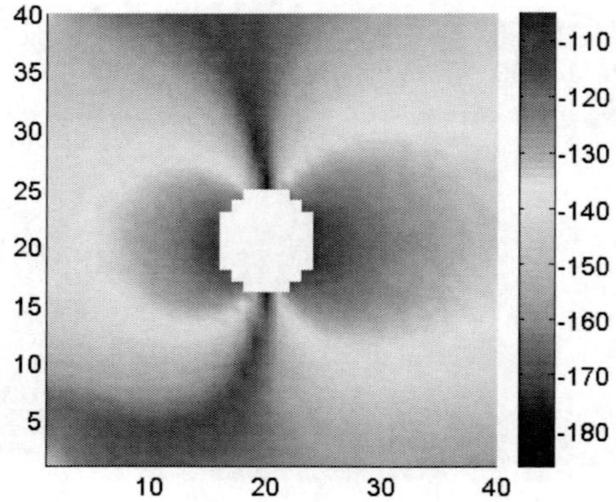

Figure 4.16. The magnetic field H_{yg} distribution in the subgrid section.

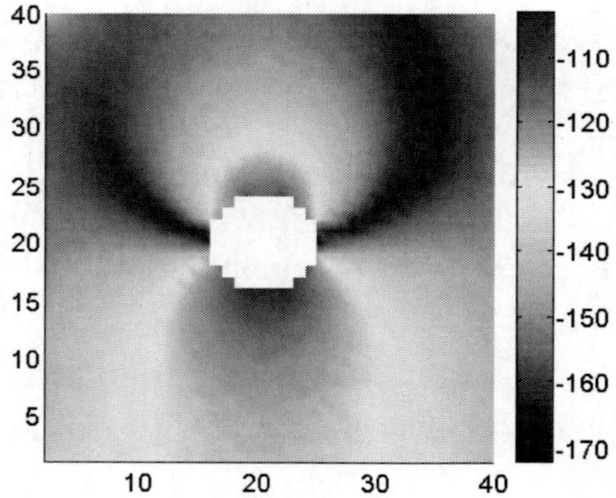

Figure 4.17. The magnetic field H_{xg} distribution in the subgrid section.

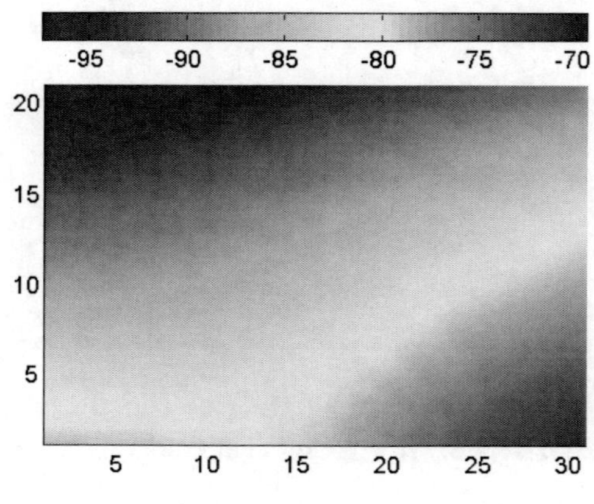

Figure 4.18. The induced electric field E_z at 1.75 m above metallic pipeline.

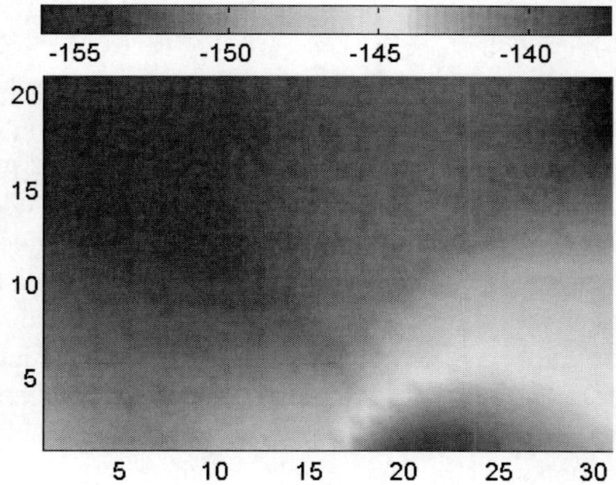

Figure 4.19. The induced magnetic field H_y at 1.75 m above metallic pipeline.

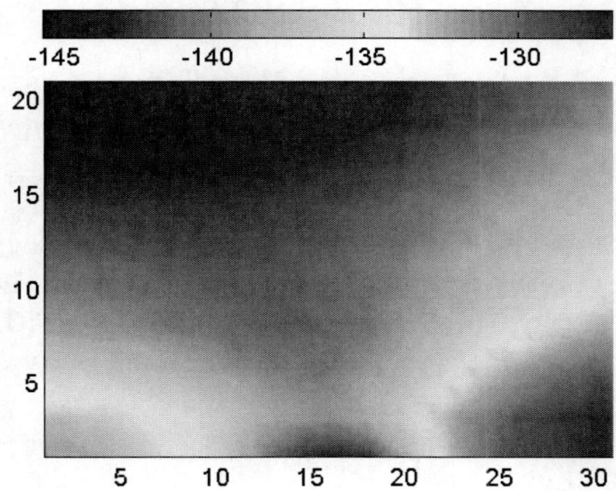

Figure 4.20. The induced magnetic field H_x at 1.75 m above metallic pipeline.

CONCLUSION

An approach to model the interaction between overhead transmission lines and underground utility pipeline at power-line frequency has been presented. This uses the FDTD technique for the whole structure of the problem combined with subgridding method at the object of interest particularly at the underground pipeline. By implementing a modified version of Berenger's

PML, the reflection on the boundary layers inside the spatial FDTD computational region has been successfully decreased, although it is surrounded by lossy penetrable media. The computational burden due to huge number of time steps in the order of tens of millions has been eased to tens of thousands by employing the method called quasi-static approximation scheme. The use of inhomogeneous soil in the common corridor permits a non-trivial proximity region of authentic ground properties to be simulated. Profound investigation of the interaction between electromagnetic fields and natural or utility arrangement with different electrical characteristics at different level of spatial resolution can be assisted by such tools. The combination of frequency scaling SGFDTD approach with arbitrary inhomogeneous dielectric volume, and the modified Berenger's PML paves a way as a good candidate model of EM fields interaction modelling for complex geometries.

REFERENCES

[1] A. W. Peabody and A. L. Verhiel, "The effects of high-voltage AC transmission lines on buried pipelines," *IEEE Transactions on Industry and General Applications*, vol. IGA-7, 1971.

[2] R. Xiong, B. Chen, Y. Yi, H.-L. Chen, and H.-F. Zhao, "Two-step method for the FDTD analysis of long apertures having depth," *IEEE Antennas and Wireless Propagation Letters*, vol. 12, pp. 39-42, 2013.

[3] S. K. Kamepally, B. P. Kumar and C. S. Paidimarry, "FDTD estimation for accurate specific absorption rate in a tumor," *Annual International Conference on Emerging Research Areas and International Conference on Microelectronics, Communications and Renewable Energy (AICERA/ICMiCR)*, pp. 1-5, 2013.

[4] W. G. Whittow, C. C. Njoku and Y. C. Vardaxoglou, "Fine scale simulations of patch antennas with heterogeneous substrates," *USNC-URSI Radio Science Meeting (Joint with AP-S Symposium)*, pp. 223, 2013.

[5] A. C. Lesina, A. Vaccari and A. Bozzoli, "A modified RC-FDTD algorithm for plasmonics in Drude dispersive media nanostructures," *6th European Conference on Antennas and Propagation (EUCAP)*, pp. 687-690, 2012.

[6] G. M. Noetscher, Y. Xu and S. N. Makarov, "Accuracy of point source models with coincident phase centers in a cubic FDTD grid for arbitrary

source orientation," *IEEE Antennas and Propagation Society International Symposium (APSURSI)*, pp. 1-2, 2012.

[7] V. Nayyeri, M. Soleimani and M. Dehmollaian, "Implementation of a PEMC boundary condition in the 2-D FDTD technique," *IEEE Antennas and Propagation Society International Symposium (APSURSI)*, pp. 1-2, 2012.

[8] D.-C. Liu and H.-C. Chang, "The second-order condition of FDTD method at sloped dielectric interfaces by averaging the permittivities along surface normal," *IEEE Transactions on Antennas and Propagation*, vol. 60, no. 11, pp. 5259-5267, 2012.

[9] M. Weiwei, S. Dong and W. Xianliang, "UPML-FDTD parallel computing on GPU," *International Conference on Microwave and Millimeter Wave Technology (ICMMT)*, vol. 4, pp. 1-4, 2012.

[10] C. Guiffaut, A. Reineix and B. Pecqueux, "New oblique thin wire formalism in the FDTD method with multiwire junctions," *IEEE Transactions on Antennas and Propagation*, vol. 60, no. 3, pp. 1458-1466, 2012.

[11] D. M. Shyroki, "Modeling of sloped interfaces on a Yee grid," *IEEE Transactions on Antennas and Propagation*, vol. 59, no. 9, pp. 3290-3295, 2011.

[12] A. Taflove and S. C. Hagness, *Computational electrodynamics: The finite-difference time-domain method,* 3rd ed. Boston, MA: Artech House, 2005.

[13] K. S. Yee, "Numerical solution of initial boundary value problems involving Maxwell's equations in isotropic media," *IEEE Transactions on Antennas and Propagation*, vol. AP-14, pp. 302-307, 1966.

[14] G. Mur, "Absorbing boundary conditions for the finite-difference approximation of the time-domain electromagnetic-field equations," *IEEE Transactions on Electromagnetic Compatibility*, vol. EMC-23, pp. 377-382, 1981.

[15] S. N. Makarov, S. D. Kulkarni, A. G. Marut, and L. C. Kempel, "Method of moments solution for a printed patch/slot antenna on a thin finite dielectric substrate using the volume integral equation," *IEEE Transactions on Antennas and Propagation*, vol. 54, pp. 1174-1184, 2006.

[16] K. Ramli, R. A. Abd-Alhameed, C. H. See, P. S. Excell, and J. M. Noras, "Hybrid Computational Scheme for Antenna-Human Body Interaction," *Progress In Electromagnetics Research (PIER)*, vol. 133, pp. 117-136, 2013.

[17] K. Xiao, D. J. Pommerenke and J. L. Drewniak, "A three-dimensional FDTD subgridding algorithm with separated temporal and spatial interfaces and related stability analysis," *IEEE Transactions on Antennas and Propagation*, vol. 55, pp. 1981-1990, 2007.

[18] T. Ohtani, K. Taguchi and T. Kashiwa, "A subgridding technique for the complex nonstandard FDTD method," *Electronics and Communications in Japan Part 2 - Electronics*, vol. 87, pp. 1-9, 2004.

[19] K. M. Krishnaiah and C. J. Railton, "A stable subgridding algorithm and its application to eigenvalue problems," *IEEE Transactions on Microwave Theory and Techniques*, vol. 47, pp. 620-628, 1999.

[20] S. M. Wang, F. L. Teixeira, R. Lee, and J. F. Lee, "Optimization of subgridding schemes for FDTD," *IEEE Microwave and Wireless Components Letters*, vol. 12, pp. 223-225, 2002.

[21] S. Kapoor, "Sub-cellular technique for finite-difference time-domain method," *IEEE Transactions on Microwave Theory and Techniques*, vol. 45, pp. 673-677, 1997.

[22] R. A. Chilton and R. Lee, "Conservative and provably stable FDTD subgridding," *IEEE Transactions on Antennas and Propagation*, vol. 55, pp. 2537-2549, 2007.

[23] R. A. Chilton and R. Lee, "Conservative subgridding for lobatto cell FDTD method," *IEEE Antennas and Propagation Society International Symposium*, pp. 1-4, 2008.

[24] T. C. Lin and C. W. Kuo, "A novel and efficient subgridding scheme in FDTD method using the Crank-Nicolson algorithm," *Microwave and Optical Technology Letters*, vol. 49, pp. 3103-3106, 2007.

[25] S. H. Sun and C. T. M. Choi, "A new multilevel subgridding scheme for two-dimensional FDTD method," *IEEE Transactions on Magnetics*, vol. 40, pp. 1025-1028, 2004.

[26] S. H. Sun and C. T. M. Choi, "A new subgridding scheme for two-dimensional FDTD and FDTD (2,4) methods," *IEEE Transactions on Magnetics*, vol. 40, pp. 1041-1044, 2004.

[27] B. Donderici and F. L. Teixeira, "Improved FDTD subgridding algorithms via digital filtering and domain overriding," *IEEE Transactions on Antennas and Propagation*, vol. 53, pp. 2938-2951, 2005.

[28] O. Podebrad, M. Clemens and T. Weiland, "New flexible subgridding scheme for the finite integration technique," *IEEE Transactions on Magnetics*, vol. 39, pp. 1662-1665, 2003.

[29] C. C. Chang and S. K. Jeng, "A multilevel subgridding scheme for two-dimensional finite-difference time-domain method," *IEEE Antennas and Propagation Society International Symposium*, vol. 1, pp. 36-39, 2001.

[30] C. C. Chang and S. K. Jeng, "Modified subgridding algorithm with multilevel scheme for two-dimensional finite-difference time-domain method," *Asia-Pacific Microwave Conference Proceedings*, vol. 1-3, pp. 831-834, 2001.

[31] W. H. Yu and R. Mittra, "A new subgridding method for the finite-difference time-domain (FDTD) algorithm," *Microwave and Optical Technology Letters*, vol. 21, pp. 330-333, 1999.

[32] K. M. Krishnaiah and C. J. Railton, "Passive equivalent circuit of FDTD: An application to subgridding," *Electronics Letters*, vol. 33, pp. 1277-1278, 1997.

[33] B. U. Musa, W. H. Siew and M. D. Judd, "Computation of transient electromagnetic fields due to switching in high-voltage substations," *IEEE Transactions on Power Delivery*, vol. 25, pp. 1154-1161, 2010.

[34] J. Dedkova and T. Kritz, "FDTD analysis of a nonlinear transmission line," *Progress in Electromagnetics Research Symposium*, Beijing, China, pp. 282-285, 2009.

[35] J. Tang, X. Cui, L. Qi, T. Lu, L. Li, P. Zhu, G. Yang, and W. Zhang, "Analysis of transient inductive interference in underground pipelines due to faults on nearby power lines," *COMPEL: The International Journal for Computation and Mathematics in Electrical and Electronic Engineering*, vol. 26, pp. 1346-1363, 2007.

[36] T. Lu and X. Cui, "Transient analysis of wave processes for multi-conductor transmission lines with branches using FDTD," *IEEE International Symposium on Electromagnetic Compatibility*, vol. 2, pp. 699-703, 2000.

[37] T. Lu, L. Qi, J. Li, and X. Cui, "Application of multi-conductor transmission lines on the transient analysis in power substation," *Asia-Pacific Conference on Environmental Electromagnetics*, pp. 198-201, 2003.

[38] T. Lu, L. Qi, L. Guo, X. Cui, and X. Gu, "Research on experiments and the FDTD Method of multi-conductor transmission lines for transient analysis," *International Symposium on Electromagnetic Compatibility*, vol. 2, pp. 708-712, 2004.

[39] C. Jiao and Y. Sun, "Progress in studies of transients analysis method of multiconductor transmission lines," *Progress in Electromagnetics Research Symposium*, Beijing, China, pp. 249-253, 2009.

[40] J. Dedkova and L. Brancik, "Laplace transform and FDTD approach applied to MTL simulation," *PIERS Online*, vol. 4, pp. 16-20, 2008.

[41] J. D. Moerloose, T. W. Dawson and M. A. Stuchly, "Application of the finite difference time domain algorithm to quasi-static field analysis," *Radio Science*, vol. 32, pp. 329-341, 1997.

[42] W. T. Kaune and M. F. Gillis, "General properties of the interaction between animals and ELF electric fields," *Bioelectromagnetics*, vol. 2, pp. 1-11, 1981.

[43] A. W. Guy, S. Davidow, G. Y. Yang, and C. K. Chou, "Determination of electric current distributions in animals and humans exposed to a uniform 60-Hz high-intensity electric field," *Bioelectromagnetics*, vol. 3, pp. 47-71, 1982.

[44] L. Golestani-Rad, B. Elahi and J. Rashed-Mohassel, "Investigating the effects of external fields polarization on the coupling of pure magnetic waves in the human body in very low frequencies," *BioMagnetic Research and Technology*, vol. 5, pp. 1-5, 2007.

[45] O. P. Gandhi and J. Chen, "Numerical dosimetry at power-line frequencies using anatomically based model," *Bioelectromagnetics*, vol. 13, pp. 43-60, 1992.

[46] M. A. Stuchly and T. W. Dawson, "Human organ and tissue induced currents by 60 Hz electric and magnetic fields," *Proceedings of the 19th Annual International Conference of the IEEE Engineering in Medicine and Biology Society*, vol. 6, pp. 2464-2467, 1997.

[47] T. W. Dawson, K. Caputa and M. A. Stuchly, "High-resolution organ dosimetry for human exposure to low-frequency electric fields," *IEEE Transactions on Power Delivery*, vol. 13, pp. 366-373, 1998.

[48] M. E. Potter, M. Okaniewski and M. A. Stuchly, "Low frequency finite difference time domain (FDTD) for modeling of induced fields in humans close to line sources," *Journal of Computational Physics*, vol. 162, pp. 82-103, 2000.

[49] C. H. See, R. A. Abd-Alhameed and P. S. Excell, "Computation of Electromagnetic Fields in Assemblages of Biological Cells Using a Modified Finite Difference Time Domain Scheme", *IEEE Transactions on Microwave Theory and Techniques*, vol. 55, no. 9, pp. 1986-1994, 2007.

[50] C. H. See, R. A. Abd-Alhameed, P. S. Excell, and D. Zhou, "Microscopic biological cell level model using modified finite-difference time-domain at mobile radio frequencies," *PIERS Online*, vol. 4, pp. 6-10, 2008.

[51] A. J. Pansini, *Power transmission and distribution*, 2nd ed. Georgia: The Fairmont Press Inc., 2005.

[52] G. A. Goulty, *Visual amenity aspects of high voltage transmission.* Somerset, England: Research Studies Press Ltd., 1990.

[53] W. M. Middleton and M. E. V. Valkenburg, *Reference data for engineers: Radio, electronics, computer, and communications*, 9th ed. Boston, 2002.

[54] G. M. Amer, "Novel technique to calculate the effect of electromagnetic field of H.V.T.L. on the metallic pipelines by using EMTP program," *18th International Conference and Exhibition on Electricity Distribution (CIRED),* Turin, Italy, pp. 1-5, 2005.

In: Development of Complex Electromagnetic … ISBN: 978-1-61122-013-1
Editors: K. N. Ramli et al.

Chapter 5

EFFECT OF ANISOTROPIC MAGNETO-CHIRALITY ON MICROSTRIP RESONATOR CHARACTERISTICS

***C. Zebiri*[1], *F. Benabdelaziz*[2] *and M. Lashab*[3]**
[1]Electronics Department, University of Ferhat Abbas, Sétif, Algeria
[2]Electronics Department, University of Constantine, Constantine, Algeria
[3]Electronics Department, University of Skikda, Skikda, Algeria

ABSTRACT

The effect of a chiral bianisotropic substrate on the complex resonant frequency of a rectangular microstrip resonator is studied on the basis of the integral equation formulation. The analysis is based on numerical resolution of the integral equation using the Galerkin procedure for the moment method in the spectral domain. This work aims to study the effect of the chirality of a bianisotropic substrate upon the resonant frequency and the half-power bandwidth. The effect of a chiral-substrate bianisotropy on the surface waves of the microstrip resonator is also studied. The effective technique used to formulate the characteristic equations of the surface waves in a medium equipped with a complex anisotropy is presented. The equations concerning an evaluation of the cut-off frequencies are given in more detailed forms. A simple approximate formula for estimating the wave number of the surface mode TM_0 and TE_1 are obtained. An estimated maximum value of chiral slab thickness without the excitation of surface waves is given. The effect of a chiral bianisotropic substrate with the complex constitutive parameters on the input impedance of a rectangular microstrip antenna is also studied on the basis of the integral equation formulation. The analysis is based on numerical solution of the integral equation using the Galerkin procedure for the moment method in the spectral domain. The effect of the uniaxial anisotropy and chirality of the superstrate on the resonant frequency and bandwidth of rectangular microstrip patch in a substrate-superstrate configuration are investigated. The work treats the asymmetrical effects on the resonant frequency and the bandwidth of a rectangular microstrip patch antenna in a complex bianisotropic substrate-superstrate configuration.

5.1. INTRODUCTION

In recent years, the method of moments (MoM) [1-6] has been adopted in the analysis of radiating elements such as resonators [7] and antennas [8, 9]. Microstrip resonators appeared during the fifties, and they were especially developed during the seventies [10, 11]. They are small in size and have simplicity, easy manufacturing and easy handling. They are robust when they are fixed on rigid surfaces, and are also very effective in terms of resonance, polarization, input impedance and radiation patterns [12-14]. The major disadvantages of microstrip resonators are their weak purity of polarization, and narrow half-power bandwidth which is typically a few percent. However, increasing substrate thickness and reduced relative permittivity make it possible to improve resonator efficiency by up to 90%, and to increase bandwidth by up to 35%, provided that the surface waves are minimized [15]. The resonant frequency of microstrip resonator is strongly related to the dielectric of the substrate, and remains a very important factor to determine in resonator design. Some works have mentioned that many uniaxial anisotropic substrates have been used for printed circuit resonators [16, 17]. However, studies concerning the resonant frequency calculation of the microstrip resonator, implemented on a substrate of uniaxial anisotropy [18], have shown that this kind of substrate does not affect the quality factor, the half-power bandwidth and obviously the resonant frequency. For these reasons, other studies have focussed on cases of positive and negative anisotropy [18-20], or bianisotropic media [21]. In the present work, the Spectral Domain Approach (SDA) is applied, and intensively used in the analysis and design of planar structures. In such an approach, the tensorial Green's spectral function which relates the tangential electric field components and the corresponding current of different conductor has to be determined. Many techniques have been used to evaluate this function [18-20, 22-26]. Chew [19] showed that the number of modes refines and improves the integral equation resolution. In the present work, only the TM_{01} mode is considered to solve the integral equation in order to highlight the effect of chirality and the permeability. The simulation of this study was carried out by a development of software using professional FORTRAN. With a PC Intel Pentium M, processor 1.86 GHz and 512 MB RAM, the determinant cancellation of the Green's tensor is obtained in about 65 s.

5.2. Theoretical Formulation

5.2.1. Maxwell's Equation

In chiral media, the geometrical character of the internal structure (the handedness) causes macroscopic effects that are observed as the rotation of the polarization of the propagating field plane [27]. Composite chiral material which exhibits the effect of chirality at microwave frequencies has been fabricated and characterized in many studies [27-29]. The chirals media considered here have a non-homogenous gyro-electric, and can be described by a set of constitutive relations [30], with an artificial magnetism which exists in chiral and bianisotropic composites [27]

$$\vec{B} = \overline{\overline{\mu}}\vec{H} + \frac{1}{c_0}\overline{\overline{\xi}}\vec{E} \tag{5.1}$$

$$\vec{D} = \overline{\overline{\varepsilon}}\vec{E} + \frac{1}{c_0}\overline{\overline{\eta}}\vec{H} \tag{5.2}$$

where the permittivity and permeability tensors are of uniaxial anisotropy and the magneto-electric elements are respectively expressed as

$$\overline{\overline{\mu}} = \begin{bmatrix} \mu_t & 0 & 0 \\ 0 & \mu_t & 0 \\ 0 & 0 & \mu_z \end{bmatrix}, \qquad \overline{\overline{\varepsilon}} = \begin{bmatrix} \varepsilon_t & 0 & 0 \\ 0 & \varepsilon_t & 0 \\ 0 & 0 & \varepsilon_z \end{bmatrix},$$

$$\overline{\overline{\xi}} = -\overline{\overline{\eta}} = j\begin{bmatrix} 0 & \xi & 0 \\ -\xi & 0 & 0 \\ 0 & 0 & 0 \end{bmatrix} \tag{5.3}$$

Assuming an $e^{i\omega t}$ time variation and starting from Maxwell's equation in the Fourier domain, we can show that the transverse fields in the chiral medium are expressed with respect to the longitudinal components $\widetilde{E}_z$ and $\widetilde{H}_z$ according to the following matrix equations

$$\overline{\overline{E}}_S(\overline{\kappa}_s, z) = \begin{bmatrix} \widetilde{E}_S^e(\overline{\kappa}_s, z) \\ \widetilde{E}_S^h(\overline{\kappa}_s, z) \end{bmatrix} = \frac{1}{\kappa_s}\overline{F}(\overline{\kappa}_s, \overline{r}_s)\begin{bmatrix} \widetilde{E}_x \\ \widetilde{E}_y \end{bmatrix}$$

$$= \begin{bmatrix} j\frac{\varepsilon_z}{\varepsilon_t}\frac{1}{\kappa_s}\left(\kappa_0\xi + \frac{\partial}{\partial z}\right) & 0 \\ 0 & \omega\frac{1}{\kappa_s}\mu_0\mu_z \end{bmatrix}\begin{bmatrix} \widetilde{E}_z \\ \widetilde{H}_z \end{bmatrix} \tag{5.4}$$

$$\overline{\overline{H}}_S(\overline{\kappa}_s, z) = \begin{bmatrix} \widetilde{H}_S^e(\overline{\kappa}_s, z) \\ \widetilde{H}_S^h(\overline{\kappa}_s, z) \end{bmatrix} = \frac{1}{\kappa_s}\overline{F}(\overline{\kappa}_s, \overline{r}_s)\begin{bmatrix} \widetilde{H}_y \\ -\widetilde{H}_x \end{bmatrix}$$

$$= \begin{bmatrix} \omega\frac{1}{\kappa_s}\varepsilon_0\varepsilon_z & 0 \\ 0 & j\frac{\mu_z}{\mu_t}\frac{1}{\kappa_s}\left(\kappa_0\xi + \frac{\partial}{\partial z}\right) \end{bmatrix}\begin{bmatrix} \widetilde{E}_z \\ \widetilde{H}_z \end{bmatrix} \tag{5.5}$$

$$\overline{F}(\overline{\kappa}_s, \overline{r}_s) = \begin{bmatrix} \kappa_x & \kappa_y \\ \kappa_y & -\kappa_x \end{bmatrix}, \ \overline{\kappa}_s = \kappa_x \hat{x} + \kappa_y \hat{y}, \ \overline{r}_s = x\hat{x} + y\hat{y} \tag{5.6}$$

where $\overline{F}(\overline{\kappa}_s, \overline{r}_s)\big|_{\overline{r}_s = 0}$ is the kernel of the vector transforms, the superscripts *e* and *h* in (5.4) and (5.5) denote the TM and TE waves respectively, $\overline{\widetilde{E}}_S$ and $\overline{\widetilde{H}}_S$ are expressed according to the TE and TM modes.

5.2.2. Transverse and Longitudinal Components

Taking into account the wave equation solution, we express the longitudinal components of the electric and magnetic fields in the chiral medium according to the following expressions

$$\widetilde{E}_z(\kappa_s, z) = A^e e^{j\kappa_z^e z} + B^e e^{-j\kappa_z^e z} \tag{5.7}$$

$$\widetilde{H}_z(\kappa_s, z) = A^h e^{j\kappa_z^h z} + B^h e^{-j\kappa_z^h z} \tag{5.8}$$

where

$$\kappa_z^{e2} = \kappa_0^2(\varepsilon_t \mu_t - \xi^2) - \frac{\varepsilon_t}{\varepsilon_z}\kappa_s^2 \tag{5.9}$$

$$\kappa_z^{h2} = \kappa_0^2(\varepsilon_t \mu_t - \xi^2) - \frac{\mu_t}{\mu_z}\kappa_s^2 \tag{5.10}$$

and the spectral coefficients A^e, A^h, B^e and B^h are functions of the variables κ_s, κ_z^e and κ_z^h. These are respectively the free space propagation for TE and TM modes. According to the previous equations, the following tangential components of the fields are obtained

$$\widetilde{E}_S(\kappa_s, z) = \begin{bmatrix} \widetilde{E}^e(\kappa_s, z) \\ \widetilde{E}^h(\kappa_s, z) \end{bmatrix} = e^{j\overline{\kappa}_z z}\overline{A}(\kappa_s) + e^{-j\overline{\kappa}_z z}\overline{B}(\kappa_s) \tag{5.11}$$

$$\widetilde{H}_S(\kappa_s, z) = \begin{bmatrix} \widetilde{H}^e(\kappa_s, z) \\ \widetilde{H}^h(\kappa_s, z) \end{bmatrix} = e^{j\overline{\kappa}_z z}\overline{g}(\kappa_s)\overline{A}(\kappa_s) + e^{-j\overline{\kappa}_z z}\overline{h}(\kappa_s)\overline{B}(\kappa_s) \tag{5.12}$$

where

$$\overline{\kappa}_z = diag\begin{bmatrix}\kappa_z^e & \kappa_z^h\end{bmatrix} \tag{5.13}$$

$$\overline{A}(\kappa_s) = \left[j\frac{1}{\kappa_s^2}\frac{\varepsilon_z}{\varepsilon_t}\left(-\kappa_0\xi + j\kappa_z^e\right)B^e \quad \frac{1}{\kappa_s^2}\omega\mu_0\mu_z A^h \right]^T \tag{5.14}$$

$$\overline{B}(\kappa_s) = \left[-j\frac{1}{\kappa_s^2}\frac{\varepsilon_z}{\varepsilon_t}\left(\kappa_0\xi + j\kappa_z^e\right)B^e \quad \frac{1}{\kappa_s^2}\omega\mu_0\mu_z B^h \right]^T \tag{5.15}$$

$$\overline{g}(\kappa_s) = diag\left[\frac{\omega\varepsilon_0\varepsilon_t}{j\left(-\kappa_0\xi + j\kappa_z^e\right)} \quad \frac{j\left(\kappa_0\xi + j\kappa_z^h\right)}{\omega\mu_0\mu_t} \right] \tag{5.16}$$

$$\overline{h}(\kappa_s) = diag\left[\frac{\omega\varepsilon_0\varepsilon_t}{j\left(-\kappa_0\xi - j\kappa_z^e\right)} \quad \frac{j\left(\kappa_0\xi - j\kappa_z^h\right)}{\omega\mu_0\mu_t} \right] \tag{5.17}$$

For an electric and non magnetic medium having biaxial anisotropy with regard to the permittivity, the previous expressions are well detailed in [18-20].

5.3. Green's Tensor Evaluation

The proposed structure is studied and the boundary conditions have been applied after which the dyadic Green's function is obtained. The Green's elements are independent and do not couple in this case, each of them regarding the medium as an equivalent isotropic medium, as can be certified by another case proved by [27]. The Green's function is expressed in compact form as the following tensor

$$\overline{\mathbf{G}}(\kappa_s) = \frac{1}{j\omega\varepsilon_0}\text{diag}\left[\frac{N^e}{D^e}\kappa_z\kappa_z^e, \frac{1}{D^h}\kappa_0^2\mu_t\right]\cdot\sin(\overline{\kappa}_z d) \tag{5.18}$$

where

$$N^e = \frac{1}{\kappa_z^e}\left(\kappa_0^2\varepsilon_t\mu_t - \frac{\varepsilon_t}{\varepsilon_z}\kappa_s^2\right) \tag{5.19}$$

$$D^e = \kappa_z^e\kappa_z\varepsilon_t\cos(\kappa_z^h d) + j\left(\kappa_0^2\varepsilon_t\mu_t - \frac{\varepsilon_t}{\varepsilon_z}\kappa_s^2 - j\kappa_z\kappa_0\xi\varepsilon_t\right)\sin(\kappa_z^e d) \tag{5.20}$$

$$D^h = \kappa_z^h\cos(\kappa_z^h d) + j(\kappa_z\mu_t - j\kappa_0\xi)\sin(\kappa_z^h d) \tag{5.21}$$

5.4. INTEGRAL EQUATION SOLUTION

It is clear from (5.18) that the Green's dyadic is diagonal in the (TM, TE) representation although a uniaxial bianisotropy is present. This results in simplification when deriving the Green's dyadic for a multilayer anisotropic media. The surface current J on the patch can be expanded into a finite series of basis functions

$$\mathbf{J}(r_s) = \sum_{n=1}^{N} a_n\begin{bmatrix} J_{xn}(r_s) \\ 0 \end{bmatrix} + \sum_{m=1}^{M} b_m\begin{bmatrix} 0 \\ J_{ym}(r_s) \end{bmatrix} \tag{5.22}$$

where a_n and b_m are the mode expansion coefficients to be sought. The integral equation describing the electric field on the patch is [18-20]

$$\iint_{-\infty}^{\infty} d\kappa_s \overline{F}(\overline{\kappa}_s, \overline{r}_s)\cdot\overline{\mathbf{G}}(\kappa_s)\cdot\tilde{\mathbf{J}}(\kappa_s) = 0 \tag{5.23}$$

Using the well-known Galerkin procedure of the moment method, the integral equation in (5.23) is discretized into the following matrix equation

$$\begin{bmatrix} (\overline{\mathbf{B}}_1)_{N\times N} & (\overline{\mathbf{B}}_2)_{N\times M} \\ (\overline{\mathbf{B}}_3)_{M\times N} & (\overline{\mathbf{B}}_4)_{M\times M} \end{bmatrix} \cdot \begin{bmatrix} (\mathbf{a})_{N\times 1} \\ (\mathbf{b})_{M\times 1} \end{bmatrix} = 0 \qquad (5.24)$$

where $(\overline{B}_1)_{N\times N}$, $(\overline{B}_2)_{N\times M}$, $(\overline{B}_3)_{M\times N}$ and $(\overline{B}_4)_{M\times M}$ are the elements of the digitalized matrix equation.

5.5. Half-power Bandwidth

Equation (5.24) has a non-trivial solution only in the case where the condition below is verified

$$\det(\overline{\mathbf{B}}(\omega)) = 0 \qquad (5.25)$$

The resonator is designed to operate near its resonant frequency, and all its characteristics are estimated around this frequency. Equation (5.25) is called the characteristic equation for the complex resonant frequency $f = f_r + if_i$, where f_r is the resonant frequency and f_i is the dampling factor that characterises the losses of the radiating antenna. The quality factor and the half-power bandwidth are defined in [31, 32] as

$$Q = f_r/(2f_i) \qquad (5.26)$$

$$BW = 1/Q \qquad (5.27)$$

5.5.1. Asymptotic Formula for Resonant Frequency

The effect of the two chiral constitutive parameters has been studied. These are first the chirality and the magnetic uniaxial anisotropy of the substrate upon the complex resonant frequency and secondly the half-power bandwidth of the microstrip resonator. In order to validate the results (in both isotropic and chiral cases), we have made a comparison with the isotropic case presented in the literature. The figures presented below show the normalized real and imaginary parts of the resonant frequency, and the half-power bandwidth, compared with the measurements in the literature with ε_x=ε_z=2.35 and ε_x=ε_z=7. The medium considered is defined as follows: η=ξ=-1, 0 and 1 respectively (magneto-electric elements), μ_t=0.8, μ_z=1 (positive anisotropy of the permeability), μ_t=1.2, μ_z=1 (negative anisotropy). Uniaxial anisotropy is obtained by changing μ_t and keeping μ_z constant.

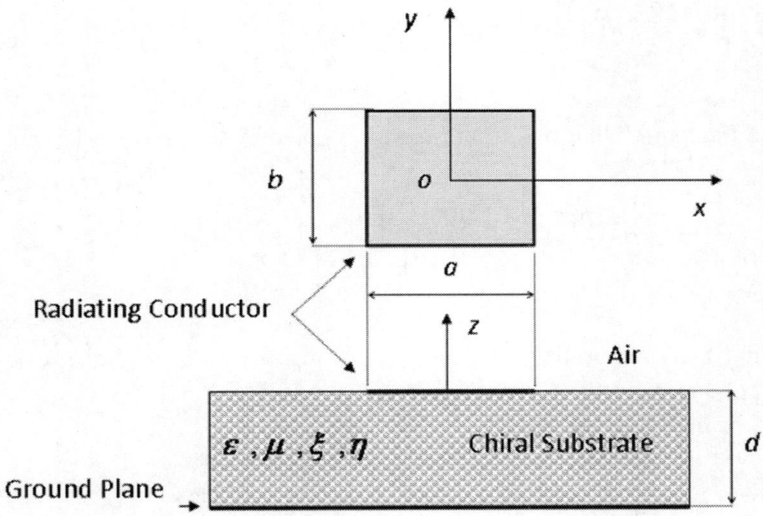

Figure 5.1. Rectangular microstrip patch antenna with chiral substrate.

To demonstrate the effect of the constitutive parameters of the dielectric chiral layer, we can consider the asymptotic form of the resonant frequency, when the layer d is electrically thin. However, the dyadic Green's function takes the following form

$$\overline{\mathbf{G}}(\mathbf{k}_s) \underset{d \to 0}{\to} \frac{d}{j\omega\varepsilon} \mathrm{diag}\left[\kappa_0^2 \mu_t \; -\frac{1}{\varepsilon_z}\kappa_s^2, \mu_t \, \kappa_0^2\right] \tag{5.28}$$

Some authors such as Chew [19], have shown that only one basis function (N=0, M=1) is needed to obtain excellent convergence of the results. In this case, the current distribution on the patch conductor is given by

$$\mathbf{J}(\mathbf{r}_s) = b_1 \begin{bmatrix} 0 \\ J_{y1}(\mathbf{r}_s) \end{bmatrix} = b_1 \begin{bmatrix} 0 \\ \sin\left[\frac{\pi}{b}\left(y + \frac{b}{2}\right)\right] \end{bmatrix} \tag{5.29}$$

and the characteristic equation for the resonant frequency driven from the fourth element of the matrix equation (5.24) is given as

$$\left(\overline{\mathbf{B}}_4\right)_{11} = \frac{d}{j\omega\varepsilon} \int\int_{-\infty}^{+\infty} d\kappa_s \left[\kappa_0^2 \mu_t - \tfrac{1}{\varepsilon_z} \kappa_y^2\right] \widetilde{J}_{y1}(-\kappa_s) \widetilde{J}_{y1}(\kappa_s) = 0 \tag{5.30}$$

The function J_{y1} has a Fourier transform given by the analytical expression

$$\widetilde{J}_{y1}(\kappa_s) = \pi b \frac{\sin\left(\kappa_x \frac{a}{2}\right)}{\kappa_x} \frac{\cos\left(\kappa_y \frac{b}{2}\right)}{\left(\frac{\pi}{2}\right)^2 - \left(\kappa_y \frac{b}{2}\right)^2} \tag{5.31}$$

Using the asymptotic expression of $\overline{\mathbf{G}}$ given by (5.28) and taking into account (5.31), equation (5.30) becomes

$$\kappa_0^2 \mu_t I_1 - \frac{4}{\varepsilon_z b^2} I_2 = 0 \tag{5.32}$$

where

$$I_1 = \int_0^{\infty} d\kappa_y \frac{\cos^2 \kappa_y}{\left[\kappa_y^2 - (\pi/2)^2\right]^2}, \quad I_2 = \int_0^{\infty} dk_y \frac{\kappa_y^2 \cos^2 \kappa_y}{\left[\kappa_y^2 - (\pi/2)^2\right]^2} \tag{5.33}$$

Using the integration along the contour of the spectral domain, the integral equations of (5.33) is resolved analytically as

$$I_1 = \frac{1}{\pi}, \quad I_2 = \frac{\pi}{4} \tag{5.34}$$

Substituting (5.34) into (5.32), the estimation of the resonant frequency can be determined, expressed by

$$f_r = \frac{c}{2b\sqrt{\varepsilon_z \mu_t}} \tag{5.35}$$

where c is the speed of light in the vacuum. The resulting formula gives an idea of the effect of the parameters ε_z and μ_t upon the antenna resonant frequency. It is clear that this depends on the permittivity along the optic axis and the perpendicular component of the permeability in the limit of a thin substrate, hence the resonant frequency depends only on the components ε_z and μ_t.

5.5.2. Numerical Results

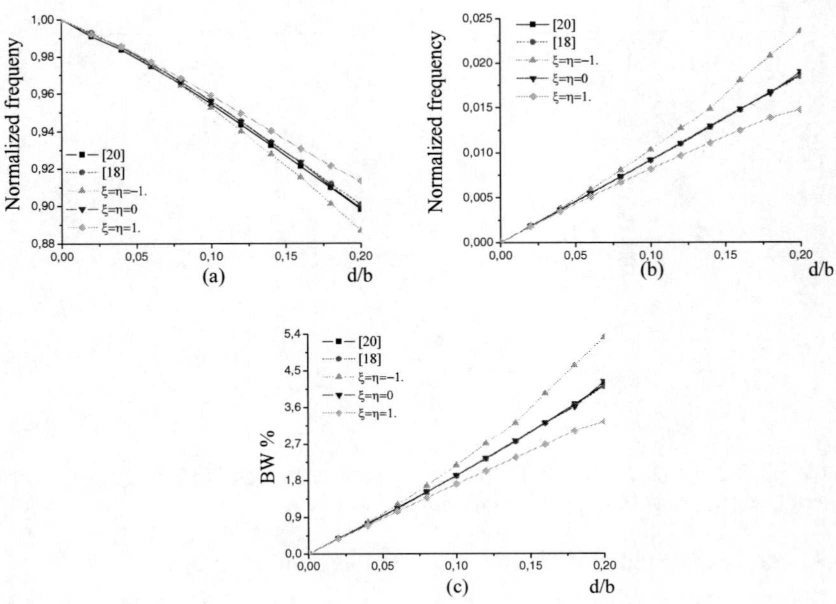

Figures 5.2. Normalized resonant frequency, a=1.5 cm, b=1 cm, ε_r=7, (a) Real part, (b) Imaginary part, (c) Half-power bandwidth.

The effect of dielectric medium, provided with a gyrotropy, upon the complex resonant frequency and the half-power bandwidth has been studied. The substrate is considered to be bianisotropic with relative permittivities such as $\varepsilon_x=\varepsilon_z=7$ and $\varepsilon_x=\varepsilon_z=2.35$. The dimensions of the rectangular patch are 1 cm × 1.5 cm. In Figure 5.1, the complex resonant frequency and the half-power bandwidth are presented with respect to the substrate thickness for different values of chirality depending on the magnetic anisotropy choice. Normalization is with respect to the frequency f_0 which is obtained from the fundamental mode $f_r = c/2b\sqrt{\varepsilon_z}$.

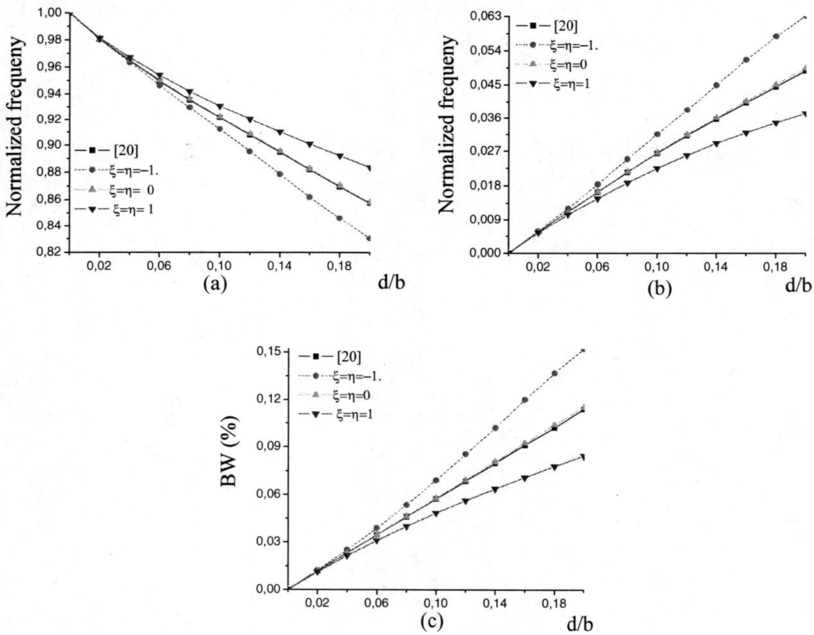

Figures 5.3. Normalized resonant frequency, a=1.5 cm, b=1 cm, ε_r=2.35, (a) Real part, (b) Imaginary part, (c) Half-power bandwidth.

According to Figures 5.2 and 5.3, it follows that

- The effect of chirality is remarkable only for thick layers, whereas for those infinitely small, no effect is perceived, which is confirmed by the asymptotic form expression of the resonant frequency given by Equation (5.35).

- In the case of a positive chirality (ξ=η=1), the real resonant frequency increases, whereas the imaginary part decreases, which also involves a reduction in the bandwidth and an increase in the factor of quality. In contrast, the negative chirality case, with ξ=η=-1, leads to opposite variations compared to the preceding ones.
- An increase or decrease in the resonator parameters of f_r, f_i or bandwidth are accentuated when we consider a low permittivity of about ε=2.35, which is not the case for the value ε=7.

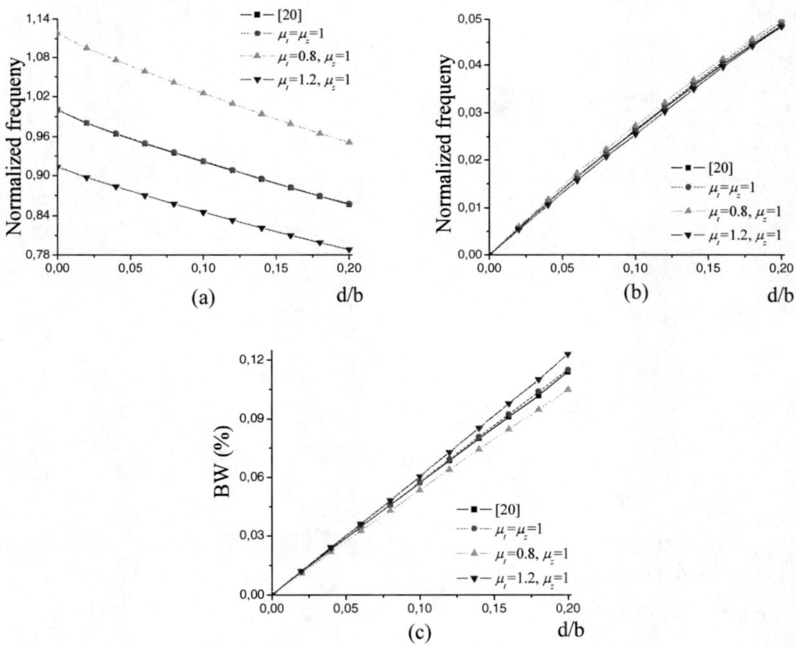

Figures 5.4. Normalized resonant frequency, a=1.5 cm, b=1 cm, ε_r=2.35, (a) Real part, (b) Imaginary part, (c) Half-power bandwidth.

It can be seen that from Figures 5.4, in the case of permeability values less than 1, for example μ_t=0.8, that the effect of uniaxial anisotropy of the permeability clearly results in a large increase in the real part of the resonance frequency, but an unimportant increase in the imaginary part of the resonance frequency. Therefore, the bandwidth undergoes a slight reduction. Permeability values greater than one, such as μ_t=1.2, cause variations in the opposite sense with respect to the preceding case. The chirality is apparent only in thick layers, whereas the effect of the permeability remains whatever the thickness of the support. According to these results one can predict the desired values of three parameters: f_r, f_i and bandwidth of the resonator, as illustrated by Figure 5.5.

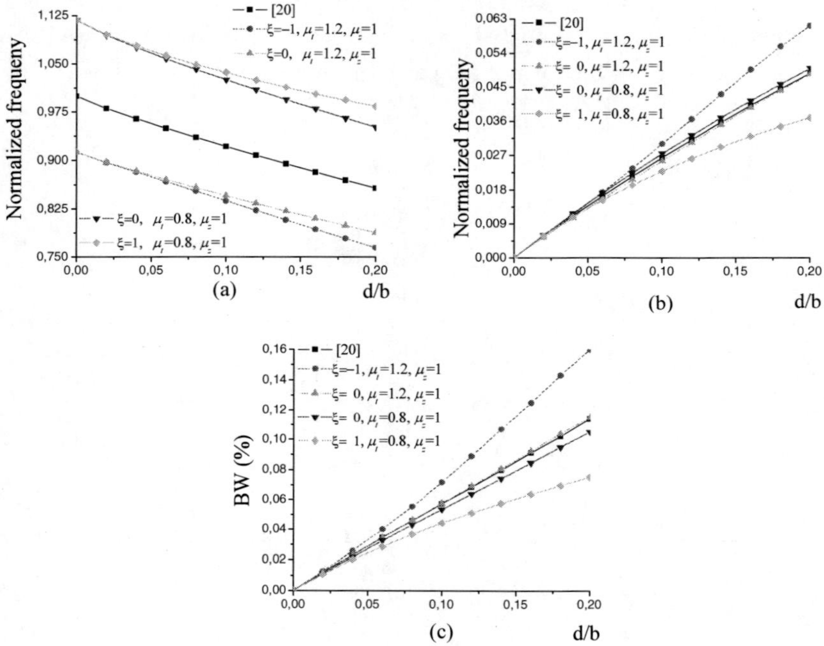

Figures 5.5. Normalized resonant frequency, a=1.5 cm, b=1 cm, ε_r=2.35, (a) Real part, (b) Imaginary part, (c) Half-power bandwidth.

5.6. Surface Waves Investigation

In the past few years, a significant interest has been directed towards the study of the effect of complex materials, such as anisotropic dielectrics, magnetized ferrites and chiral materials on the realization of Microwave Integrated Circuits (MIC) and Antenna Printed Circuits (APC). Certain standard anisotropic media are used as substrates for microwave printed resonators [16]. Pozar proved that the substrate anisotropy should always be taken into account when designing planar microstrip resonators, otherwise, the structure is likely to operate out of the predicted frequency band [17].

Magnetized ferrites belong to the complex material class which have proved to have potential applications as substrates in the field of MIC and APC. Measurements have confirmed that the resonant frequencies of the microstrip structures printed on ferrite substrates can be fixed to suit a range of applications, simply by adjusting the polarization of an external magnetic field [33]. However, ferrites can be employed to reduce the microstrip resonator [34, 35] and to design circularly polarized antennas by the application of a simple feed [36].

The possibility of employing chiral materials as substrates for the design of MIC and APC, was reported by Lindell [27]. On the other hand, Pozar [37] underlined the serious disadvantages of the use of these materials as substrates, because of the losses due to the surface wave excitation and the significant appearance of poles during numerical calculations. However, Toscano *et al.* [38] and Zebiri *et al.* [39, 40] have recently proved that chiral substrates can advantageously be employed to increase the bandwidth of microstrip antennas. This result indicates that more attention must be paid to detailed studies of the surface wave excitation in the microstrip antennas over chiral substrates. This study evaluates the cut-off frequencies of TE_0 and TM_1 modes of a resonator printed on an anisotropic medium in order to optimize these structures according to their applications.

These resonators appeared during the 1950s and especially were developed during the 1970s [10, 11]. They combine small size, simplicity, facility of manufacture and practical implementation. Moreover, they are easily matched to plane and non-plane surfaces, and they exhibit a great robustness when assembled on rigid surfaces. They are also very effective in terms of resonance, polarization, input impedance and radiation diagram [12, 41].

The major disadvantages of microstrip resonators lie in their low purity of polarization. As regards antennas we can consider the reduction of their bandwidth which is typically a few percent [41]. However, increasing substrate thickness and reduced relative permittivity make it possible to improve resonator efficiency by up to 90%, and to increase bandwidth by up to 35%, provided that the surface waves are minimized [15]. The idea of using chiral materials as substrates and superstrates in the design of printed antenna was first presented by Engheta [42] and the term "chirostrip" was then invented. In the literature, it is shown that the power of the surface wave can generally be reduced when a chiral substrate is employed for antennas intended for printed circuitry [43].

5.6.1. Surface Waves in Chiral Media

The surface wave modes are guided by the interface separating two different dielectric media [44]. In the design of microstrip circuits as well as in their applications, the study of surface waves is justified by one of the two following reasons; to minimize their effects or to use them in certain applications [45]. It is initially essential to take into account the propagation of modes in the chiral layer. We study the propagation of these modes along *z*-axis. For such propagation, the interface of the structure can be considered as a particular case of a chiro-waveguide [46].

In [47] it is proved that in all chiro-waveguides the proper modes are hybrid [46], but for the case of the selected chiral, decomposition into TE and TM modes is possible [39, 40, 48, 49]. Their study would be also made during the evaluation of the improper integrals appearing in the formulation of the problem of a microstrip structure by the integral method. It would be also necessary, in certain situations, to determine as a preliminary the propagating modes of the surface waves which appear in a microstrip structure. The spectral method and the tensorial Green's function are adopted as an analysing tool. We can mention as an example the case where the integration of the characteristic matrix elements deduced from the Galerkin's method is performed on the wave numbers real axis [50, 51].

Moreover, if we are interested in the fields radiated by a microstrip structure without neglecting the radiation due to the surface waves, the determination of their wave numbers is necessary [52]. However, the majority of the studies relating to this subject deal with the problem of the surface waves as being a procedure of solving two transcendent equations for which the expressions appear in the denominators of the Green's function [32, 52, 53].

The study in the case of simple structures, made up of one or two dielectric layers, does not present difficulties [54, 55]. However, for the structures that consist of more than two layers, the study becomes increasingly complicated, requiring tedious calculations. As a result, the resolution of the resulting characteristic equations and the asymptotic behavior of the solutions for low operating frequencies and low dielectric-layer thicknesses is more complicated. Traditional methods do not lead, in general, to simple expressions without significant algebraic calculation.

5.6.1.1. Isotropic Case

Surface waves induce unwanted radiation at the non-radiant ends of the structure. Moreover, they can lead to the undesirable coupling between the radiant elements in the case of antenna arrays. The phase velocity of the surface waves strongly depends on the permittivity ε_r and the substrate thickness h. Surface wave excitation in a dielectric layer established on a ground plane was well studied in [56]. The dominant mode with the lowest cut-off frequency is TM_0 (f_c=0), and the cut-off frequencies of higher modes (TM_n and TE_n) are presented as

$$f_c = \frac{nc_0}{4h\sqrt{\varepsilon_r - 1}}, \; n=0, 1, 2, \ldots \tag{5.36}$$

where c_0 is the speed of light in vaccuo, and h and ε_r are respectively the thickness and the permittivity of the substrate. Odd values of n give the cut-off frequencies for the TE_n modes, and even values of n give cut-off frequencies for the TM_n modes. For the TE_1 mode, the computed values of the ratio $h/\lambda_c^{(1)}$ are 0.217 for Duroid (ε_r=2.32) and 0.0833 for alumina (ε_r=10), using the following equation

$$\frac{h}{\lambda_c^{(1)}} = \frac{n}{4\sqrt{\varepsilon_r - 1}} \tag{5.37}$$

Thus, the low order mode TE_1 is excited at 41 GHz for a Duroid substrate of thickness 1.6 mm, and at approximately 39 GHz for an alumina substrate of thickness 0.635 mm. The thickness of the substrate is chosen so that the h/λ_0 ratio is lower than the $h/\lambda_c^{(1)}$ (λ_0 is the wavelength in open space at the operating frequency), which gives [57]

$$h < \frac{c}{4f_u\sqrt{\varepsilon_r - 1}} \tag{5.38}$$

where f_u is the highest frequency in the operation band. Note that h should be chosen as high as possible, under the constraint indicated in (5.38), so that the maximum effectiveness is achieved. Moreover, h must be in conformity with the substrates available in the market. Another practical formula for h is given in [58] by

$$h < \frac{0.3 \cdot c}{2\pi f_u \sqrt{\varepsilon_r}} \tag{5.39}$$

The TM_0 mode has no cut-off frequency and is always present to a certain extent. The excitation of the surface wave TM_0 mode becomes appreciable when $h/\lambda > 0.09$ ($\varepsilon_r \cong 2.3$) and when $h/\lambda > 0.03$ ($\varepsilon_r \cong 10$). Generally, suppressing the TM_0 mode requires lower permittivity and smaller substrate size. Unfortunately, decreasing ε tends to increase the antenna size, whereas decreasing h thus leads to a limitation of the antenna effectiveness, hence a reduction of its bandwidth. The reduction of surface waves, for microstrip antennas printed on ferrites, has been discussed by many researchers [59-63]. In [64-66], electromagnetic band gap (EBG) structures, also known under the name of photonic crystals, were found to reduce and in certain cases to eliminate the surface waves, leading to increases in directivity and bandwidth. In this context, Yang [67] was the first who proposed antennas with large gain, which could be carried out with a unique radiant element printed on a 2-D photonic bandgap (PBG) material.

5.6.1. Bianisotropic Case

The problem of surface waves, in simple anisotropic monolayer microstrip structures, is rigorously formulated and solved. The advantages of the method are highlighted, and comparative analyses are carried out between our results and those published in the literature. Here we consider the monolayer structure represented in Figure 5.1. The radiant conducting plate is suppressed, the structure is a chiral layer without an excitation source established on a ground plane (Figure 5.6). The surface waves which occur and propagate in this structure have been studied.

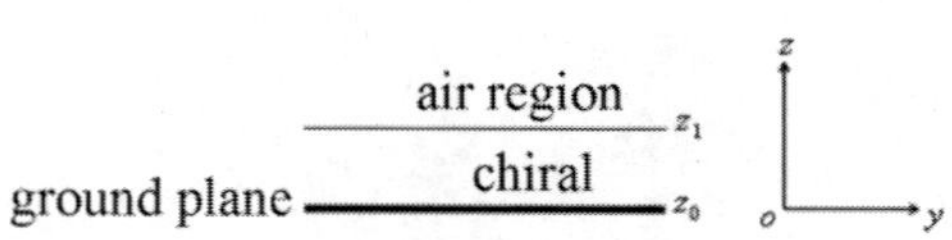

Figure 5.6. Monolayer chiral substrate configuration.

In previous work [39], we calculated the denominators of the Green tensor elements. We point out here the modal equations of the surface waves. For the TM modes

$$D^e = \kappa_z^e \kappa_z \varepsilon_t \cos(\kappa_z^e d) + j\left(\kappa_0^2 \varepsilon_t \mu_t - \tfrac{\varepsilon_t}{\varepsilon_z}\kappa^2 - j\kappa_z \kappa_0 \xi \varepsilon_t\right)\sin(\kappa_z^e d) = 0 \tag{5.40}$$

For TE modes

$$D^h = \kappa_z^h \cos(\kappa_z^h d) + j(\kappa_z \mu_t - j\kappa_0 \xi)\sin(\kappa_z^h d) = 0 \tag{5.41}$$

where the wavenumbers κ_z^e, κ_z^h, κ_z, κ_0^2 and κ^2 are given in [39, 40].

We develop asymptotic limits at low frequencies for TM_0 modes, where the cut-off frequencies of the surface waves are characterized by $\kappa_s = \kappa$. In the borderline of low frequency, $\kappa \to 0$. By taking into account the preceding Equations (5.40) and (5.41), the asymptotic limits at low frequencies for the surface waves are obtained as follows. For the TM_0 mode, and at low frequencies, by letting $\lim_{\kappa_0 \to 0} \kappa_s^2 = \kappa_0^2$, and according to [39], κ_z^e becomes

$$\lim_{\kappa_0 \to 0} \kappa_z^{e2} = \kappa_0^2\left(\varepsilon_t \mu_t - \xi^2 - \tfrac{\varepsilon_t}{\varepsilon_z}\right) \tag{5.42}$$

Knowing that in this case, $\lambda \to \lambda_0$ since $\kappa_s \to \kappa_0$, by substituting (5.42) into (5.40)

$$\kappa_z = -j\kappa_0^2 \frac{(\varepsilon_z \mu_t - 1)}{\varepsilon_z (1 + \kappa_0 \xi d)} d \tag{5.43}$$

For the isotropic case, Equation (5.43) is reduced to [55]

$$\kappa_z = -j\kappa_0^2 d \frac{(\varepsilon_z - 1)}{\varepsilon_z} \tag{5.44}$$

For the TE_1 mode, at low frequencies, where $\kappa_0^2 \rightarrow 0$ and $\lim_{\kappa_0 \rightarrow 0} \kappa_s^2 = \kappa_0^2$, and according to [39], κ_z^h becomes

$$\lim_{\kappa_0 \rightarrow 0} \kappa_z^{h2} = \kappa_0^2 \left(\varepsilon_t \, \mu_t - \xi^2 - \frac{\mu_t}{\mu_z} \right) \tag{5.45}$$

Knowing that in this case, $\lambda \rightarrow \lambda_0$ since $\kappa_s \rightarrow \kappa_0$, by substituting (5.45) into (5.41)

$$\kappa_z = (1 - \kappa_0 \xi d) \frac{1}{\mu_t \, d} \tag{5.46}$$

The wavenumbers in Equations (5.43) and (5.46) are general formulas for the anisotropic case with chiral, they can be shown to equal those obtained by Peixeiro et al. [55]. The development of an approximate formula for the wavenumbers of TM_0 and TE_1 modes gives for the TM_0 mode

$$\kappa_z = -j \kappa_0^2 \mu_t \frac{1}{(1 + \kappa_0 \xi d)} \left(1 - \frac{1}{\varepsilon_z \mu_t} \right) d \tag{5.47}$$

and for the TE_1 mode

$$\kappa_z = (1 - \kappa_0 \xi d) \frac{1}{\mu_t \, d} \tag{5.48}$$

The chirality effect on the two wavenumbers is similar, for the case of very thin layers, while bringing the ratio $1/1+\kappa_0 \xi d$ closer to $1-\kappa_0 \xi d$. According to the equation $\kappa_z^2 = \kappa_0^2 - \kappa_s^2$. By replacing κ_z in the Equation (5.42), an approximate expression can be found as

$$\kappa_{sp}^2 = \kappa_0^2 \left(1 + \frac{\kappa_0^2 d^2}{\varepsilon_z^2} \frac{(\varepsilon_z \mu_t - 1)^2}{(1 + \kappa_0 \xi d)^2} \right) \tag{5.49}$$

$$\lambda_{TM_0} = \kappa_0 \left(1 + \frac{\kappa_0^2 d^2}{\varepsilon_z^2} \frac{1}{(1 + \kappa_0 \xi d)^2} (\varepsilon_z \mu_t - 1)^2 \right)^{1/2} \tag{5.50}$$

and for the TE$_1$ mode

$$\kappa_{sp}^2 = \kappa_0^2 \left(1 - \left(\frac{(1 - \kappa_0 \xi d)}{\kappa_0 \mu_t d} \right)^2 \right) \tag{5.51}$$

$$\lambda_{TE_1} = \kappa_0 \left(1 - \left(\frac{(1 - \kappa_0 \xi d)}{\kappa_0 \mu_t d} \right)^2 \right)^{1/2} \tag{5.52}$$

Equations (5.50) and (5.52) are useful as a good initial estimate for the pole location, in a standard routine, to seek the true solution, as shown in Figure 5.7. The approximation of the pole location in the integration path is harmful for effective numerical evaluations of the integrals. It is advantageous to subtract the pole singularity and to reinstate it thereafter analytically. In the case of a thick substrate, several poles may exist, and the analytical evaluation of the integrals around the half-circles becomes quite complicated if two or several poles are very close to each other [68]. The total number of the poles is determined by the operating frequency and the dimensional parameters of the substrate [52, 68]. In the isotropic case, it is possible to predict the number of poles, such as in [69], if we have $\kappa_0 d\sqrt{\varepsilon_r - 1} < \pi/2$ then the denominator of $G^h(\kappa_s)$ does not have any zero, while that of $G^e(\kappa_s)$ has only one.

In the case of a structure having only one isotropic dielectric layer with $\varepsilon_z=\varepsilon_t=\varepsilon_r$, $\mu_z=\mu_t=1$ and $\xi=0$ the equality (5.51) reduces to the simple expression given by [63, 70]

$$\kappa_{sp}^2 = \kappa_0^2 \left(1 + \kappa_0^2 d^2 \frac{(\varepsilon_r - 1)^2}{\varepsilon_r^2} \right) \tag{5.53}$$

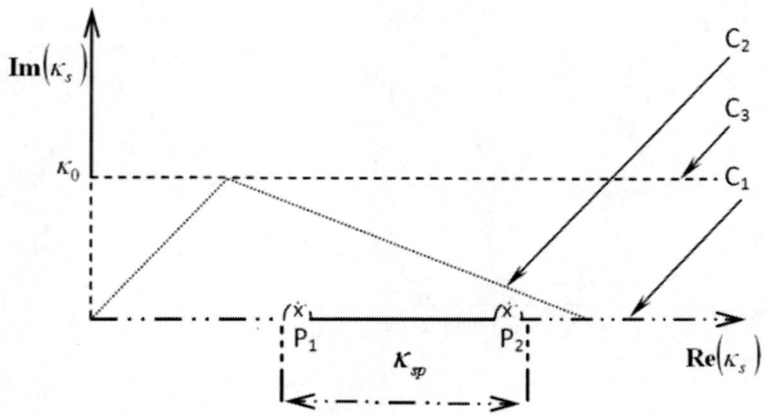

Figure 5.7. Surface waves approximate wavenumber in the integration path.

By replacing (5.42) in (5.40), the TM_0 mode cut-off frequency is obtained as follows

$$\kappa^2 = \kappa_0^2 \frac{\kappa_0 \xi d}{(1+\kappa_0 \xi d)} \tag{5.54}$$

It is sufficient to put κ_z=0 in Equations (5.40) and (5.41), to deduce the cut-off frequencies of TE and TM modes. For TM modes

$$f_c^{TM} = \frac{n \cdot c}{4 \cdot d} \frac{1}{\sqrt{\left(\varepsilon_t \mu_t - \frac{\varepsilon_z}{\varepsilon_t} - \xi^2\right)}}, \; n \text{ is even.} \tag{5.55}$$

For TE modes

$$f_c^{TE} = \frac{c}{d} \frac{1}{\sqrt{\left(\varepsilon_t \mu_t - \xi^2 - \frac{\mu_t}{\mu_z}\right)}} \left(\frac{n}{2} - \frac{1}{2\pi} \arctan \frac{\sqrt{\left(\varepsilon_t \mu_t - \xi^2 - \frac{\mu_t}{\mu_z}\right)}}{\xi} \right), \quad n \text{ is odd.} \tag{5.56}$$

In the absence of chirality, Equations (5.55) and (5.56) become

$$f_c^{TM} = \frac{n \cdot c}{4 \cdot d} \frac{1}{\sqrt{\left(\varepsilon_t \mu_t - \frac{\varepsilon_z}{\varepsilon_t} \right)}} \tag{5.57}$$

$$f_c^{TE} = \frac{2n+1}{4} \frac{c}{d} \frac{1}{\sqrt{\left(\varepsilon_t \mu_t - \frac{\mu_t}{\mu_z} \right)}} \tag{5.58}$$

In the case of nonmagnetic anisotropic dielectrics, Equations (5.57) and (5.58) become

$$f_c^{TM} = \frac{n \cdot c}{4 \cdot d} \frac{1}{\sqrt{\left(\varepsilon_t - \frac{\varepsilon_z}{\varepsilon_t} \right)}} \tag{5.59}$$

$$f_c^{TE} = \frac{2n+1}{4} \frac{c}{d} \frac{1}{\sqrt{\left(\varepsilon_t - 1 \right)}} \tag{5.60}$$

In the case of isotropic dielectrics, Equations (5.59) and (5.60) become

$$f_c^{TM} = \frac{n \cdot c}{4 \cdot d} \frac{1}{\sqrt{\left(\varepsilon_r - 1 \right)}} \tag{5.61}$$

$$f_c^{TE} = \frac{2n+1}{4} \frac{c}{d} \frac{1}{\sqrt{\left(\varepsilon_r - 1 \right)}} \tag{5.62}$$

The previous equations have the same forms of those developed in [55, 57]. The mode of the low order TE_1 is excited at 40.77 GHz in a duroid (ε_r=2.32) substrate of thickness 1.6 mm without the effect of the chirality [57]. On the other hand, using the same substrate with the consideration of a positive (ξ=+1) or negative (ξ=-1) coefficient of chirality, the TE_1 mode for the same structure is excited at 138.40 GHz or 192.75 GHz respectively. Similarly, the low order TE_1 is excited at approximately 39.33 GHz for an isotropic 0.635 mm thick alumina (ε_r=10) substrate [57], but at 50.75 GHz or

116.16 GHz respectively with a positive (ξ=+1) or negative (ξ=-1) chirality coefficient.

The practical expression of d is no more that given by (5.38), but rather that given by (5.63), and we can in this case increase the thickness of the substrate without exciting surface waves [71]

$$d < \frac{c}{f} \frac{1}{\sqrt{\left(\varepsilon_t \mu_t - \xi^2 - \frac{\mu_t}{\mu_z}\right)}} \left(\frac{n}{2} - \frac{1}{2\pi} \arctan \frac{\sqrt{\left(\varepsilon_t \mu_t - \xi^2 - \frac{\mu_t}{\mu_z}\right)}}{\xi} \right) \tag{5.63}$$

5.6.1. Numerical Examples

According to the results obtained for different expressions of the surface modes cut-off frequencies illustrated in Figures 5.8 and 5.9, chirality moves the surface waves away from the operating frequency of the structure, which constitutes an apparent advantage for the antennas. However, for filter structures it is preferable to use the pure imaginary form of the magneto-electric elements, formulated from more complex expressions. This case of study of TM and TE surface waves cut-off frequencies according to the medium constitutive parameters, ended in results which show clearly that the thicker the dielectric, the more favoured is excitation of surface waves.

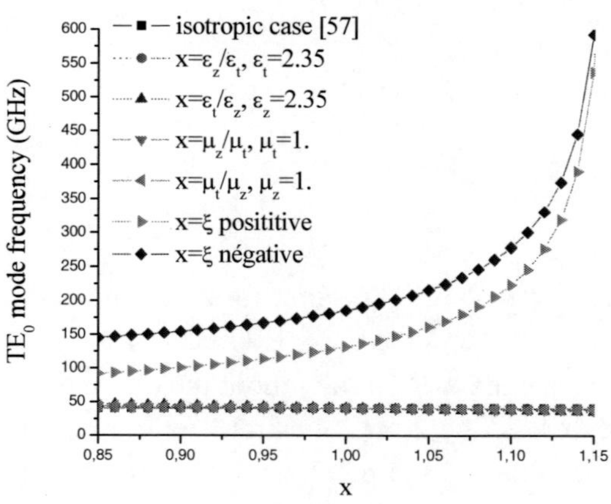

Figure 5.8. Effect of the chiral constitutive parameters on the TE_0 mode cut-off frequency versus different ratios. TE_0 is excited at 40.77 GHz in a Duroid (ε_r=2.32) substrate of thickness 1.6 mm [57].

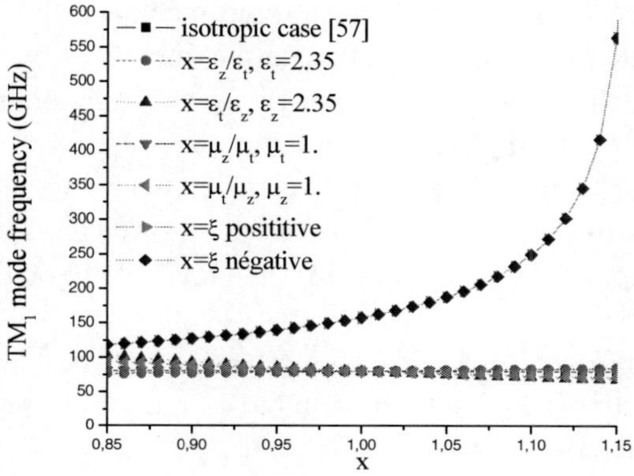

Figure 5.9. Effect of the chiral constitutive parameters on the TM_1 mode cut-off frequency versus different ratios, in a Duroid (ε_r=2.32) substrate of thickness 1.6 mm [57]. isotropic again.

5.7. Complex Media Parameter Effect on the Input Impedance

The effect of a chiral bianisotropic substrate with complex constitutive parameters on the input impedance of a rectangular microstrip antenna has been studied on the basis of numerical solution of the integral equation using the Galerkin procedure for the moment method in the spectral domain. The input impedance of the resonator studied in Figure 5.1 is one of the problems that must be rigorously formulated in the spectral domain [72]. The method of moments using the Galerkin procedure, where the test functions are the same basis functions, can be used to solve numerically these integrals. The forms most commonly used for the elementary cells are the triangle and the rectangle [73]. The spectral domain approach (SDA) is used for the analysis and design of planar structures. In this approach, the tensorial Green's spectral function which relates the tangential electric field components and the corresponding currents on different conductors has to be determined. A great many techniques have been used to evaluate this function [10, 22]. The fundamental difference between an isotropic medium and an isotropic chiral one is that the latter is characterized by a more complex form of the wave equation. In electromagnetism, the chirality is an additional parameter for the design of materials. However, chiral materials are supposed to have particular absorbing properties [74]. The effect of the bianisotropic media on the propagation in waveguides and antennas has also stimulated the interest of many researchers. Thus, much research work on electromagnetic propagation in waveguides filled with bi-isotropic/chiral materials is reported in the literature, such as in [75, 76]. The use of chiral media in microstrip antennas is also proposed in [76-78] specifically in order to reduce losses due to surface wave radiation in a microstrip antenna printed on a chiral substrate [76, 78], although this matter is discussed in [79].

The input impedance of an antenna case can be calculated as a combination of electric field dispersed with the current on the probe, according to the following expression

$$Z_{in} = -\left(\sum_{m=1}^{M} I_{xm} V_{xm} + \sum_{n=1}^{N} I_{yn} V_{yn}\right) \tag{5.64}$$

where I_{xm} and I_{yn} are the coefficients to be determined, and m and n define the number of basis functions along the directions x and y, respectively.

$$V_{xm} = \frac{1}{4\pi^2} \int_{-\infty}^{+\infty}\!\!\int J_{xm} e^{-i\left(\kappa_x x_f + \kappa_y y_f\right)} \int_0^d G_{zx} dz d\kappa_x d\kappa_y \tag{5.65}$$

$$V_{yn} = \frac{1}{4\pi^2} \int_{-\infty}^{+\infty}\!\!\int J_{yn} e^{-i\left(\kappa_x x_f + \kappa_y y_f\right)} \int_0^d G_{zy} dz d\kappa_x d\kappa_y \tag{5.66}$$

$G_{zx}=G_{zx}(\kappa_x,\kappa_y,z)$ and $G_{zy}=G_{zy}(\kappa_x,\kappa_y,z)$ are the spectral Green functions for the chiral substrate case, where $\widetilde{J}_{xm} = \widetilde{J}_{xm}\left(\kappa_x, \kappa_y, z\right)$ and $\widetilde{J}_{yn} = \widetilde{J}_{yn}\left(\kappa_x, \kappa_y, z\right)$ are the Fourier transforms of the basis functions in the spectral domain. After some algebra, it can be written as

$$V_m^p = \frac{1}{4\pi^2 \omega \varepsilon_0} \frac{\varepsilon_t}{\varepsilon_z} \iint \frac{\kappa_z \sin\left(\kappa_z^e d\right) \kappa_y \widetilde{J}_y}{\kappa_z^e T} d\kappa_x d\kappa_y \tag{5.67}$$

$$V_n^p = \frac{1}{4\pi^2 \omega \varepsilon_0} \frac{\varepsilon_t}{\varepsilon_z} \iint \frac{\kappa_z \sin(\kappa_z^e d) \kappa_x \widetilde{J}_x}{\kappa_z^e T} d\kappa_x d\kappa_y \tag{5.68}$$

$$T = \varepsilon_t \kappa_z \cos(\kappa_z^e d) + j \varepsilon_z \kappa_z^e \sin(\kappa_z^e d) \left(1 - \frac{\kappa_0 \xi}{\kappa_z^e} \tan(\kappa_z^e d) \right) \tag{5.69}$$

where the surface currents J on the patch are expanded into a finite series of basis function J_{xn} and J_{ym}. For digital convergence of Equations (5.67)-(5.69), the steps discussed in [57, 80, 81] are followed.

Figures 5.9 and 5.10 show that the input impedance is directly related to the chirality, which leads us to consider this effect in seeking a better adaptation. The amplitude parameters raised vary depending on the chirality, while in [72], where the case of chiral bi-isotropic is considered, these parameters are constant in amplitude, which is advantageous for the circuit design.

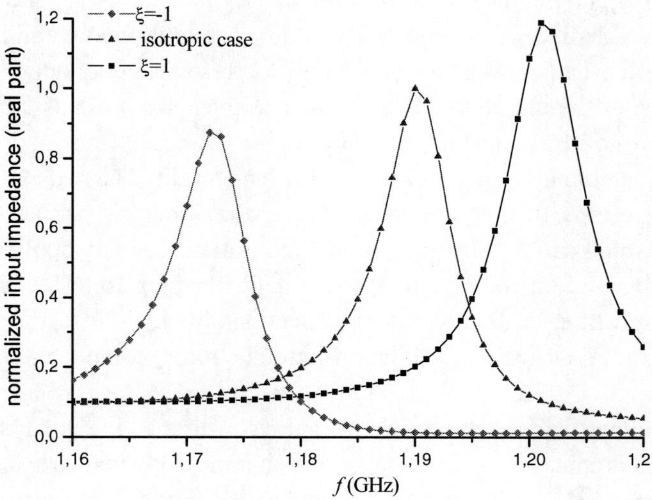

Figure 5.9. Chirality effect on the real part of the input impedance (a=7.62 cm, b=11.43 cm, x_0=1.52 cm, y_0=0.385 cm, ε_x=ε_z=2.64).

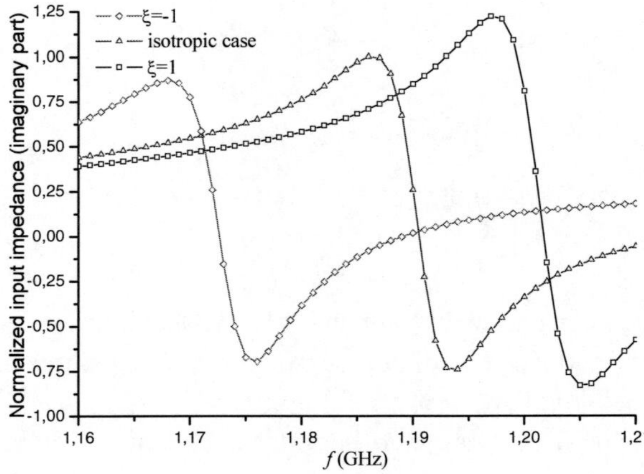

Figure 5.10. Chirality effect on the imaginary part of the input impedance (a=7.62 cm, b=11.43 cm, x_0=1.52 cm, y_0=0.385 cm, $\varepsilon_x = \varepsilon_z$=2.64).

According to the literature, positive and negative uniaxial anisotropy both introduce a slight change in the half-power bandwidth, whereas in this work, the structure is implemented on a chiral substrate. The results show that the effect of the constitutive parameters (chirality and axial permeability) is very apparent on the half-power bandwidth and the resonant frequency, according to the choice of the chirality element. The positive case presents a diminution of the real part of the resonant frequency and an increase of the imaginary part. The inverse is obtained for the case of negative chirality. The advantage of the uniaxial anisotropy of the permeability is on the widening of the half-power bandwidth which can be very important for a radiating cavity application. The inverse is obtained for the case of a cavity built for filter use. The introduction of a chiral substrate leads to various applications. Its main effect is the ability of structure miniaturization, which makes electronic component integration easier.

The adopted approach can be employed during the analysis of the microstrip to predict the surface waves which appear in such structures. It can also be used for locating integrand singularities that can be encountered in solutions using the method of moments via the Galerkin procedure. The introduction of chiral leads to a diversity of applications. One appreciable effect, which is the capacity of miniaturization of the structures, facilitates the electronic components integration. The chirality also moves away the surface waves from the operating frequency of the structure, which is an advantage for antennas. Furthermore, for filters, it is preferable to use the pure imaginary shape of the magneto-electric elements of the chiral. This case of study of the TM or TE surface waves cut-off frequencies, with respect to the constitutive parameters of the medium, ended in results which show clearly that the possibility of excitation of the surface waves is more favoured when the dielectric become increasingly thick, compared to components with traditional dielectrics. Thus chirality presents interesting effects on the resonant frequency, bandwidth and input impedance.

5.8. Rectangular Microstrip Antenna with Uniaxial Bianisotropic Chiral Substrate-Superstrate

Structures with substrate-superstrate are employed, in general, to improve parameters such as the efficiency, effectiveness and polarization of the antenna [82], where the superstrate is used as a protective structure for the aerial elements against bad climatic conditions such as rain, fog, and snow [83], which form in unexpected ways, and affect the performance of the printed antennas [20, 84, 85]. Therefore, for better characterization and precise design, these effects must be taken into account. The major undesirable effect of the considered structure is resonant frequency shifting, which can force the antenna to work at frequencies other than the predefined ones [83]. Another possible disadvantage could be the excitation of a large number of surface waves, due to large thicknesses. Lifting the superstrate off the antenna substrate would help to alleviate this problem [17]. The use of complex materials, such as biaxial or bianisotropic materials, or the use of an anisotropic superstrate, can have interesting effects [17]. The effects of uniaxial anisotropy on the resonant frequency and the bandwidth of a monolayer microstrip antenna, have been studied in several works, for a rectangular patch in [18, 86, 87], and for a circular patch in [83, 85, 88], where the positive and negative uniaxial anisotropy were considered. On the other hand, the effect of an isotropic dielectric superstrate on the resonant frequency and the bandwidth of a radiating rectangular patch microstrip was evaluated experimentally by Bahl [89], and theoretically by Bouttout [20]. They deduced that the bandwidth decreases for a high permittivity thin superstrate.

On the basis of previous work for the monolayer case [39], where magnetic uniaxial anisotropy and chirality effects are studied, the effects of a superstrate having uniaxial electric, magnetic anisotropies and chiral elements on the resonant frequency and the bandwidth of the rectangular microstrip are investigated. The idea of using chiral materials as a substrate and superstrate in printed antenna circuits was first introduced by Engheta [42]. It has been shown that the power of the surface waves can be generally reduced when a chiral substrate is employed in a printed microstrip circuit [90]. The theoretical concept of the complex resonant frequency considered here is formulated in terms of an integral equation, by using the vector Fourier transform [91, 92]. This part of work is presented as follows. Firstly, the bianisotropic material is presented with its magneto-electric tensors, with a chiral material as substrate and superstrate, Green's tensors are deduced and the integral equation is obtained in the spectral domain. Secondly, the numerical results are presented for different values of the magneto-electric elements. In this contribution, the intention is to show and observe the effects of chiral elements on the resonant frequency and half-bandwidth of a microstrip antenna.

5.8.1. Constitutive Equations for Bianisotropic Media

The geometry under consideration is shown in Figure 5.11. A rectangular patch with dimensions (a, b) along the two axes (ox, oy) respectively, is printed on a grounded dielectric slab of thickness d_1. Above the radiating patch is the superstrate layer of thickness d_2. All the dielectric materials are assumed to be bianisotropic. The chiral media considered in this study can be described by the following constitutive relations [30]

$$\vec{B} = \mu\vec{H} + \frac{1}{c_0}\xi\vec{E} \tag{5.70}$$

$$\vec{D} = \varepsilon\vec{E} + \frac{1}{c_0}\eta\vec{H} \tag{5.71}$$

where $\vec{E}$ and $\vec{H}$ are the electric and magnetic field vectors, $\vec{D}$ and $\vec{B}$ are the electric and magnetic induction vectors, c_0 is the speed of light in vacuum and ε, μ, ξ and η are the constitutive tensors, expressed as

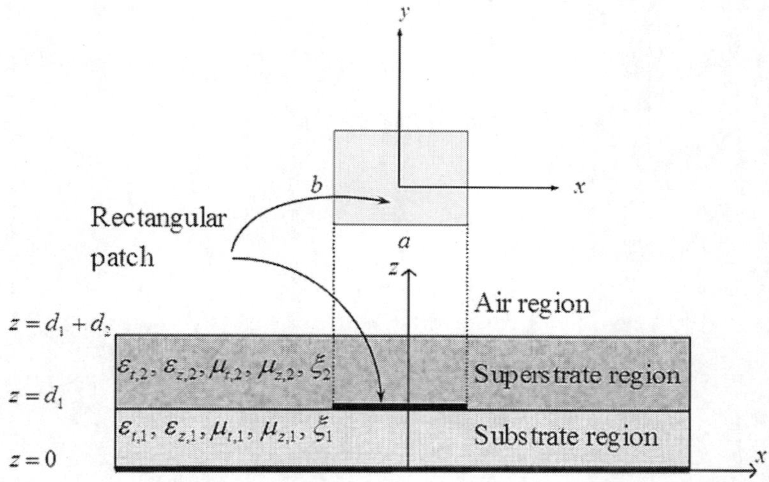

Figure 5.11. Geometrical structure of rectangular patch trapped between bianisotropic substrate-superstrate.

$$\mu = \mu_0 \begin{bmatrix} \mu_t & 0 & 0 \\ 0 & \mu_t & 0 \\ 0 & 0 & \mu_z \end{bmatrix}, \varepsilon = \varepsilon_0 \begin{bmatrix} \varepsilon_t & 0 & 0 \\ 0 & \varepsilon_t & 0 \\ 0 & 0 & \varepsilon_z \end{bmatrix}, \ \xi = \eta = j \begin{bmatrix} 0 & \xi & 0 \\ -\xi & 0 & 0 \\ 0 & 0 & 0 \end{bmatrix} \tag{5.72}$$

A more detailed classification of the bianisotropic materials in terms of reciprocity and some other properties can be found in [30, 39, 90, 93-95].

5.8.2 Green's Tensor Evaluation

Starting from Maxwell's equations, the transverse magnetic (e) and transverse electric (h) counterparts of the tangential electric and magnetic fields in the Fourier domain for a bianisotropic bounded region with an anisotropy tensor of type (5.72) can be expressed in compact matrix form [39]. Keeping in mind the boundary conditions for the tangential field components in the proximity of the conducting patch, for the structure shown in Figure 5.11, the relation between the electric source and the electric field can be derived in terms of the spectral Green's dyad [96].

$$G(\kappa_s) = \frac{1}{j\omega\varepsilon}\begin{bmatrix} \dfrac{(N_i^e + N_c^e)}{(D_i^e + D_c^e)} N^e \kappa_{z,1}^e & 0 \\ 0 & \dfrac{(N_i^h + N_c^h)}{(D_i^h + D_c^h)} \kappa_0^2 \end{bmatrix} \sin(\kappa_{z,1} d_1) \tag{5.73}$$

$$N^e = \frac{\kappa_0^2 \varepsilon_{t,1} \mu_{t,1} - \kappa_s^2 \dfrac{\varepsilon_{t,1}}{\varepsilon_{z,1}}}{\kappa_{z,1}^{e2}} \tag{5.74}$$

$$N_i^e = \kappa_z \cos(\kappa_{z,2}^e d_2) + j \frac{\kappa_{z,2}^e}{\varepsilon_{t,2}} \sin(\kappa_{z,2}^e d_2) \tag{5.75}$$

$$N_c^e = \frac{\kappa_0 \xi_2}{\kappa_{z,2}^e} (\kappa_z \varepsilon_{t,2} + j\kappa_0 \xi_2) \sin(\kappa_{z,2}^e d_2) \tag{5.76}$$

$$N_i^h = \cos(\kappa_{z,2}^h d_2) + j \frac{\kappa_z}{\kappa_{z,2}^h} \mu_{t,2} \sin(\kappa_{z,2}^h d_2) \tag{5.77}$$

$$N_c^h = -j \frac{\kappa_0 \xi_2}{\kappa_{z,2}^h} \sin(\kappa_{z,2}^h d_2) \tag{5.78}$$

$$\begin{aligned} D_i^e &= (\kappa_z \varepsilon_{t,1} \cos(\kappa_{z,1}^e d_1) + j\kappa_{z,1}^e \sin(\kappa_{z,1}^e d_1)) \cos(\kappa_{z,2}^e d_2) \\ &\quad + j\left(\frac{\varepsilon_{t,1}}{\varepsilon_{t,2}} \kappa_{z,2}^e \cos(\kappa_{z,1}^e d_1) + j \frac{\kappa_{z,1}^e \kappa_z}{\kappa_{z,2}^e} \varepsilon_{t,2} \sin(\kappa_{z,1}^e d_1) \right) \sin(\kappa_{z,2}^e d_2) \end{aligned} \tag{5.79}$$

$$\begin{aligned} D_c^e &= \frac{\kappa_0 \xi_1}{\kappa_{z,1}^e} (\kappa_z \varepsilon_{t,1} + j\kappa_0 \xi_1) \sin(\kappa_{z,1}^e d_1) \cos(\kappa_{z,2}^e d_2) \\ &+ \frac{\varepsilon_{t,1}}{\varepsilon_{t,2}} \frac{\kappa_0 \xi_2}{\kappa_{z,2}^e} (\kappa_z \varepsilon_{t,2} + j\kappa_0 \xi_2) \sin(\kappa_{z,2}^e d_2) \cos(\kappa_{z,1}^e d_1) \\ &+ \kappa_0 \left(\left(\frac{\varepsilon_{t,1}}{\varepsilon_{t,2}} - \frac{\xi_1}{\xi_2} \right) \frac{\kappa_0 \xi_1 \xi_2}{\kappa_{z,1}^e \kappa_{z,2}^e} (\kappa_z \varepsilon_{t,2} + j\kappa_0 \xi_2) - j\xi_2 \frac{\kappa_{z,1}^e}{\kappa_{z,2}^e} + j\xi_1 \frac{\varepsilon_{t,1}}{\varepsilon_{t,2}} \frac{\kappa_{z,2}^e}{\kappa_{z,1}^e} \right) \sin(\kappa_{z,2}^e d_2) \sin(\kappa_{z,1}^e d_1) \end{aligned} \tag{5.80}$$

$$D_i^h = \frac{1}{\mu_{t,1}} \left[\begin{array}{l} \left(\kappa_{z,1}^h \cos\left(\kappa_{z,1}^h d_1\right) + j\kappa_z \mu_{t,1} \sin\left(\kappa_{z,1}^h d_1\right)\right) \cos\left(\kappa_{z,2}^h d_2\right) \\ \quad + j\left(\mu_{t,2} \frac{\kappa_z \kappa_{z,1}^h}{\kappa_{z,2}^h} \cos\left(\kappa_{z,1}^h d_1\right) + j\frac{\mu_{t,1}}{\mu_{t,2}} \kappa_{z,2}^h \sin\left(\kappa_{z,1}^h d_1\right)\right) \sin\left(\kappa_{z,2}^h d_2\right) \end{array} \right] \tag{5.81}$$

$$D_c^h = \frac{\kappa_0}{\mu_{t,1}} \left[\begin{array}{l} j\frac{\left(\xi_2 \mu_{t,1} - \xi_1 \mu_{t,2}\right)\left(j\kappa_0 \xi_2 - \kappa_z \mu_{t,2}\right)}{\kappa_{z,2}^h \mu_{t,2}} \sin\left(\kappa_{z,1}^h d_1\right) \sin\left(\kappa_{z,2}^h d_2\right) \\ \quad + \frac{\kappa_{z,1}^h}{\kappa_{z,2}^h} \xi_2 \sin\left(\kappa_{z,2}^h d_2\right) \cos\left(\kappa_{z,1}^h d_1\right) + \xi_1 \cos\left(\kappa_{z,2}^h d_2\right) \sin\left(\kappa_{z,1}^h d_1\right) \end{array} \right] \tag{5.82}$$

$$\kappa_z = \sqrt{\kappa_0^2 - \kappa_s^2} \tag{5.83}$$

The indices i, c, 1 and 2 respectively represent the uniaxial anisotropic case, chiral case, first layer and second layer.

5.8.3. Galerkin's Method

It is clear from (5.73) that the Green's dyadic is diagonal in the (TM, TE) representation in the presence of uniaxial anisotropic chirality can be simplified when deriving Green's dyadic for a multilayer anisotropic media. Furthermore, if the substrate and superstrate layers are isotropic the coefficients with 'c' indices are cancelled. The dyadic elements given in (5.73) are the same as given in [20] and [97]. The surface current J on the patch can be expanded into a finite series of basis functions

$$J(r_s) = \sum_{n=1}^{N} a_n \begin{bmatrix} J_{xn}(r_s) \\ 0 \end{bmatrix} + \sum_{m=1}^{M} b_m \begin{bmatrix} 0 \\ J_{ym}(r_s) \end{bmatrix} \tag{5.84}$$

where a_n and b_m are the mode expansion coefficients to be sought. The integral equation describing the electric field on the patch is [98]

$$\iint_{-\infty}^{\infty} d\kappa_s F(\kappa_s, r_s) \cdot G(\kappa_s) \cdot \tilde{J}(\kappa_s) = 0 \tag{5.85}$$

$F(\kappa_s, r_s)\big|_{r_s=0}$ is the kernel of the vector transforms, where $F(\kappa_s, r_s)_{r_s=0} = \begin{bmatrix} \kappa_x & \kappa_y \\ \kappa_y & -\kappa_x \end{bmatrix}$, and $\widetilde{J}(\kappa_s)$ is the Fourier transformation of the surface current J. Using the well known Galerkin procedure of the method of moments, the integral equation in (5.85) is discretized and brought into the following matrix equation

$$\begin{bmatrix} (B_1)_{N\times N} & (B_2)_{N\times M} \\ (B_3)_{M\times N} & (B_4)_{M\times M} \end{bmatrix} \cdot \begin{bmatrix} (a)_{N\times 1} \\ (b)_{M\times 1} \end{bmatrix} = 0 \tag{5.86}$$

where $(B_1)_{N\times N}$, $(B_2)_{N\times M}$, $(B_3)_{M\times N}$ and $(B_4)_{M\times M}$ are the elements of the digitalized matrix equation. $(a)_{N\times 1}$ and $(a)_{M\times 1}$ are respectively vectors of the unknown expansion coefficients of patch surface currents $\widetilde{J}_{xn}(\kappa_s)$ and $\widetilde{J}_{ym}(\kappa_s)$. A nontrivial solution of (5.86) is derived by seeking a complex frequency $f=f_r+if_i$, where f_r and f_i are respectively the real and imaginary parts of the complex resonant frequency and $2f_i/f$ is the half-power bandwidth of the antenna [18, 99, 100]. Owing to the fact that the narrow bandwidth of the resonator should be exact, the obtained results are validated with those in [20] and [89]. Table 5.1 presents a comparison between the calculated frequencies and those measured by Bahl [89] and calculated in [20]. The superstrate and the substrate are considered in the isotropic case. The values are closer to those given in [89] than to those given in [20]. Table 5.1 shows a better convergence of our results.

In Figure 5.12, the effect of the electric uniaxial anisotropy is presented by comparison with the isotropic case [20] (a rectangular patch trapped between a superstrate and a substrate; a=6 cm, b=5 cm, d_1=0.1 cm, $\varepsilon_{t,1}=\varepsilon_{z,1}=2.35$), for various values of the permittivity ($\varepsilon_{t,2}=\varepsilon_{z,2}=1.5$, 2.35, 4, 10) and superstrate thickness.

Table 5.1. Comparison between the real resonant frequencies calculated and measured

N°	$\varepsilon_{t,2}$	d_2 (cm)	Resonant Frequencies (GHz)		
			Measured [89]	Calculated [20]	Our results
1	1.0	∞	4.104	4.123	4.10121
2	2.32	0.0800	4.008	4.033	4.01228
3		0.1590	3.934	3.983	3.96249
4		0.3180	3.895	3.924	3.90398
5	2.6	0.1120	3.952	3.991	3.96994
6		0.1590	3.912	3.958	3.93791
7		0.3180	3.874	3.887	3.86747
8		0.6360	3.806	3.822	3.80391
9	3.0	0.0064	4.070	4.108	4.08679
10		0.0128	4.058	4.095	4.07364
11		0.0384	4.010	4.051	4.02978
12	10.2	0.0635	3.640	3.782	3.76112
13	10	0.1540	3.482	3.518	3.49265
14		0.3120	3.260	3.222	3.27192

The normalized real frequency is plotted according to the thickness of the superstrate, and the normalization is with respect to the real resonant frequency f_{r0}, related to the monolayer antenna ($d_2=0$, $\varepsilon_{r,1}=2.35$). This is the case for all the following sections. According to Figure 5.12(a) and 5.12(b), it is noticed that for lower values of the permittivity, the effect of the two axial components of the permittivity ($\varepsilon_{t,2}$,$\varepsilon_{z,2}$) becomes more apparent for low values, but in fact it is very important for high values of the permittivity. These values of the permittivity mask the effect of the axial components. The component $\varepsilon_{t,2}$ has a more marked effect on the real part of the resonant frequency than the component $\varepsilon_{z,2}$. These observations are valid for the ratio d_2/d_1 greater than unity and for high values for the permittivity. In Figure 5.12(a)-(c), the influence of the component $\varepsilon_{z,2}$ on the real resonant frequency is almost constant (±0,15%-±0,2%). Consequently, the most advantageous of these anisotropies is the positive uniaxial anisotropy case, where the real resonant frequency is subject to a great increase.

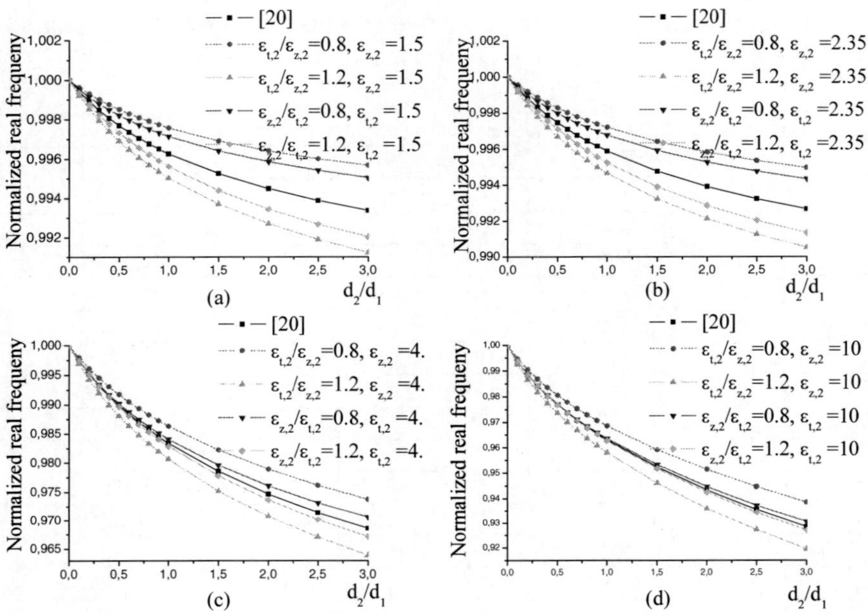

Figure 5.12. The effect of uniaxial anisotropic permittivity of a superstrate on the normalized real resonant frequency; a=6 cm, b=5 cm, d_1=0.1 cm, $\varepsilon_{r,1}$=2.35, (a) $\varepsilon_{t,2}/\varepsilon_{z,2}$=1.5, (b) $\varepsilon_{t,2}/\varepsilon_{z,2}$=2.35, (c) $\varepsilon_{t,2}/\varepsilon_{z,2}$=4, (d) $\varepsilon_{t,2}/\varepsilon_{z,2}$=10.

In Figure 5.13, the normalized imaginary frequency is plotted according to the thickness of the superstrate. The numerical results show that the influencing factor is $\varepsilon_{t,2}$, and this effect is maximum in the region ($1.5d_1 > d_2 > d_1$). For a positive anisotropy ($\varepsilon_{z,2}$=1.5 and $\varepsilon_{t,2}=0.8\varepsilon_{z,2}$) and $d_2=1.5d_1$, the imaginary resonant frequency f_i reaches the monolayer case, whereas for the isotropic case the resonant frequency f_i is reached for $d_2=2.6d_1$. We deduce that the first case leads to superstrate miniaturisation (superstrate thickness reduction of 42.30%) and a slight increase in bandwidth. In [18], it has been found that the bandwidth is only weakly affected by the uniaxial anisotropy of the mono layer.

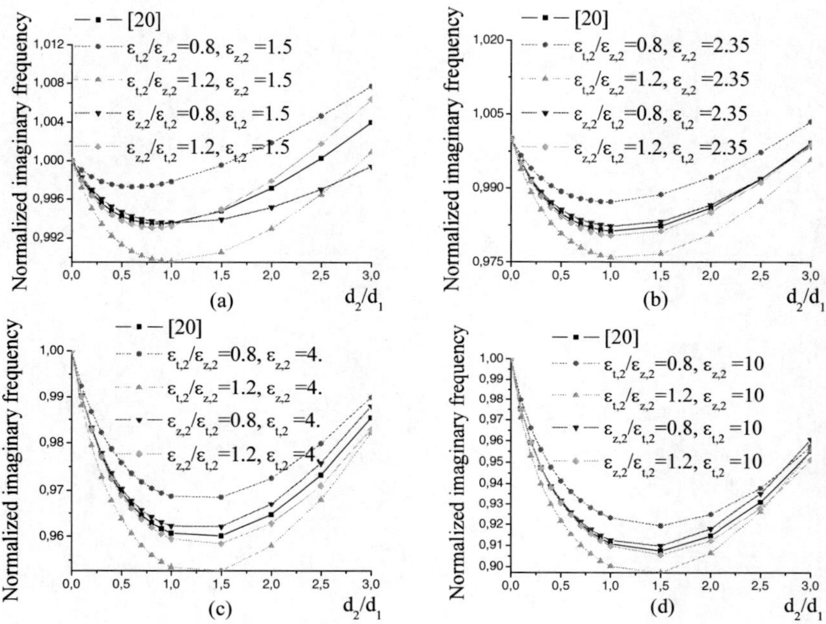

Figure 5.13. The effect of uniaxial anisotropic permittivity of a superstrate on the normalized imaginary resonant frequency; a=6 cm, b=5 cm, d_1=0.1 cm, $\varepsilon_{r,1}$=2.35, (a) $\varepsilon_{t,2}/\varepsilon_{z,2}$=1.5, (b) $\varepsilon_{t,2}/\varepsilon_{z,2}$=2.35, (c) $\varepsilon_{t,2}/\varepsilon_{z,2}$=4, (d) $\varepsilon_{t,2}/\varepsilon_{z,2}$=10.

The effect of a dielectric superstrate on the bandwidth of a rectangular antenna microstrip is also studied. In Figure 5.14, the normalized bandwidth is plotted versus the thickness of the superstrate. Numerical results show that for low permittivity ($1<\varepsilon_{z,2}<\varepsilon_{z,1}$) and low thicknesses ($d_2<d_1$) of the protective dielectric layer, the bandwidth is appreciably constant for the isotropic case [89], but it is more sensitive to the variations of the permittivity of the superstrate $\varepsilon_{t,2}$, whereas the component $\varepsilon_{z,2}$ does not have any effect on the bandwidth. For higher thicknesses ($d_2>d_1$), one can see two regions. In the first one ($d_1<d_2<2d_1$), the bandwidth increases slowly and the component $\varepsilon_{t,2}$ remains more influential than $\varepsilon_{z,2}$, and in of the second region ($2d_1<d_2$), the component $\varepsilon_{z,2}$ becomes the most important factor. In the case of an isotropic superstrate, ($\varepsilon_{z,2}$, $\varepsilon_{t,2}>\varepsilon_{t,1}=\varepsilon_{z,1}$) and relatively lower thicknesses ($d_2<2.6d_1$), the bandwidth is always lower than BW_0 (related to the monolayer antenna d_2=0, $\varepsilon_{r,1}$=2.35) which is the bandwidth of the monolayer antenna [20].

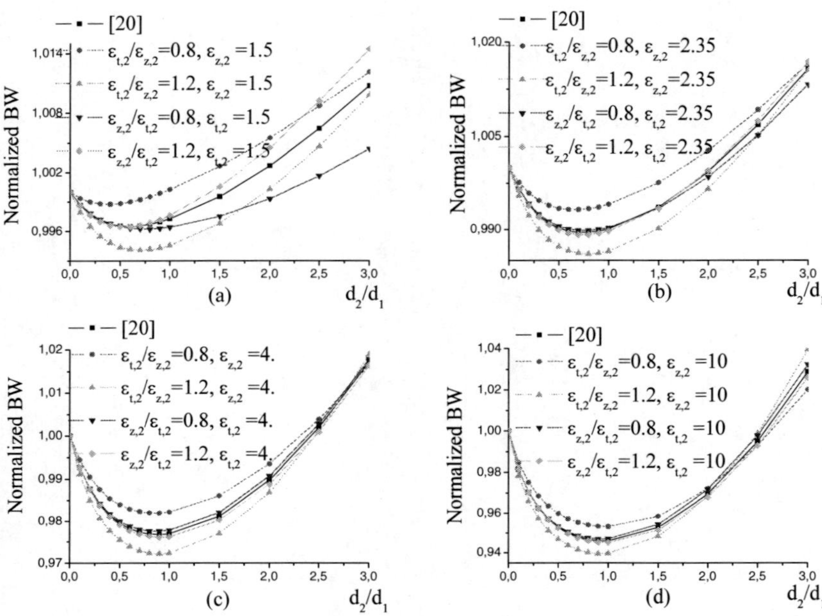

Figure 5.14. The effect of uniaxial anisotropic permittivity of a superstrate on the normalized bandwidth; a=6 cm, b=5 cm, d_1=0.1 cm, $\varepsilon_{r,1}$=2.35, (a) $\varepsilon_{t,2}/\varepsilon_{z,2}$=1.5, (b) $\varepsilon_{t,2}/\varepsilon_{z,2}$=2.35, (c) $\varepsilon_{t,2}/\varepsilon_{z,2}$=4, (d) $\varepsilon_{t,2}/\varepsilon_{z,2}$=10.

In this region, the components $\varepsilon_{t,2}$ hold the bandwidth close to BW_0 or higher. For dielectric permittivity ($\varepsilon_{z,2}$, $\varepsilon_{t,2}>\varepsilon_{t,1}=\varepsilon_{z,1}$) and higher thicknesses ($3d_1>d_2>2.6d_1$), the components $\varepsilon_{z,2}$ and $\varepsilon_{t,2}$ have a tiny effect on the bandwidth, but for much higher values ($\varepsilon_{z,2}$ or $\varepsilon_{t,2}$=10), the component $\varepsilon_{t,2}$ has an important effect. In Figure 5.15, the effect of the uniaxial permeability is presented in comparison with the isotropic case, for various values of the permeability and thickness of the superstrate. According to Figure 5.15, while increasing the value of the permittivity ($\varepsilon_{t,2},\varepsilon_{z,2}>\varepsilon_{r,1}$), the effect of the two axial components of the permeability ($\mu_{t,2}$, $\mu_{z,2}$) becomes more negligible, whereas for weak permittivity ($1<\varepsilon_{t,2},\varepsilon_{z,2}\leq\varepsilon_{r,1}$), the component $\mu_{z,2}$ has a stronger effect on the real part of the resonant frequency than the component $\mu_{t,2}$.

Figure 5.16 shows that for low permittivity ($1<\varepsilon_{t,2},\varepsilon_{z,2}\leq\varepsilon_{r,1}$), the component of the permeability $\mu_{z,2}$ for ($d_2<1.5d_1$) is the most influential factor on the imaginary resonant frequency, whereas for thick layers ($d_2>2.6d_1$), the component $\mu_{t,2}$ becomes more important than $\mu_{z,2}$. For higher permittivity ($\varepsilon_{t,2},\varepsilon_{z,2}>\varepsilon_{r,1}$), one can distinguish two regions. For the first one ($d_2<1.5d_1$), the anisotropy in this case has a negligible effect (0.5% maximum), while in the second region (d_2>1.5d_1), the component $\mu_{t,2}$ has a visible effect (up to 2.5%). Figure 5.17 shows that for (d_2<d_1), the influencing component on the bandwidth is $\mu_{z,2}$. Unfortunately, this effect remains weak in this region, whereas for thick layers (d_2>d_1), the component $\mu_{t,2}$ becomes more important than $\mu_{z,2}$, and the bandwidth in this case reaches a maximum of 3%. For higher thicknesses of the superstrate (d_2>d_1), the effect of permeability on the imaginary frequency and bandwidth is more important than the permittivity, whereas for real frequency the latter constitutive elements have a reverse effect.

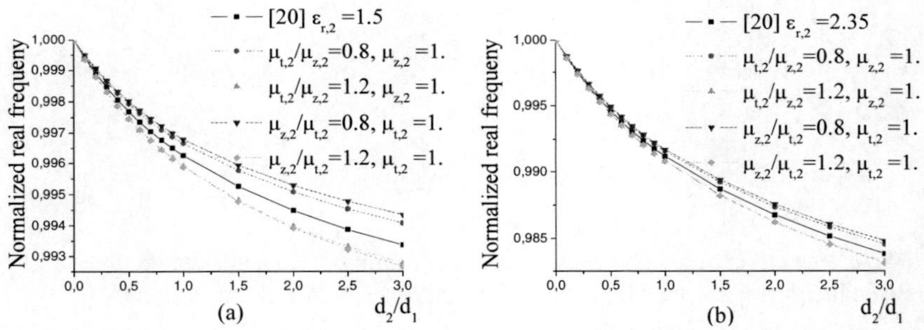

Figure 5.15. Continued on next page.

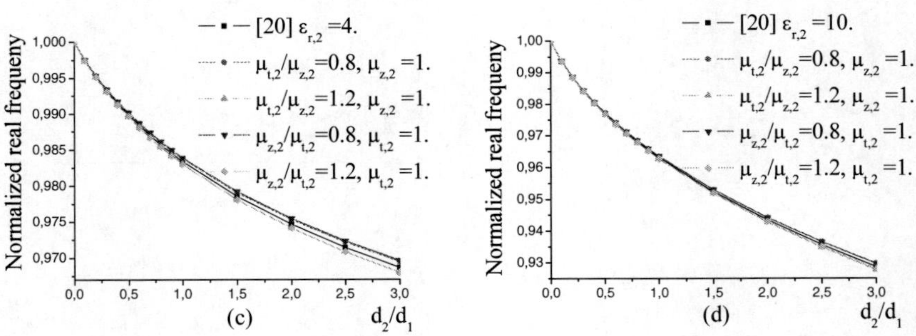

Figure 5.15. The effect of uniaxial anisotropic permeability of a superstrate on the normalized real resonant frequency; a=6 cm, b=5 cm, d_1=0.1 cm, $\varepsilon_{r,1}$=2.35, (a) $\varepsilon_{r,2}$=1.5, (b) $\varepsilon_{r,2}$=2.35, (c) $\varepsilon_{r,2}$=4, (d) $\varepsilon_{r,2}$=10.

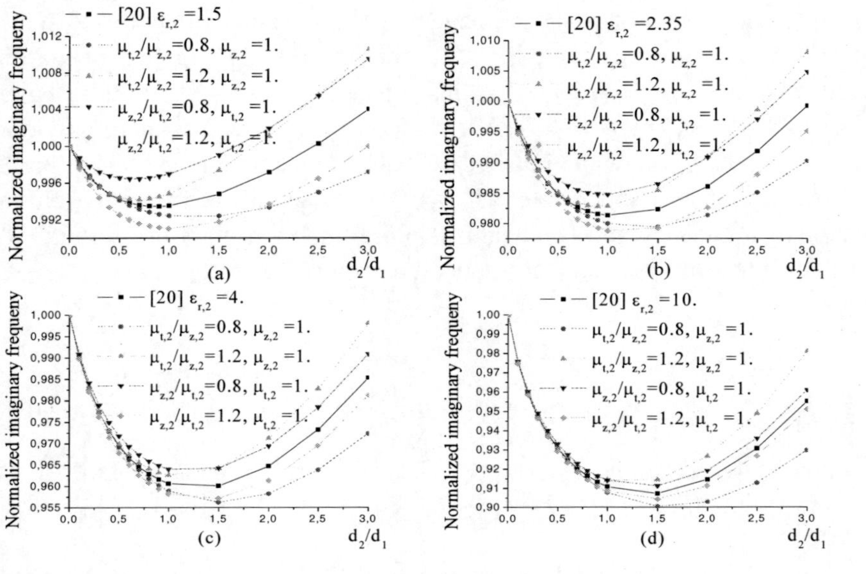

Figure 5.16. The effect of uniaxial anisotropic permeability of a superstrate on the normalized imaginary resonant frequency; a=6 cm, b=5 cm, d_1=0.1 cm, $\varepsilon_{r,1}$=2.35, (a) $\varepsilon_{r,2}$=1.5, (b) $\varepsilon_{r,2}$=2.35, (c) $\varepsilon_{r,2}$=4, (d) $\varepsilon_{r,2}$=10.

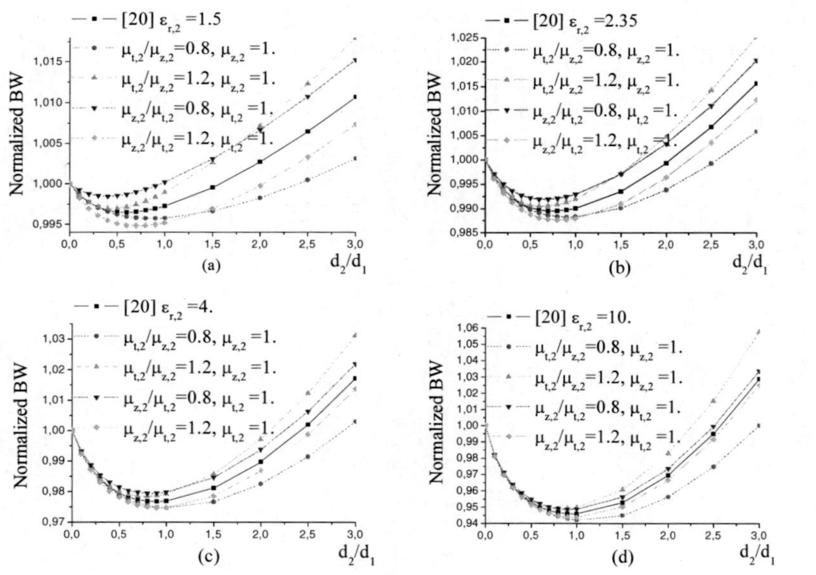

Figure 5.17. The effect of uniaxial anisotropic permeability of a superstrate on the normalized bandwidth; a=6 cm, b=5 cm, d_1=0.1 cm, $\varepsilon_{r,1}$=2.35, (a) $\varepsilon_{r,2}$=1.5, (b) $\varepsilon_{r,2}$=2.35, (c) $\varepsilon_{r,2}$=4, (d) $\varepsilon_{r,2}$=10.

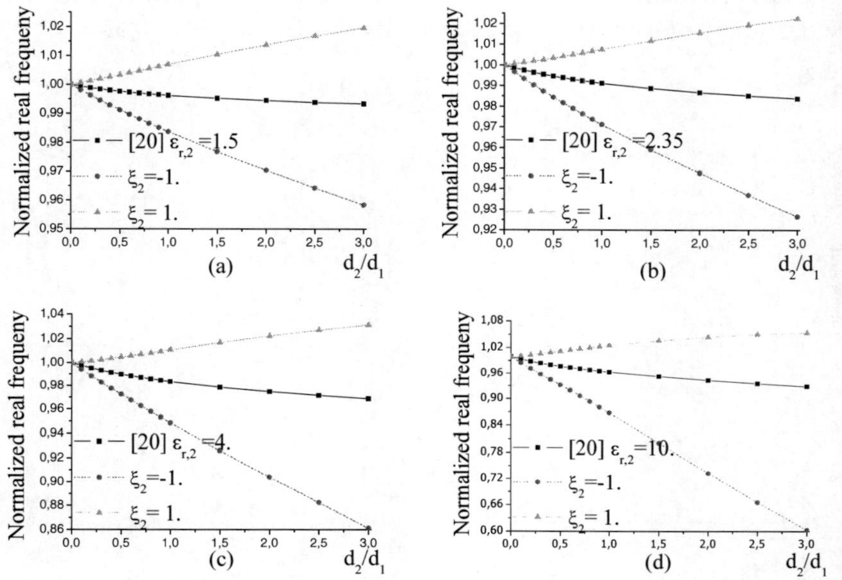

Figure 5.18. The chirality effect of a superstrate on the normalized real resonant frequency; a=6 cm, b=5 cm, d_1=0.1 cm, $\varepsilon_{r,1}$=2.35, (a) $\varepsilon_{r,2}$=1.5, (b) $\varepsilon_{r,2}$=2.35, (c) $\varepsilon_{r,2}$=4, (d) $\varepsilon_{r,2}$=10.

The chirality effect on the resonant frequency is shown in Figure 5.18. In this case, the effect of the chirality is that the real resonant frequency monotonously increases with respect to superstrate thickness for a positive magneto-electric element compared to the previous case (uniaxial anisotropies), whereas for negative elements the real resonant frequency is subject to an accentuated decrease. In addition, by increasing the permittivity, the normalized real resonant frequency increases or decreases further, according to the chirality parameter. In Figure 5.19, the chirality effect on the imaginary resonant frequency always remains the same as the real part, with the exception that this effect is more important for the imaginary part than for the real part. The normalized imaginary frequency is three times the monolayer case of imaginary frequency for a positive chirality, and six times for a negative chirality parameter. The effect of the negative chirality element on the bandwidth is more important than the positive chirality element for d_2>1.5d_1 and for higher permittivity ($\varepsilon_{t,2}$,$\varepsilon_{z,2}$>$\varepsilon_{r,1}$). This is illustrated in Figure 5.20(d), where an increase of eleven times in BW_0 is obtained. For lower permittivity ($\varepsilon_{t,2}$,$\varepsilon_{z,2}$<$\varepsilon_{r,1}$) the effect of the positive chirality element on the bandwidth is more important than the negative chirality. It should be noted that the negative chirality element is advantageous over the positive one, due to the possibility of bandwidth cancellation for the case when $d_2 \approx 0.5 d_1$.

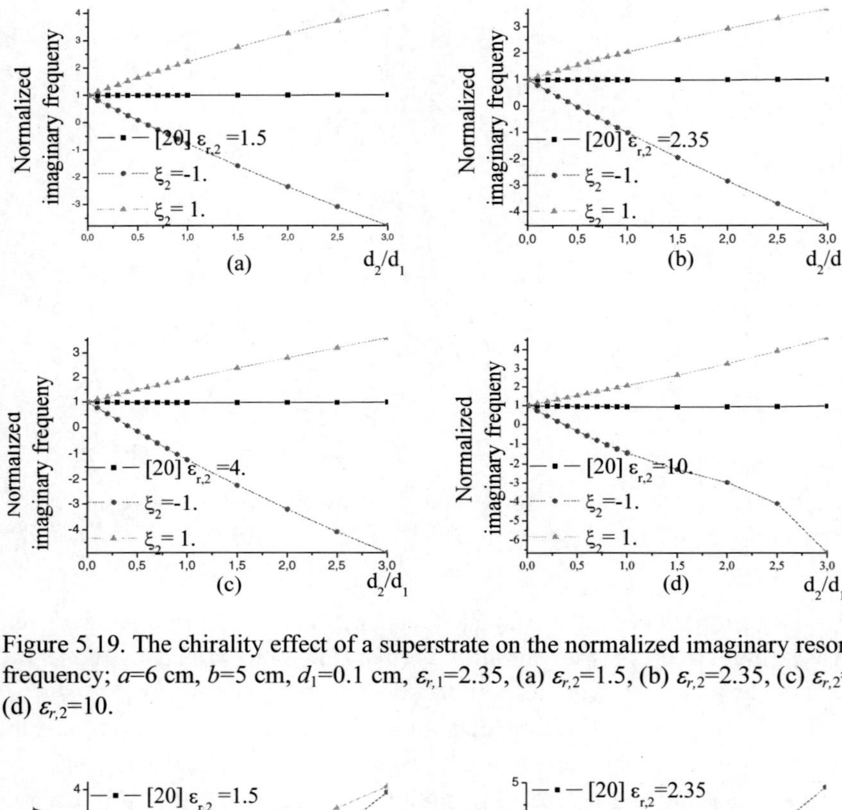

Figure 5.19. The chirality effect of a superstrate on the normalized imaginary resonant frequency; a=6 cm, b=5 cm, d_1=0.1 cm, $\varepsilon_{r,1}$=2.35, (a) $\varepsilon_{r,2}$=1.5, (b) $\varepsilon_{r,2}$=2.35, (c) $\varepsilon_{r,2}$=4, (d) $\varepsilon_{r,2}$=10.

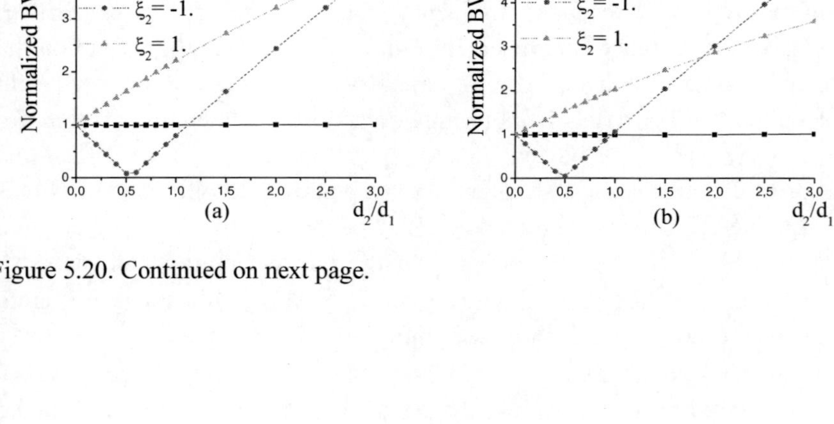

Figure 5.20. Continued on next page.

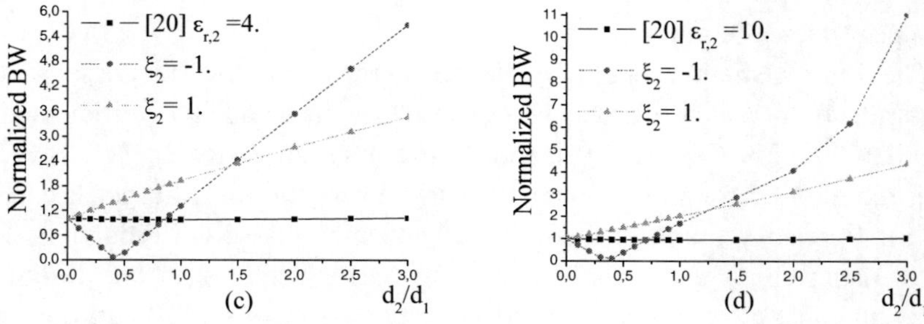

Figure 5.20. The chirality effect of a superstrate on the normalized bandwidth; a=6 cm, b=5 cm, d_1=0.1 cm, $\varepsilon_{r,1}$=2.35, (a) $\varepsilon_{r,2}$=1.5, (b) $\varepsilon_{r,2}$=2.35, (c) $\varepsilon_{r,2}$=4, (d) $\varepsilon_{r,2}$=10.

5.9. Asymmetrical Effects of Bianisotropic Substrate-Superstrate Sandwich Structure

Earlier publications focused on understanding how the superstrate affects the resonant frequency and subsequent power matching concerns [101-103]. Also other general types of antenna are treated - multilayered stacked geometry [104], several patch forms, such as the triangular one [103, 104], rectangular type [104], circular patch [105] and circular annular ring microstrip [106], where the structures with substrate-superstrate, were generally employed to improve the characteristics of the resonator and especially those of the antenna [82, 102, 103, 105-109]. Several particular cases of such a medium have found practical applications in recent years (chiral materials, biased ferrites and biaxial optical fibres) and additional applications are likely to occur as a consequence of the introduction of novel synthetic electronic materials [48]. Pozar quoted some disadvantages posed by the superstrate, and suggested that the most complex mediums, such as biaxial, bianisotropic, or the use of an anisotropic superstrate, can have more interesting effects [17]. This suggestion motivated studies of several configurations of the superstrates, used to improve antenna radiation properties, such as dielectric slabs [110, 111] and electromagnetic bandgap (EBG) structures [66, 112, 113]. Yang [67] was the first who proposed the antennas with high profit could be obtained by a radiating element printed on EBG, and the use of highly-reflective surfaces [109], the artificial magnetic [102, 114], effects of self-biased magnetic films on patch antennas with substrate-superstrate sandwich structure [59-63], and bianisotropic substrate and superstrates [39, 40].

The improvement brought by the anisotropy and the thickness of the superstrate is to reduce surface waves resulting in a radiation efficiency of more than 90% [59-63, 101, 115], and radar cross section (RCS) [116, 117]. In [118], the authors found the superstrate loaded on the metallic patch can be used for frequency tuning, bandwidth enhancement and RCS reduction. The idea of using chiral materials as substrates and superstrate in the design of printed antennas was first introduced in 1988 by Engheta [42], and the term 'chirostrip' was then invented. Later in 1989, chiroptical guides were invented (and named) by Engheta and Pelet in [46]. These discoveries have opened up a whole new field of theoretical studies on the properties of these guides or resonator-based chiral: new polarizations, crossing patterns, coupling TE/TM wave generation 'slow' resonant frequency, bandwidth and surface wave [90]. It was not until 2001 that the resolution of Maxwell's equations in these structures was fully completed [119, 120]. Over the years, the interaction of electromagnetic fields with chiral matters has been studied, with chiral media is used in many applications involving arrays antennas, radome antennas, microstrip substrates and waveguides [121]. Therefore, it is important to predict the electromagnetic behaviour of a general bianisotropic medium, aside from the fundamental studies contained in [122-124] and related publications. The dispersion relation of such media has been studied in [125] and an application to layered structures has been fully made in [48, 126].

Chiral media are optically active, a behaviour caused by their asymmetrical molecular structure that enables a substance to rotate the plane of incident polarized light, where the amount of rotation in the plane of polarization is proportional to the thickness of the traversed medium as well as to the light wavelength [47, 121, 127-129]. The media can be divided basically into reciprocal [130] and nonreciprocal magneto-electric materials [122, 131, 132]. The most general linear medium is described by bianisotropic constitutive relations [122-124]. Note that the constitutive tensors can define whether the medium is reciprocal or not [133], and the effect of this media in electronic components parameters can be symmetrical or asymmetrical [134, 135]. It should be pointed out that the paper [136] is specifically focused on chiral media. Those reciprocal bi-isotropic media whose existence is out of controversy-nonreciprocal bi-isotropic media have neither been found nor manufactured, and the possibility of their existence is currently under discussion in the open literature [97, 137]. The nonreciprocal behaviour of particular classes of such materials has also been studied [48, 123, 138]. In the following section, the combined effects of asymmetric chiral substrate and superstrate on the resonant frequency and bandwidth of rectangular microstrip are considered.

5.9.1. Combined Effect of Substrate and Superstrate Chirality (Thin Layer Case)

Figure 5.21 shows a rectangular microstrip patch antenna in a bianisotropic substrate-superstrate configuration. In the monolayer case [39], the asymmetry effect is more visible when the substrate permittivity ε_r=7 than in a second case with ε_r=2.35; it is also more apparent on the bandwidth than on the resonant frequency. An increase of layer thickness leads to an increase of the asymmetry effect of the chiral (?) [39]. Figures 5.22(a)-5.22(d) and 5.23(a)-5.23(d) show the combined effect of the two chiral substrates. Knowing that the substrate thickness d_1=1 mm, in the configuration (substrate/superstrate) case, the asymmetry effect of the substrate is negligible both on the resonant frequency or on the bandwidth, this case concord with the last one [39] and in [40].

The effect of asymmetry on an increase or decrease of the resonant frequency, for a positive or negative magneto-electric element, is not the same as compared to the isotropic case. The chiral superstrate becomes more important with the increasing of these two factors. The thickness and permittivity (i.e. the asymmetry reaches 1% up to 33% for respectively ε_r=1.5% to ε_r=10 on the resonant frequency, and on the bandwidth, the asymmetry effect can reach 700% for ε_r=10). In the case of the resonant frequency, the asymmetry of the chiral superstrate is important (the effect of the substrate chirality is always negligible due to the weak thickness of this layer). This occurs according to the thickness layer ratio, d_2/d_1. For a positive magneto-electric, there is a monotonis increase of the bandwidth. However, in the negative magneto-electric case, three zones are observed, the first one located in $0<d_2/d_1<0.5$, when the bandwidth undergoes an increase reaching 100% for all cases. In the second region, $0.5 <d_2/d_1<1$, the bandwidth decreases with the same slope. In the last region $d_2/d_1> 1$, a monotonic increase of the bandwidth is visible. As stated above in the previous section, it is preferable to have the substrate layer thick enough for better observing the combined effect of the both media, indeed the value of d_1=5 mm is considered. The dimensions of the studied structure and the ratio d_2/d_1, are preserved as in previous cases and the normalization is compared to the monolayer case (d_2=0).

5.9.2. Combined Effect of Substrate and Superstrate Chirality (Thick Layer Case)

Figures 5.24(a-d) present the associated effect of the two magneto-electric components of the various layers (substrate and superstrate). The behaviour of the real resonant frequency is as studied previously. Even with d_1=5 mm, the asymmetry effect of the chiral is not visible, but the combination between the chirality of the substrate and superstrate is asymmetric. By increasing the thickness of the superstrate and its permittivity ε_r, the substrate chirality effect with a negative magneto-electric element of the superstrate becomes weaker, whereas for a positive chirality of the superstrate, the chirality effect of the substrate is important. The asymmetry effect of the superstrate chirality on the resonant frequency is also remarkable in this case.

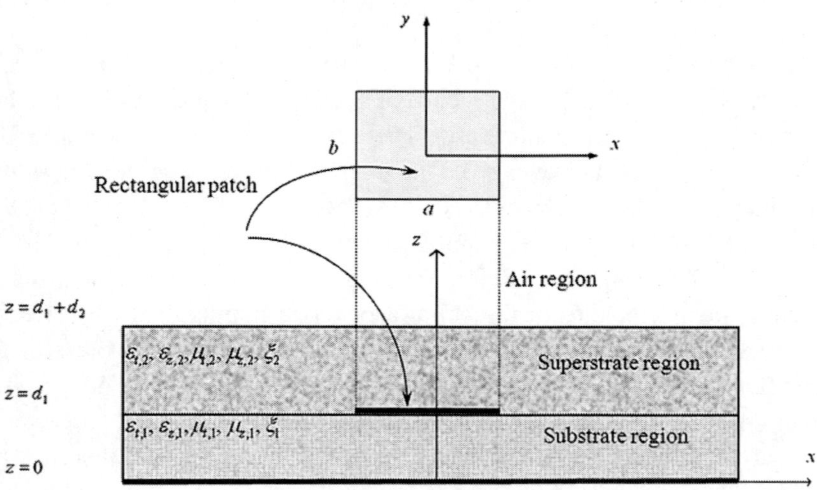

Figure 5.21. Rectangular microstrip patch antenna in a bianisotropic substrate-superstrate configuration.

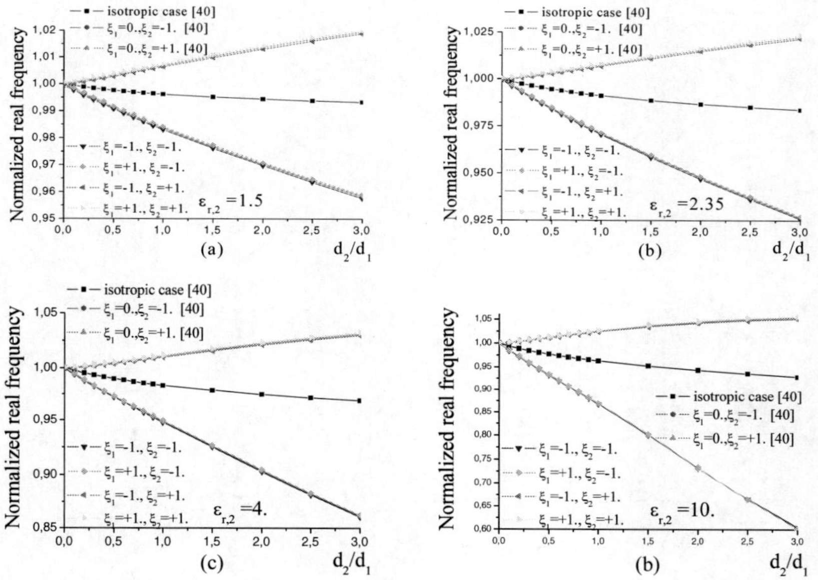

Figure 5.22. Normalized resonant frequency of a rectangular microstrip patch antenna in a bianisotropic substrate-superstrate configuration versus superstrate thickness; a=6 cm, b=5 cm, d_1=0.1 cm, $\varepsilon_{t,1}$=$\varepsilon_{z,1}$=2.35, $\varepsilon_{z,2}$=1.5, 2.35, 4, 10.

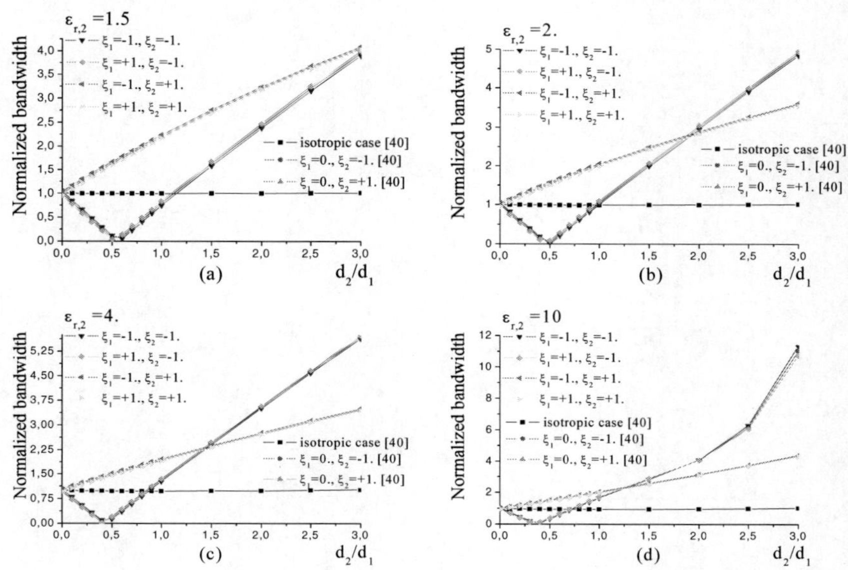

Figure 5.23. Normalized bandwidth of a rectangular microstrip patch antenna in a bianisotropic substrate-superstrate configuration versus superstrate thickness; a=6 cm, b=5 cm, d_1=0.1 cm, $\varepsilon_{t,1}$=$\varepsilon_{z,1}$=2.35, $\varepsilon_{z,2}$=1.5, 2.35, 4, 10.

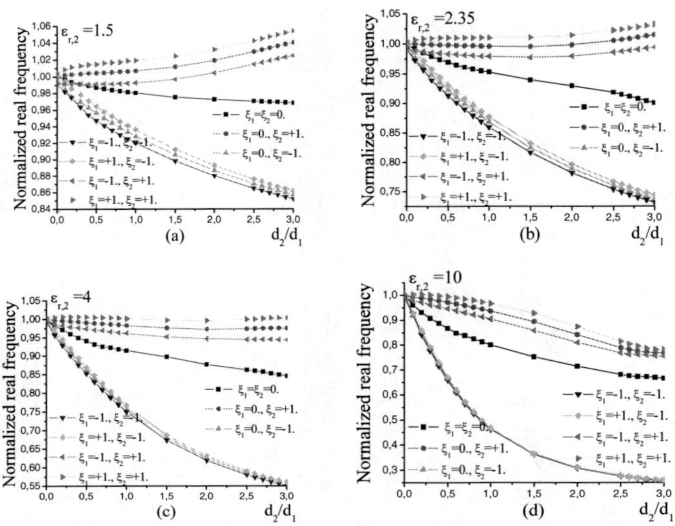

Figure 5.24. Combined effect of substrate and superstrate on the normalized resonant frequency versus superstrate thickness; a=6 cm, b=5 cm, d_1=0.5 cm, $\varepsilon_{t,1}=\varepsilon_{z,1}$=2.35, $\varepsilon_{z,2}$=1.5, 2.35, 4, 10.

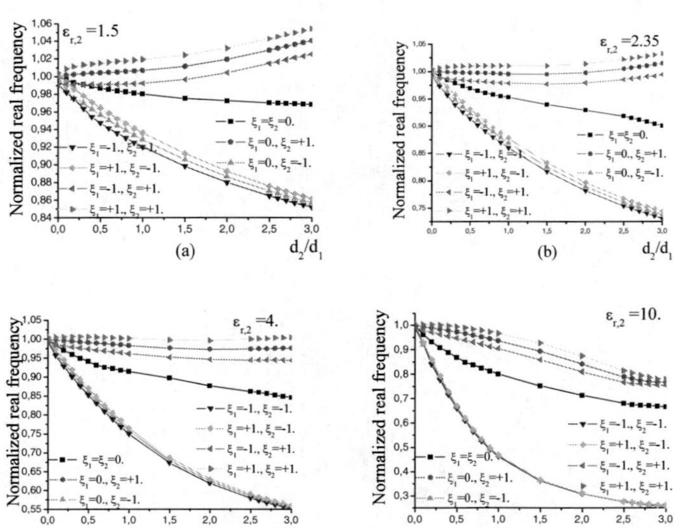

Figure 5.25. Combined effect of substrate and superstrate on the normalized bandwidth versus superstrate thickness; a=6 cm, b=5 cm, d_1=0.5 cm, $\varepsilon_{t,1}=\varepsilon_{z,1}$=2.35, $\varepsilon_{z,2}$=1.5, 2.35, 4, 10.

Figures 5.25(a-d) show that the chirality effect of the substrate on the bandwidth is symmetric in this case, but the combined effect (ξ_1 and ξ_2) is asymmetric on the bandwidth. This is more apparent for low permittivity ε_r, important ratio d_2/d_1 and for a positive magneto-electric element of the superstrate, ξ_2=+1, Figures 5.25(a-b). The asymmetry effect of the superstrate chirality on the bandwidth becomes important when increasing both the permittivity ε_r and the thickness of this layer, with a negative magneto-electric element of the superstrate ξ_2=-1, Figure 5.25(d), whereas in the case of a positive magneto-electric element of the superstrate, the chirality of the substrate is negligible.

The symmetry and asymmetry of the bi-isotropic chiral media, having scalar constitutive parameters, have been reported by some authors [134, 135]. In [134], the authors found that the chiral substrate produces additional longitudinal asymmetric and transverse symmetric field components, this is the case of isotropic study. This effect could significantly change the properties of microwave devices built on a chiral substrate, and with respect to the asymmetry, it is necessary to evaluate the asymptotic form of the Green tensor. This form is obtained analytically just below, when $d_1 \rightarrow 0$, and d_2 are maintained constant to show the asymmetry effect of this chiral layer, according to the Green tensor given in (5.73) as the following

$$G(\kappa_s) \rightarrow \frac{d_1}{j\omega\varepsilon}\begin{bmatrix} \dfrac{(\kappa_0^2\varepsilon_{t,1}-\kappa_s^2)\left(\kappa_z\cos(\kappa_{z,2}^e d_2)+\left(j\dfrac{\kappa_{z,2}^e}{\varepsilon_{t,2}}+\dfrac{\kappa_0\xi_2}{\kappa_{z,2}^e}(\kappa_z\varepsilon_{t,2}+j\kappa_0\xi_2)\right)\sin(\kappa_{z,2}^e d_2)\right)}{\left(\kappa_z\varepsilon_{t,1}\cos(\kappa_{z,2}^e d_2)+j\left(\kappa_{z,2}^e\sin(\kappa_{z,2}^e d_2)-j\dfrac{\varepsilon_{t,1}}{\varepsilon_{t,2}}\dfrac{\kappa_0\xi_2}{\kappa_{z,2}^e}(\kappa_z\varepsilon_{t,2}+j\kappa_0\xi_2)\right)\sin(\kappa_{z,2}^e d_2)\right)} & 0 \\ 0 & \kappa_0^2 \end{bmatrix} \tag{5.87}$$

It can be noticed from the previous expression that the chirality anomaly is observed as

$$G^e(\kappa_s) \rightarrow f\left(\frac{\kappa_0}{\kappa_{z,2}^e}(\kappa_z\varepsilon_{t,2}\xi_2+j\kappa_0\xi_2^2)\sin(\kappa_{z,2}^e d_2)\right) \tag{5.88}$$

The term $\left(\varepsilon_{t,2}\xi_2\sin(\kappa_{z,2}^e d_2)\right)$ is significant for important values of $\varepsilon_{t,2}$, and this last result can explain some previous results. The anomaly can be explained also by the fact that the fields are divided into the symmetric and asymmetric field components. The symmetric field components were only slightly affected by the chirality of the media, but the asymmetric field components were entirely due to the chirality of the media. This can be clearly noticed by rewriting the equations, TE and TM immitance in symmetric and asymmetric forms

$$g(\kappa_s) = diag\left[\underbrace{j\frac{\omega\varepsilon_0\varepsilon_t\kappa_0\xi}{\left(\kappa_0^2\xi^2+\kappa_z^{e2}\right)}}_{\text{asymmetric part}} + \underbrace{j\kappa_z^e\frac{\omega\varepsilon_0\varepsilon_t}{\left(\kappa_0^2\xi^2+\kappa_z^{e2}\right)}}_{\text{symmetric part}} \quad \underbrace{\frac{j\left(\kappa_0\xi+j\kappa_z^h\right)}{\omega\mu_0\mu_t}}_{\text{asymmetric part}} \right] \quad (5.89)$$

$$h(\kappa_s) = diag\left[\underbrace{j\frac{\omega\varepsilon_0\varepsilon_t\kappa_0\xi}{\left(\kappa_0^2\xi^2+\kappa_z^{e2}\right)}}_{\text{asymmetric part}} - \underbrace{j\kappa_z^e\frac{\omega\varepsilon_0\varepsilon_t}{\left(\kappa_0^2\xi^2+\kappa_z^{e2}\right)}}_{\text{symmetric part}} \quad \underbrace{\frac{j\left(\kappa_0\xi-j\kappa_z^h\right)}{\omega\mu_0\mu_t}}_{\text{asymmetric part}} \right] \quad (5.90)$$

The asymmetric part occurs in the TE and TM immitance equations (5.89) and (5.90) according to the sign of the magneto-electric element ξ. Contrary to the case studied in [134], and taking into account Equation (5.72), which shows a gyrotropic form of the bianisotropic media, the non-diagonal shape of the magneto-electric element produces an asymmetry in the longitudinal and transverse components.

In addition, according to the observations, it was found that the layer thickness has an effect on the chiral asymmetry, where the variation of the layer leads to a variation of the transverses components.

CONCLUSION

Numerical results for resonant frequency and bandwidth were presented for different values of constitutive parameters. The asymmetric effect of a thin chiral substrate is negligible in the first case examined, a monolayer microstrip antenna, but the effect of the superstrate is always important. Another original result was obtained, not previously discussed in the literature, namely the relationship between the permittivity (exactly $\varepsilon_{t,2}$) and the chiral constitutive parameter. It is shown that by increasing the permittivity, the asymmetrical effect of the chiral becomes high, and for complex values of the magneto-electric parameters ξ, η (in this case a non-diagonal matrix) has shown that the transverse component occurs in the chiral asymmetry, which is not the case for bi-isotropic case study where ξ, η have a scalar form. In addition to what is found in the literature, it is shown that the layer plays an important role in the chiral asymmetry. Bianisotropy and specially the asymmetry of the substrates should always be taken into account in the design of the microstrip resonators. Furthermore, to improve the resonator parameters the chiral is probably most convenient for this task, since this artificial medium can be modelled as required. The effects of a uniaxial permittivity, uniaxial permeability and chirality of the superstrate on the complex resonant frequency and bandwidth have been studied using the integral equation formulation. Fast numerical convergence is obtained using sinusoidal basis functions to expand the current on the patch, with the calculation being carried out on a PC with 1.86 GHz Intel Pentium M and 1,024 GB RAM. The determinant cancellation of the Green's tensor is obtained in about 42 s. Numerical results for the resonant frequency and bandwidth have been presented for various constitutive parameter configurations. The most important parameters for a substrate-superstrate are $\varepsilon_{t,2}$, $\mu_{z,2}$, in contrast to the monolayer case where they are $\varepsilon_{z,1}$, $\mu_{t,1}$, whereas the chirality elements have an important effect for any case. The advantage of positive uniaxial permittivity anisotropy is summarized by the fact that the resonant frequency and the bandwidth are subject to an increase. For all presented behaviour of the chiral medium in a rectangular microstrip antenna substrate-superstrate configuration, it can be concluded that the material is suitable for the conception of radiating resonators or selective filter resonators.

REFERENCES

[1] X. -M. Pan and X. -Q. Sheng, "Improved algebraic preconditioning for MoM solutions of large-scale electromagnetic problems," *IEEE Antennas and Wireless Propagation Letters*, vol. 13, pp. 106-109, 2014.

[2] M. A. Echeverri Bautista, M. A. Francavilla, F. Vipiana and G. Vecchi, "A hierarchical fast solver for EFIE-MoM analysis of multiscale structures at very low frequencies," *IEEE Transactions on Antennas and Propagation*, vol. 62, no. 3, pp. 1523-1528, 2014.

[3] M. Shafieipour, I. Jeffrey, J. Aronsson and V. I. Okhmatovski, "On the equivalence of RWG method of moments and the locally corrected nyström method for solving the electric field integral equation," *IEEE Transactions on Antennas and Propagation*, vol. 62, no. 2, pp. 772-782, 2014.

[4] A. Schroder, G. A. Rasek, H. -D. Bruns, Z. Reznicek, J. Kucera, S. E. Loos and C. Schuster, "Analysis of high intensity radiated field coupling into aircraft using the method of moments," *IEEE Transactions on Electromagnetic Compatibility*, vol. 56, no. 1, pp. 113-122, 2014.

[5] N. Hendijani, J. Cheng and R. J. Adams, "Combined field integral equation using a constraint-based Helmholtz decomposition," *IEEE Transactions on Antennas and Propagation*, vol. 62, no. 3, pp. 1500-1503, 2014.

[6] J. Alvarez, L. D. Angulo, M.R. Cabello, A. Rubio Bretones and S. G. Garcia, "An analysis of the leap-frog discontinuous Galerkin method for Maxwell's equations," *IEEE Transactions on Microwave Theory and Techniques*, vol. 62, no. 2, pp. 197-207, 2014.

[7] K. Aliane, N. Benahmed, R. Bouhmidi, S. M. Meriah, F. T. Bendimerad and N. Benabdallah, "Analysis and design of a 12-element coupled-microstrip-line quasi-TEM resonator for MRI," *Spring Congress on Engineering and Technology (SCET)*, pp. 1-4, 2012.

[8] M. Amir, S. Bedra, S. Benkouda and T. Fortaki, "Bacterial foraging optimisation and method of moments for modelling and optimisation of microstrip antennas," *IET Microwaves, Antennas and Propagation*, vol. 8, no. 4, pp. 295-300, 2014.

[9] S. R. Zang and J. R. Bergmann, "Analysis of omnidirectional dual-reflector antenna and feeding horn using method of moments," *IEEE Transactions on Antennas and Propagation*, vol. 62, no. 3, pp. 1534-1538, 2014.

[10] C. A. Balanis, Antenna theory, analysis and design, 2nd ed., New York: John Wiley & Sons, Inc., pp. 722-783, 1997.

[11] G. A. Deschamps, "Microstrip microwave antennas," presented in the third symposium on resonators, 1953.

[12] W. F. Richards, Y. T. Lo and D. D. Harrison, "An improved theory of microstrip antennas with applications," *IEEE Transaction on Antennas and Propagation*, vol. AP-29, no. 1, pp. 38-46, 1981.

[13] D. H. Schaubert, F. G. Farrar, A. Sindoris, and S. T. Hayes, "Microstrip antennas with frequency agility and polarisation diversity," *IEEE*

Transaction on Antennas Propagation, vol. AP-29, no. 1, pp. 118-123, 1981.

[14] M. P. Purchine, J. T. Aberle and C. R. Birtcher, "A tunable L-band circular microstrip patch antenna," *Microwave Journal*, vol. 36, no. 10, pp. 80-88, 1993.

[15] D. M. Pozar, "Microstrip Antennas," *Proceeding of the IEEE*, vol. 80, no. 1, pp. 79-81, 1992.

[16] N. G. Alexopoulos, "Integrated-circuit structures on anisotropic substrate," *IEEE Transaction on Microwave Theory and Techniques*, vol. MTT-33, pp. 847-881, 1985.

[17] D. M. Pozar, "Radiation and scattering from a microstrip patch on a uniaxial substrate," *IEEE Transaction on Antennas and Propagation*, vol. AP-35, pp. 613-621, 1987.

[18] K. -L. Wong, J. -S. Row, C. -W. Kuo and K. -C. Huang, "Resonance of a rectangular microstrip patch on a uniaxial substrate," *IEEE Transaction on Microwave Theory and Techniques*, vol. 41, pp. 698-701, 1993.

[19] W. C. Chew and Q. Liu, "Correction to resonance frequency of a rectangular microstrip patch," *IEEE Transaction on Antennas and Propagation*, vol. 36, pp. 1827, 1988.

[20] F. Bouttout, F. Benabdelaziz, T. Fortaki and D. Khedrouche, "Resonant frequency and bandwidth of a superstrate-loaded rectangular patch on a uniaxial anisotropic substrate," *Communications in Numerical Methods in Engineering, John Wiley & Sons*, vol. 16, no. 7, pp. 459-473, 2000.

[21] F. Bilotti and L. Vegni, "Chiral cover effects on microstrip antennas," *IEEE Transactions on Antennas and Propagation*, vol. 51, no. 10, pp. 2891-2898, 2003.

[22] T. Itoh, "Spectral domain immitance approach for dispersion characteristics of generalized printed transmission lines," *IEEE Transaction on Microwave Theory and Techniques*, vol. MTT-28, pp. 733-736, 1980.

[23] N. K. Das and D. M. Pozar, "A generalized spectral-domain Green's function for multilayer dielectric substrates with application to multilayer transmission lines," *IEEE Transaction on Microwave Theory and Techniques*, vol. MTT-35, pp. 326-335, 1987.

[24] F. L. Mesa, R. Marques, and M. Horno, "A general algorithm for computing the bi-dimensional spectral Green's dyad in multilayered complex bianisotropic media: The equivalent boundary method," *IEEE*

Transaction on Microwave Theory and Techniques, vol. 39, pp. 1640-1649, 1991.

[25] P. Bernardi and R. Cicchetti, "Dyadic Green's function for conductor-backed layered structures excited by arbitrary tri-dimensional sources," *IEEE Transaction on Microwave Theory and Techniques*, vol. 42, pp. 1474-1483, 1994.

[26] A. Dreher, "A new approach to dyadic Green's function in spectral domain," *IEEE Transaction on Antennas and Propagation*, vol. 43, pp. 1297-1302, 1995.

[27] V. Lindell, A. H. Sihvola, S. A. Tretyskov, and A. J. Vitanen, Electromagnetic waves in chiral and bi-isotropic media, Norwood, MA: Artech House, 1994.

[28] K. W. Whites and C. Y. Chung, "Composite uniaxial bianisotropic chiral materials characterization: Comparison of predicted and measured scattering," *Journal of Electromagnetic Waves and Applications*, vol. 11, pp. 371-394, 1997.

[29] G. Busse, J. Reinert, and, A. F. Jacob, "Waveguide characterization of chiral material: Experiments," *IEEE Transactions on Microwave Theory and Techniques*, vol. 47, pp. 297-301, 1999.

[30] V. Dmitriev; "Tables of the second rank constitutive tensors for linear homogeneous media described by the point magnetic groups of symmetry," *Progress In Electromagnetics Research*, PIER 28, pp. 43-95, 2000.

[31] D. M. Pozar, "General relations for a phased array of printed antennas derived from infinite current sheets," *IEEE Transaction on Antennas and Propagation*, vol. AP-33, pp. 498-504, 1985.

[32] W. P. Harokopus, L. P. B. Katehi, W. Y. Ali-Ahmed, and G. M. Rebiez, "Surface wave excitation from open microstrip discontinuities," *IEEE Transaction on Microwave Theory and Techniques*, vol. 39, pp. 1098-1107, 1991.

[33] D. M. Pozar and V. Sanchez, "Magnetic tuning of a microstrip antenna on a ferrite substrate," *Electronic Letters,* vol. 24, no. 12, pp. 729-731, 1988.

[34] D. M. Pozar, "Radar cross-section of microstrip antenna on normally biased ferrite substrate," *Electronic Letters,* vol. 25, no. 16, pp. 1079-1080, 1989.

[35] H. Y. Yang. J. A. Castaneda, and N. G. Alexopoulos, "Multifunctional and low RCS nonreciprocal microstrip antennas," *Electromagnetics*, vol. 12, no. 1, pp. 17-31, 1992.

[36] D. M. Pozar, "Radiation and scattering characteristics of microstrip antennas on normally biased ferrite substrates," *IEEE Transaction on Antennas and Propagation*, vol. 40, no. 9, pp. 1084-1092, 1992.

[37] D. M. Pozar, "Microstrip antennas and arrays on chiral substrates," *IEEE Transaction on Antennas and Propagation,* vol. 40, no. 10, pp. 1260-1263, 1992.

[38] A. Toscano and L. Vegni, "A new efficient moment method formulation for the design of microstrip antennas over a chiral grounded slab," *Journal of Electromagnetic Waves and Application*, vol. 11, no.5, pp. 567-592, 1997.

[39] C. Zebiri, M. Lashab and F. Benabdelaziz, "Effect of anisotropic magneto-chirality on the characteristics of a microstrip resonator," *IET Microwaves, Antennas and Propagation*, vol. 4, no. 4, pp. 446-452, 2010.

[40] C. Zebiri, M. Lashab and F. Benabdelaziz, "Rectangular microstrip antenna with uniaxial bi-anisotropic chiral substrate-superstrate," *IET Microwaves, Antennas and Propagation*, vol. 3, no. 1, pp. 17-29, 2011.

[41] J. R. James and P. S. Hall, Handbook of microstrip antennas, Institution of Electrical Engineers, 1989.

[42] N. Engheta, "The theory of chirostrip antennas," Proceedings of the 1988 URSI International Radio Science Symposium, Syracuse, New York, pp. 213, 1988.

[43] N. Engheta and P. Pelet, "Reduction of surface waves in chirostrip antennas," *Electronic Letters*, vol. 27, no. 1, pp. 5-7, 1991.

[44] R. F. Harrington, Time-harmonic electromagnetic fields, New York: McGraw-Hill, 1961.

[45] A. J. M. Soares, S. B. D. A. Fonseca, and A. J. Giarola, "Surface wave excitation on a microstrip ring antenna," *IEEE Transaction on Antennas and Propagation*, vol. 37, no. 10, pp. 1310-1312, 1989.

[46] N. Engheta and P. Pelet, "Modes in chirowaveguides," *Optics Letters*, vol. 14, no. 11, pp. 593-595, 1989.

[47] N. Engheta and D. L. Iaggaro, "Electromagnetic chirality and its applications," *IEEE Antennas Propagation Society Newsletter*, vol. 30, no. 5, pp. 6-12, 1988.

[48] P. L. E. Uslenghi, "TE TM decoupling for Guided propagation in bi-anisotropic media," *IEEE Transaction on Antennas and Propagation*, vol. 45, no. 2, pp. 284-286, 1997.

[49] L. Vegni and A. Toscano, "Shielding and radiation characteristics of cylindrical layered bianisotropic structures," *Radio Engineering*, vol. 14, no. 4, pp 68-74, 2005.

[50] W. C. Chew and J. A. Kong, "Resonance of non-axial symmetric modes in circular microstrip disk antenna," *Journal of Mathematical Physics*, vol. 21, no. 10, pp. 2590-2598, 1980.

[51] D. M. Pozar, "Input impedance and mutual coupling of rectangular microstrip antennas," *IEEE Transaction on Antennas and Propagation*, vol. AP-30, pp. 1191-1196, 1982.

[52] J. M. Laheurte, L. P. B. Katehi and G. M. Rebeiz, "CPW-fed slot antennas on multilayer dielectric substrates," *IEEE Transaction on Antennas and Propagation*, vol. 44, no. 8, pp. 1102-1111, 1996.

[53] J. Zehentner and J. Machac, "Upper cutoff frequency of the bound wave and new leaky wave on the slotline," *IEEE Transaction on Microwave Theory and Techniques*, vol. 46, no. 4, pp. 378-386, 1998.

[54] J. P. Damiano, "Computation of input impedance in microstrip antennas. Graphic representation and numerical integration of oscillating function," *IEE Proceedings on Microwave, Antennas and Propagation*, vol. 134, no. 5, pp. 456-466, 1987.

[55] C. Peixeiro and A. M. Barbosa, "Leaky and surface waves in anisotropic printed antenna structures," *IEEE Transaction on Antennas and Propagation*, vol. 40, no. 5, pp. 566-569, 1992.

[56] R. E. Collin, Field theory of guided waves, 2nd ed., New York: Pression de Wiley, 1990.

[57] P. Bhartia, K. V. S. Rao and R. S. Tomar, Millimeter-wave microstrip and printed circuit antennas, Artech House, 1991.

[58] R. A. Sainati, CAD of microstrip antennas for wireless applications, Artech House, 1996.

[59] L. Dixit and P. K. S. Pourush, "Radiation characteristics of switchable ferrite microstrip array antenna," *IEE Proceedings on Microwave, Antennas and Propagation*, vol. 147, no. 2, pp. 151-155, 2000.

[60] J. C. Batchelor and R. J. Langley, "Beam scanning using microstrip line on biased ferrite," *Electronic Letters*, vol. 33, no. 8, pp. 645-646, 1997.

[61] P. Y. Ufimtsev, R. T. Ling and J. D. Scholle, "Transformation of surface waves in homogenous absorbing layers," *IEEE Transaction on Antennas and Propagation*, vol. 48, no. 2, pp. 214-222, 2000.

[62] B. Horsfield and J. A. R. Ball, "Surface wave propagation on a grounded dielectric slab covered by a high-permittivity material," *IEEE Microwave and Guided Wave Letters*, vol. 10, no. 5, pp. 171-173, 2000.

[63] N. K. Saxena, N. Kumar and P. K. S. Pourush, "Microstrip rectangular patch antenna printed on liti ferrite with perpendicular DC magnetic biasing," *The Journal of American Science*, vol. 6, no. 3, pp 46-52, 2010.

[64] K. C. Huie, Microstrip antennas: Broadband radiation patterns using photonic crystal substrates, Thesis Master of Science, Electrical Engineering, Faculty of the Virginia Polytechnic Institute and State University, 2002.

[65] A. De, N. S. Raghava, S. Malhotra, P. Arora and R. Bazaz, "Effect of different substrates on compact stacked square microstrip antenna," *Journal of Telecommunications*, vol. 1, no. 1, pp 63-65, 2010.

[66] H. Boutayeb and T. A. Denidni, "Gain enhancement of a microstrip patch antenna using a cylindrical electromagnetic crystal substrate," *IEEE Transaction on Antennas and Propagation*, vol. 55, no. 11, pp. 3140-3145, 2007.

[67] H. Y. D. Yang, N. G. Alexopoulos and E. Yablonovitch, "Photonic band gap materials for high gain printed circuit antennas," *IEEE Transaction on Antennas and Propagation*, vol. 45, no. 1, pp.185-187, 1997.

[68] N. Kumprasert and W. Kiranon, "Simple and accurate formula for the resonant frequency of the circular microstrip disk antenna," *IEEE Transaction on Antennas and Propagation*, vol. 43, no. 11, pp. 1331-1333, 1995.

[69] T. Itoh, Numerical techniques for microwave and millimeter-wave passive structures, New York: A Wiley-Interscience Publication, Wiley, 1989.

[70] W. C. Chew and J. A. Kong, "Resonance of the axial-symmetric modes in microstrip disk resonators," *Journal of Mathematical Physics*, vol. 21, no. 3, pp. 582-591, 1980.

[71] D. R. Jackson and N. G. Alexopoulos, "Simple approximate formulas for input resistance, bandwidth, and efficiency of a resonant rectangular patch," *IEEE Transaction on Antennas and Propagation*, vol. 39, no. 3, pp. 407-410, 1991.

[72] J. C. S. Lacava, I. Bianchi and F. Lumini, "A dynamic model for printed apertures in open chirostrip structures," *Proceeding of the XXVIth General Assembly of the International Union of Radio Science, Toronto, Canada*, pp. 201, 1999.

[73] J. R. Mosig, Integral equation technique in numerical techniques for microwave and millimeter-wave passive structures, New York: John Wiley, 1989.

[74] H. Guan, S. Liu, Y. Zhao and Y. Duan, "Electromagnetic characteristics of nanometer manganese dioxide composite materials," *Journal of Electronic Materials*, vol. 35, no. 5, pp 892-896, 2006.

[75] C. R. Paiva and A. M. Barbosa, "A linear-operator formalism for the analysis of inhomogeneous biisotropic planar waveguides," *IEEE Transaction on Microwave Theory and Techniques*, vol. 40, no. 4, pp. 672-678, 1992.

[76] H. Cary and S. Waxman, "Wave propagation along a fully or partially loaded parallel plate chiro-waveguide," *IEE Proceedings on Microwaves, Antennas and Propagation*, vol. 141, no. 4, pp. 209-306, 1994.

[77] C. Zebiri and F. Benabdelaziz, "Evaluation du tenseur de Green pour une microbande avec substrat d'une anisotropie tensorielle," 4th International Conference: Sciences of Electronic, Technologies of Information and Telecommunications (SETIT), Tunisia, pp. 1-4, 2007.

[78] C. Zebiri, F. Benabdelaziz and D. Sayad, "Surface waves investigation of a bianisotropic chiral substrate resonator," *Progress In Electromagnetics Research B*, vol. 40, pp. 399-414, 2012.

[79] H. Cory, "Chiral devices an overview of canonical problems," *Journal of Electromagnetic Waves and Applications*, vol. 9, no. 5-6, pp. 805-829, 1995.

[80] G. Wang and G. W. G. Pan, "Full wave analysis of microstrip floating line structures by wavelet expansion method," *IEEE Transaction on Antennas and Propagation*, vol. MTT-43, pp. 131-142, 1995.

[81] G. W. G. Pan, J. Tan and J. D. Murphy, "Full wave analysis of microstrip floating-line discontinuities," *IEEE Transaction on Electromagnetic Compatibility*, vol. 36, no. 1, pp. 49-59, 1994.

[82] N. G. Alexopoulos and D. R. Jackson, "Fundamental superstrate cover effects on printed circuit antennas," *IEEE Transaction on Antennas and Propagation*, vol. 32, no. 8, pp. 807-816, 1984.

[83] V. Losada, R. R. Boix and M. Horno, "Full-wave analysis of circular microstrip resonators in multilayered media containing uniaxial anisotropic dielectrics, magnetized ferrites, and chiral materials," *IEEE Transaction on Microwave Theory and Techniques*, vol. 48, no. 6, pp. 1057-1164, 2000.

[84] J. T. Bernhard and C. J. Tousignant, "Resonant frequencies of rectangular microstrip antennas with flush and spaced dielectric superstrates," *IEEE Transaction on Antennas and Propagation*, vol. 47, no. 2, pp. 302-308, 1999.

[85] V. Losada, R. R. Boix and M. Horno, "Resonant modes of circular microstrip patches in multilayered substrates," *IEEE Transaction on Microwave Theory and Techniques*, vol. 47, no. 4, pp. 488-498, 1999.

[86] R. M. Nelson, D. A. Rogers and A. G. D'Assuncao, "Resonant frequency of a rectangular microstrip patch on several uniaxial substrates," *IEEE Transaction on Antennas and Propagation*, vol. 38, no. 7, pp. 973-981, 1990.

[87] F. Bouttout, F. Benabdelaziz, A. Benghalia, D. Khedrouche and T. Fortaki, "Uniaxially anisotropic substrate effects on resonance of rectangular microstrip patch antenna," *Electronic Letters*, vol. 35, no. 4, pp. 255-256, 1999.

[88] C. S. Gurel and E. Yazga, "Characteristics of a circular patch microstrip antenna on uniaxially anisotropic substrate," *IEEE Transaction on Antennas and Propagation*, vol. 52, no. 10, pp. 2532-2537, 2004.

[89] I. J. Bahl, P. Bhartia and S. Stuchly, "Design of microstrip antennas covered with a dielectric layer," *IEEE Transaction on Antennas and Propagation*, vol. 30, no. 2, pp. 314-318, 1982.

[90] N. Engheta and P. Pelet, "Reduction of surface waves in chirostrip antennas," Electronic Letters, vol. 27, no. 1, pp. 5-7, 1991.

[91] W. C. CHEW and T. M. Habashy, "The use of vector transforms in solving some electromagnetic scattering problems," *IEEE Transaction on Antennas and Propagation*, vol. 34, no. 7, pp. 871-879, 1986.

[92] H. -Y. Yang, A. Nakatani and J. A. Castaneda, "Efficient evaluation of spectral integrals in the moment method solution of microstrip antennas and circuits," *IEEE Transaction on Antennas and Propagation*, vol. 38, no. 7, pp. 1127-1130, 1990.

[93] A. H. Sihvola and I. V. Lindell, "Material effects in bi-anisotropic electromagnetic," *IEICE Transaction Electronics (Japan)*, vol. E78-C, no. 10, pp. 1383-1390, 1995.

[94] S. A. Tretyakov, A. H. Sihvola, A. A. Sochava and C. R. Simovski, "Magnetoelectric interactions in bi-anisotropic media," *Journal of Electromagnetic Waves and Applications*, vol. 12, no. 4, pp. 481-497, 1998.

[95] A. Sihvola, "Metamaterials in electromagnetic," *ELSEVIER, Metamaterials*, vol. 1, no. 1, pp. 2-11, 2007.

[96] A. Toscano and L. Vegni, "Analysis of printed-circuit antennas with chiral substrates with the method of lines," *IEEE Transaction on Antennas and Propagation*, vol. 49, no. 1, pp. 48-54, 2001.

[97] K. M. Luk, W. Y. Tam and C. L. Yip, "Analysis of circular microstrip antennas with superstrate," *IEE Proceedings H Microwaves, Antennas and Propagation*, vol. 136, no. 3, pp. 261-262, 1989.
[98] E. Newman and D. Forrai, "Scattering from a microstrip patch," *IEEE Transaction on Antennas and Propagation*, vol. 35, no. 3, pp. 245-251, 1987.
[99] W. C. Chew and Q. Liu, " Resonance frequency of a rectangular microstrip patch," *IEEE Transaction on Antennas and Propagation*, vol. 36, no. 8, pp. 1045-1056, 1988.
[100] K. -L. Wong, Design of non planar microstrip antennas and transmission lines, New York: Wiley Inter-Science Publication, John Wiley & Sons, Inc., 1999.
[101] O. M. Ramahi and Y. T. Lo, "Superstrate effect on the resonant frequency of microstrip antennas," *Microwave and Optical Technology Letters*, vol. 5, no. 6, pp. 254-257, 1992.
[102] H. Attia, L. Yousefi and O. M. Ramahi, "Analytical model for calculating the radiation field of microstrip antennas with artificial magnetic superstrates: Theory and experiment," *IEEE Transaction on Antennas and Propagation*, vol. 59, no. 5, pp. 1438-1445, 2011.
[103] M. Biswas and D. Guha, "Input impedance and resonance characteristic of superstrate loaded triangular microstrip patch," *IET Microwaves, Antennas and Propagation*, vol. 3, no. 1, pp. 92-98, 2009.
[104] D. Mirshekar-Syankal and H. R. Hassani, "Characteristics of stacked rectangular and triangular patch antennas for dual band application," *8th IEE International Conference Antennas and Propagation*, vol. 2, pp. 728-731, 1993.
[105] D. Guha and J. Y. Siddiqui, "Resonant frequency of circular microstrip antenna covered with dielectric superstrate," *IEEE Transaction on Antennas and Propagation*, vol. 51, no. 7, pp. 1649-1652, 2003.
[106] J. Shinde, P. Shinde, R. Kumar, M. D. Uplane and B. K. Mishra, "Resonant frequencies of a circularly polarized nearly circular annular ring microstrip antenna with superstrate loading and airgaps," *ITU-T Kaleidoscope: Beyond the Internet? Innovations for Future Networks and Services*, pp. 1-7, 2010.
[107] A. Foroozesh and L. Shafai, "Effects of artificial magnetic conductors in the design of low-profile high-gain planar antennas with high permittivity dielectric superstrate," *IEEE Antennas and Wireless Propagation Letters*, vol. 8, pp. 10-13, 2009.

[108] C. A. Valagiannopoulos and N. L. Tsitsas, "On the resonance and radiation characteristics of multi-layered spherical microstrip antennas," *Electromagnetics*, vol. 28, no. 4, pp. 243-264, 2008.

[109] A. Foroozesh and L. Shafai, "Investigation into the effects of the patch type FSS superstrate on the high-gain cavity resonance antenna design," *IEEE Transaction on Antennas and Propagation*, vol. 58, no. 2, pp. 258-270, 2010.

[110] H. Vettikalladi, O. Lafond and M. Himdi, "High-efficient and high-gain superstrate antenna for 60-GHz indoor communication," *IEEE Antennas and Wireless Propagation Letters*, vol. 8, pp. 1422-1425, 2009.

[111] R. Mittra, Y. Li and K. Yoo, "A comparative study of directivity enhancement of microstrip patch antennas with using three different superstrates," *Microwave and Optical Technology Letters*, vol. 52, no. 2, pp. 327-331, 2010.

[112] Y. J. Lee, J. Yeo, R. Mittra and W. S. Park, "Application of electromagnetic bandgap (EBG) superstrates with controllable defects for a class of patch antennas as spatial angular filters," *IEEE Transaction on Antennas and Propagation*, vol. 53, no. 1, pp. 224-235, 2005.

[113] H. Attia and O. M. Ramahi, "EBG superstrate for gain and bandwidth enhancement of microstrip array antennas," *IEEE Antennas and Propagation Society International Symposium*, San Diego, CA, pp. 1-4, 2008.

[114] H. Attia, L. Yousefi, O. Siddiqui and O. M. Ramahi, "Analytical formulation of the radiation field of printed antennas in the presence of artificial magnetic superstrates," *Journal of Applied Physics A: Materials Science and Processing, Springer Berlin / Heidelberg*, vol. 103, no. 3, pp. 877-880, 2011.

[115] D. Liu, H. C. Chen and B. Floyd, "An LTCC superstrate patch antenna for 60-GHz package applications," *IEEE Antennas and Propagation Society International Symposium (APSURSI)*, pp. 1-4, 2010.

[116] S. Wang, X. Guan, D. Wang, X. Ma and Y. Su, "Electromagnetic scattering by mixed conducting/dielectric objects using higher order MoM," *Progress in Electromagnetic Research*, vol. 66, pp. 51-63, 2006.

[117] D. R. Jackson, "The RCS of a rectangular microstrip patch in a substrate-superstrate geometry," *IEEE Transaction on Antennas and Propagation*, vol. 38, no. 1, pp. 2-8, 1990.

[118] Y. B. Thakare and Rajkumar, "Design of fractal patch antenna for size and radar cross-section reduction," *IET Microwaves, Antennas and Propagation*, vol. 4, no. 2, pp. 175-181, 2010.

[119] W. -N. Herman, "Polarization eccentricity of the transverse field for modes in chiral core planar waveguides," *Journal of the Optical Society of America A: Optics, Image Science and Vision*, vol. 18, no.11, pp. 2806-2818, 2001.

[120] E. Bahar, "Mueller matrices for waves reflected and transmitted through chiral materials: Waveguide modal solutions and applications," *Journal of the Optical Society of America B*, vol. 24, no. 7, pp. 1610-1619, 2007.

[121] M. Al-Sharkawy, A. Z. Elsherbeni and S. F. Mahmoud, "Electromagnetic scattering from parallel chiral cylinders of circular cross sections using an iterative procedure," *Progress In Electromagnetics Research*, vol. 47, pp. 87-110, 2004.

[122] J. A. Kong, Electromagnetic waves theory, Cambridge, MA: EMW Publishing, USA, 2005.

[123] H. -Y. Yang and P. L. E. Uslenghi, "Planar bianisotropic waveguides," *Radio Science*, vol. 28, no. 5, pp. 919-927, 1993.

[124] T. H. O'DELL, The electrodynamics of magnetoelectric media, North-Holland Publication Corporation: Amsterdam, 1970.

[125] R. D. Graglia, P. L. E. Uslenghi and R. E. Zich, "Dispersion relation for bianisotropic materials and its symmetry properties," *IEEE Transaction on Antennas and Propagation*, vol. 39, no. 1, pp. 83-90, 1991.

[126] R. D. Graglia, P. L. E. Uslenghi and R. E. Zich, "Reflection and transmission for planar structures of bianisotropic media," *Electromagnetics*, vol. 11, pp. 193-208, 1991.

[127] M. A. Al-Kanhal and E. Arvas, "Electromagnetic scattering from a chiral cylinder of arbitrary cross section," *IEEE Transaction on Antennas and Propagation*, vol. AP-44, no. 7, pp. 1041-1048, 1996.

[128] S. F. Mahmoud, "Mode characteristics in chirowaveguides with constant impedance walls," *Journal of Electromagnetic Waves and Applications (JEMWA)*, vol. 6, no. 5-6, pp. 625-640, 1992.

[129] R. D. Graglia, P. L. E. Uslenghi and C. L. Yu, "Electromagnetic oblique scattering by a cylinder coated with chiral layers and anisotropic jump-admittance sheets," *Journal of Electromagnetic Waves and Applications (JEMWA)*, vol. 6, no.1, pp. 695-719, 1992.

[130] X. Q. Sheng and Y. Ekn, "Analysis of microstrip antennas on finite chiral substrates," *International Journal of RF and Microwave Computer-Aided Engineering*, vol. 14, no. 1, pp. 49-56, 2004.

[131] E. O. Kamenetskii, "Nonreciprocal microwave bianisotropic materials: Reciprocity theorem and network reciprocity", *IEEE Transaction on Antennas and Propagation*, vol. 49, no. 3, pp. 361-366, 2001.

[132] A. Serdyukov, I. Semchenko, S. Tretyakov and A. Sihvola, Electromagnetics of bi-anisotropic materials: Theory and applications, Taylor & Francis, 2001.

[133] X. Cheng, J. A. Kong and L. Ran, "Polarization of waves in reciprocal and nonreciprocal uniaxially bianisotropic media," *PIERS Online*, vol. 4, no. 3, pp. 331-335, 2008.

[134] M. S. Kluskens and E. H. Newman, "A microstrip line on a chiral substrate," *IEEE Transaction on Microwave Theory and Techniques*, vol. 39, no. 11, pp. 1889-1891, 1991.

[135] M. S. Kluskens and E. H. Newman, Method of moments analysis of scattering by chiral media, Ft. Belvoir: Defense Technical Information Center, 1991.

[136] G. Plaza, F. Mesa and M. Horno, "Study of the dispersion characteristics of planar chiral lines," *IEEE Transaction on Microwave Theory and Techniques*, vol. 46, no. 8, pp. 1150-1157, 1998.

[137] A. Lakhtakia and W. Weiglhofer, "Constraint on linear, spatiotemporally nonlocal, spatiotemporally nonhomogenous constitutive relations," *International Journal of Infrared and Millimeter Waves*, vol. 17, no. 11, pp. 1867-1878, 1996.

[138] P. L. E. Uslenghi, "Theory of certain bianisotropic waveguides," *Proceedings of the URSI International Symposium on Electromagnetic Theory, Sydney, Australia*, pp. 293-295, 1992.

In: Development of Complex Electromagnetic … ISBN: 978-1-61122-013-1
Editors: K. N. Ramli et al.

Chapter 6

INTERACTION OF EM FIELDS TO THE HUMAN BODY USING HYBRID COMPUTATIONAL METHOD

***K. N. Ramli*[1]*, R. A. Abd-Alhameed*[2] *and P. S. Excell*[3]**

[1]Faculty of Electrical and Electronics Engineering,
Universiti Tun Hussein Onn Malaysia,
Parit Raja, Batu Pahat, Johor, Malaysia
[2]Mobile and Satellite Communications Research Centre,
Bradford University, Bradford, UK
[3]Institute for Arts, Science and Technology,
Glyndwr University, Wrexham, UK

ABSTRACT

A new hybrid method of moments (MoM)/finite-difference time-domain (FDTD), with a subgridded finite-difference time-domain (SGFDTD) approach is presented. The method overcomes the drawbacks of homogeneous MoM and FDTD simulations, and so permits accurate analysis of realistic applications. As a demonstration, it is applied to the short-range interaction between an inhomogeneous human body and a small UHF RFID antenna tag, operating at 900 MHz. Near-field and far-field performance for the antenna are assessed for different placements over the body. The cumulative distribution function of the radiation efficiency and the absorbed power are presented and analyzed. The algorithm has a five-fold speed advantage over fine-gridded FDTD.

6.1. INTRODUCTION

Rigorous determination of electromagnetic fields within arbitrary, anisotropic, and inhomogeneous dielectric bodies is important for researchers exploring the effects of microwaves upon living tissue. Electromagnetic fields are induced inside any biological system, such as the human body, when it is illuminated by an electromagnetic wave, which is also scattered externally. The human body is an irregularly shaped heterogeneous conducting medium whose permittivity and conductivity vary with the frequency of the incident wave, so the distributions of the internal and scattered electromagnetic fields depend on the body's physiological parameters and geometry, as well as on the polarization and frequency of operation. Due to its mathematical simplicity and ease of implementation, FDTD [1-12] numerical solutions have been widely adopted to solve complex stratified dielectric objects, such as human bodies [13-15] and biological tissues [16-17].

The conventional uniform mesh FDTD becomes unfavourable due to excessive computation time and memory when it is applied to electrically small objects which require high spatial resolution. To circumvent this problem, the non-uniform mesh and subgridding schemes in FDTD were proposed. However, these methods may produce spurious solutions or suffer from instability [18], so the FDTD method has been used in conjunction with other numerical techniques in order to tackle this limitation. In 1993, Aoyagi et al. [19] used the Yee algorithm with the scalar wave equation to reduce the computations needed to model a vivaldi antenna, while Cangellaris et al. [20] used a hybrid spectral-FDTD method to analyze propagation in anisotropic, inhomogeneous periodic structures. Wang [21] introduced a hybrid ray-FDTD method and used it to investigate scattering from a cavity with a complex termination and wave penetration through inhomogeneous walls. In 1994, Mrozowski [22] introduced a hybrid FDTD-PEE (partial eigenfunction expansion) method to speed up the FDTD method when solving shielded structure problems. In addition, finite element and finite volume methods have recently been combined with FDTD, [23, 24], for accuracy in handling curved geometries and systems with fine features. Hybrid methods operating entirely in the time domain have been reported in the literature [25, 26], but the time-domain MoM is not at the state of maturity and flexibility of the frequency-domain version. Time-domain MoM does have the advantage of generating information over a wide frequency band, and does not need an iterative procedure to couple with FDTD, but it requires very long run-times when treating a junction with more than two wires [27], unlike the frequency domain version in which complex metallic structures may be modeled accurately with lower run-times and with more flexibility in treating different complex geometries.

The hybrid MoM/FDTD method may be said to have been first investigated in 1982, when Taflove and Umashankar [28] used a hybrid FDTD/MoM approach to investigate electromagnetic (EM) coupling problems and aperture penetration into complex geometries and loaded cavities, In 1987, the same authors used an equivalent surface fully enclosing equivalent wire bundles (concept of equivalent radius) to replace them with a single wire in the FDTD model [29]. Later, the concepts of [28] and [29] were deployed in the computer software GEMACS [30] which was developed using method of moments/uniform geometrical theory of diffraction/finite-difference frequency-domain (MoM/UTD/FDFD) hybrids to allow users to model problems with more than one region. The work on hybrid MoM/FDTD was continued by Bo et al. in 2010 [31] to model HF antennas located on the ground. A year later, Xin *et al.* [32] employed the same technique for RF coil to determine the open vertical-field of the propagated EM wave.

In 2000, Mangoud proposed a hybrid combination of frequency domain MoM and FDTD [33], thus overcoming the drawbacks of homogeneous FDTD and MoM to solve a wide variety of electromagnetic interaction scattering problems. In 2004, a new hybrid method brought together the FDTD, finite-element time-domain (FETD) and method of moments time-domain (MoMTD) methods to analyze problems of thin-wire antennas radiating in the vicinity of arbitrarily-shaped inhomogeneous bodies [34]. From [35], stair case errors [36] from FDTD can be mitigated by using a finite element method (FEM), but this method requires high computational resources. The method in [33] was extended to include the analysis of wide band [37] and dual-band [38] antenna responses using an impedance interpolation method [38] to minimize the computation time on the MoM side.

Recently, the works [33, 37-38] have been extended by embedding subgridding inside an inhomogeneous human body model when the source excitation is a short-range radio frequency identification (RFID) antenna at 900 MHz in the UHF band [39]. The use of subgridding in the hybrid MoM-FDTD method ensures that transitions are as reflection-less as possible in order to provide specific animated display movements of the electric and magnetic fields in both the subgridded region and main grids. However, curved objects and small geometric features cannot be accurately modelled due to inherent staircase errors caused by the Cartesian grid. To circumvent this weakness, a variable time and space increment has been employed which provides for subgridding and consequently minimizes numerical dispersion. The main grid is divided into subgrids, and the boundary fields are coupled using temporal and spatial interpolations. Thus, the MoM-FDTD-SGFDTD hybridization method is well-suited to electromagnetic radiation and scattering problems as it overcomes the drawbacks of homogeneous MoM and FDTD simulations, this being appropriate for realistic electromagnetic analysis. In this paper, a new 3D hybrid MoM-FDTD-SGFDTD is developed and used to study the state of the art in body-antenna interaction.

6.2. Hybrid MoM-FDTD-SGFDTD Theoretical Formulation

Consider the electromagnetic hybrid geometry illustrated in Figure 6.1. The figure shows two regions, with the source in region *A*, and the scatterer in region *B*. The source region is bounded by a closed Huygens surface S_c. The method starts by computing the fields due to the real currents of the source region on the surface S_c, excluding region *B*. These fields are computed by applying Galerkin's method with a set of variable polynomial basis functions [33, 37].

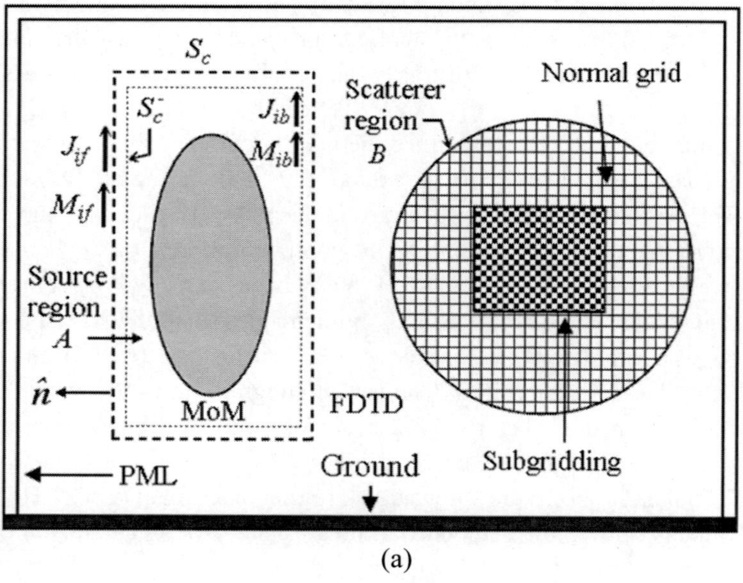

(a)

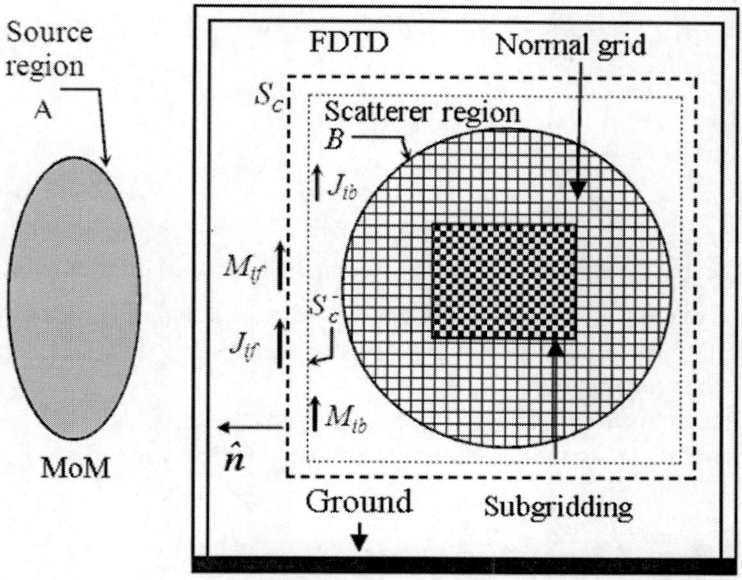

(b)

Figure 1. Hybrid MoM-FDTD-SGFDTD configuration for single source and scatterer geometries: (a) Near-field, (b) Far-field.

The equivalent surface currents on the surface S_c represent the outward travelling wave-fields from the source to the scatterer, due to the fields of the source. These may be expressed as:

$$J_{if} = \hat{n} \times H_{if} \tag{6.1}$$

$$M_{if} = E_{if} \times \hat{n} \tag{6.2}$$

Here $\hat{n}$ is the outwardly directed unit vector normal to the surface from the source region. H_{if} and E_{if} are equivalent to the forward-scattered magnetic and electric fields respectively from the source region on the equivalent surface S_c, and J_{if} and M_{if} are the corresponding electric and magnetic source currents respectively on this surface. These currents are then treated as sources in the FDTD computational province, propagating fields to the scatterer by using the E and H curl equations given by the expression:

$$\nabla \times E = -\frac{\partial B}{\partial t} - M_{if} \tag{6.3}$$

$$\nabla \times H = -\frac{\partial D}{\partial t} + J_{if} \tag{6.4}$$

The back-scattered fields were computed by FDTD at S_c^- (the closed surface interior to the surface S_c and bounding the region *A*). This surface is closed in the scattered field region, so that the calculated surface currents are due to the scattered fields only.

The equivalent surface currents due to these fields, representing an additional source to the MoM domain (region *A*), are given by:

$$J_{ib} = H_{ib} \times \hat{n} \tag{6.5}$$

$$M_{ib} = \hat{n} \times E_{ib} \tag{6.6}$$

where H_{ib} and E_{ib} are the back-scattered fields computed at S_c^-. Note that $\hat{n}$ is as above, directed outwards from the source region. J_{ib} and M_{ib} are the electric and magnetic equivalent surface currents at S_c^-.

Now, the voltage back-scattered in the source region (the excitation for the MoM) can be evaluated using either of the following equations, defined by the reciprocity theorem in the same way as in [33, 37]:

$$V_b = \iint_{S_a} dS_a (J_{ms} \cdot E_{ib}) \tag{6.7}$$

$$V_b = \left\langle J_{ib} \cdot E_{ms} - M_{ib} \cdot H_{ms}, dS_c^- \right\rangle \tag{6.8}$$

$$V_b = \iint_{S_c^-} dS_c^- (J_{ib} \cdot E_{ms} - M_{ib} \cdot H_{ms}) \tag{6.9}$$

where

$$E_{ib} = -j\omega A(r) - \nabla V(r) - \frac{1}{\varepsilon}\nabla \times F(r) \tag{6.10}$$

$$A(r) = \mu \iint_{S_c^-} dS_c^- \{J_{ib} G(r,r')\} \tag{6.11}$$

$$V(r) = -\frac{j}{\omega\varepsilon} \iint_{S_c^-} dS_c^- \left\{\nabla'_s \cdot J_{ib} G(r,r')\right\} \tag{6.12}$$

$$F(r) = \varepsilon \iint_{S_c^-} dS_c^- \{M_{ib} G(r,r')\} \tag{6.13}$$

$G(r, r')$ is the free space Green function, given by the expression:

$$G(r,r') = \frac{e^{-jk|r-r'|}}{|r-r'|} \tag{6.14}$$

The vectors r and r' apply to the source and observation points respectively and S_a is the conducting surface area of the structure within region A. J_{ms} is the electric test-function used on the wire. E_{ms} and H_{ms} are the electric and magnetic fields respectively for the test-function J_{ms}. Equation (6.7) explicitly requires a double integral to evaluate E_{ib} and integrate over the surface of the antenna, with the condition that the FDTD discretisation be very small compared to the operating wavelength. It can be simplified by ignoring the surface integral and evaluating the voltage back-scattered corresponding to the centre of the cell surface, by a summation over grid cell surfaces, to get the following equation for the hybrid case:

$$V_b = \sum_{k=1}^{n_{S_c^-}} (J_{ib_k} \cdot E_{ms}(r_k, r') - M_{ib_k} \cdot H_{ms}(r_k, r'))a_k \tag{6.15}$$

where r_k is the position vector of the centre of the cell surface and a_k is the surface area of the cell surface. Therefore J_{ib_k} and M_{ib_k} are the equivalent surface currents at the centre of the surface cell *n*. Since the excitation voltages are known, the MoM can be executed to compute the new currents and the procedure can be repeated until the steady state solution is achieved.

6.3. VALIDATION OF RADIATED FIELD

A 900 MHz centre-fed half-wavelength dipole of 0.0025λ radius was defined as a transmitter source, for example presenting the RFID reader, whereas the RFID tag was considered as a small half-wavelength meander antenna [40], as shown in Figure 6.2. The tag was designed in a zig-zag pattern with 17 turns in which subgridding cells of 6×6×6 volumetric were imposed. The dotted bold line on the left and right portray the Huygens surface for near and far field respectively. The spatial resolution of the computational domain was terminated by six faces of perfectly matched layer (PML) in which one of them describes the ground plane (represented by grey colouring in the figure).

Two different distances between the source and the tag were studied and discussed, for the far-field and near-field cases, with separate simulations for each. It should be noted that the centres of the source and the sensor lie on the *y* axis and they are separated by the distance D_t. For both simulations, a thin wire subroutine code inside the subgridded FDTD was used to include the effects of the wire radius of the meander antenna inside the new FDTD subgrid. For simplicity in this example the medium surrounding the source and the scatterer was modelled as free space; however, in general the scatterer can be placed on dielectric objects in which part of its volume can be subgridded.

A sinusoidal excitation voltage was applied at the centre of the antenna source. The antenna wires were assumed to be perfectly conducting. The near-field method as shown in Figure 6.1(a) is equivalent to the far-field configuration for short distances, such as D_t = 16.8 cm. In contrast, the far-field technique as illustrated in Figure 6.1(b) is analogous to the near-field configuration for much higher distances, such as D_t = 33.6 cm. The magnitude of E_y electric field components along the *z* axis at y = 7.2 cm were compared for near-field and far-field simulations, with good agreement as shown in Figure 6.3. The E_z components were also assessed and found to be identical to each other (see Figure 6.3).

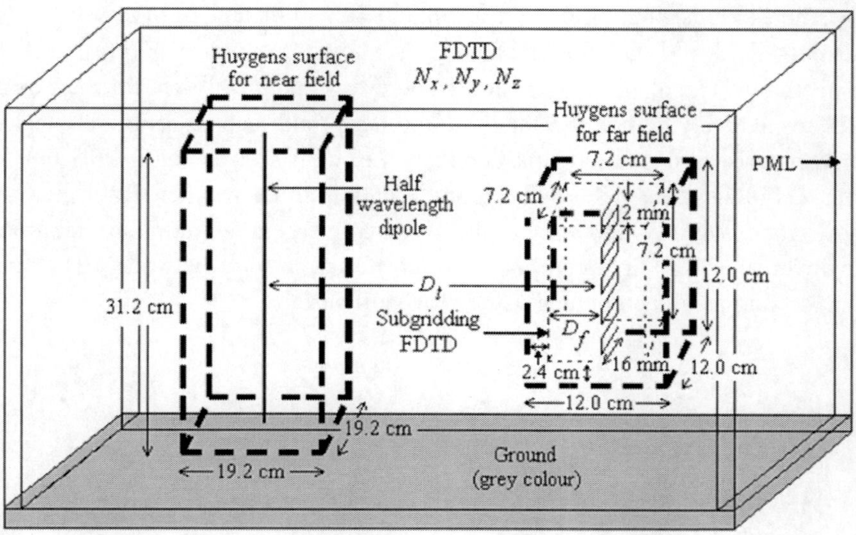

Figure 6.2. A basic geometry of FDTD-SGFDTD for near-field and far-field validation.

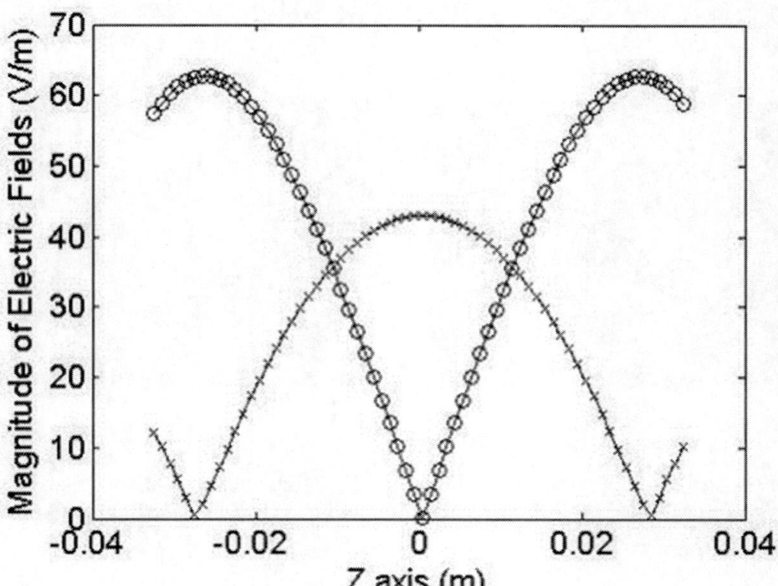

Figure 6.3. Magnitude of E_y and E_z electric field components along z axis at $y = 7.2$ cm: Near-field E_y('ooo'), E_z('xxx'), Far-field E_y and E_z ('—— ').

The field distribution over an x-z plane 7.2 cm distant from the sensor for the near-field and far-field technique is shown in Figure 6.4 (a) and (b) respectively. The plane size considered here was 20 cm × 16 cm for the x and z axes respectively. When the far-field and near-field techniques were checked for comparison of one antenna geometry i.e. with D_t was set at 16.8 cm for both techniques, the fields were found identical to each other. Both methods show good stability and the results were convergent within four iterations. However, the total field components were found to vary between ±2% when compared to results computed using NEC software.

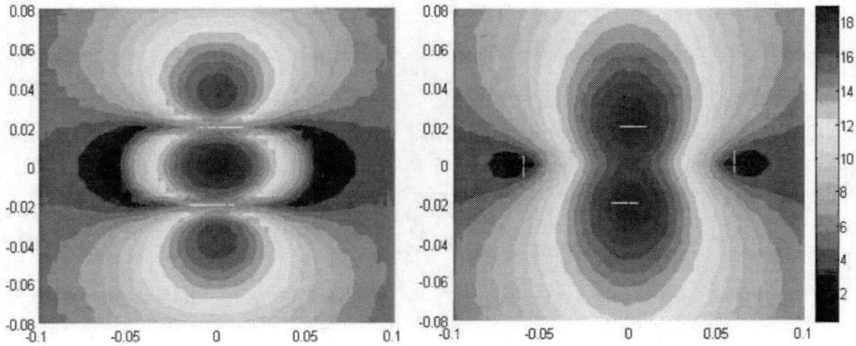

(a)

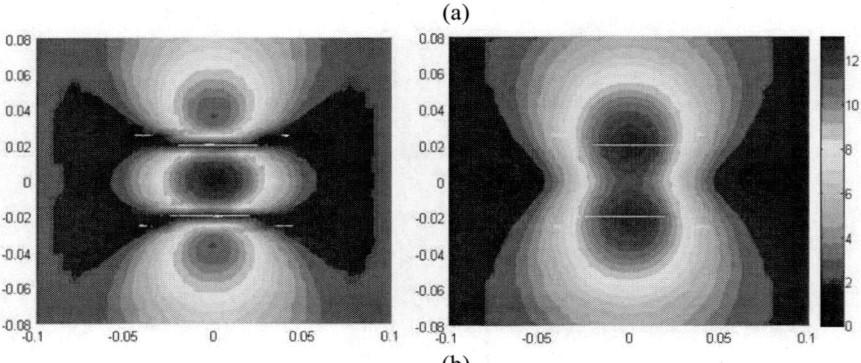

(b)

Figure 6.4. A basic. Distribution of the E_z and E_{total} field components in dB at 7.2 cm away from the sensor, (a) using near field method and (b) using far field method. (left: E_z, right: E_{total}).

To validate the accuracy of the method [40], where Figure 6.5 depicts the calculation model used, and to compare the result with existing published results [15, 41], a COST244 model [15, 41] was adopted as a human phantom. This model represents a homogeneous cubical human head of size 200 × 200 ×

200 mm^3, with relative permittivity $\varepsilon_r = 41.5$, conductivity $\sigma = 0.95$ S/m and volume density $\rho = 1000$ kg/m^3.

A 160 mm long half-wavelength dipole, operating at 900 MHz, was used as a radiation source, with a continuous wave with an input power of 250 mW exciting the dipole. The distance between the antenna and the human head was 15 mm and the origin of coordinates was set at the centre of the surface of the human head model, on the same horizontal level as the antenna feeding point. The FDTD cell size and time step in the analysis area were 2.5 mm and 3.3 ps respectively. The problem space, cell sizes and number of PML layers were (127×127×127 cells), 2.5 mm and 6, respectively. The size of the equivalent Huygens surface is 6×6×68 cells, equivalent to 15 mm × 15 mm × 170 mm. The subgridded volume is 16×16 ×16 cells, equivalent to 40 mm × 40 mm × 40 mm. It should be noted that this subgridded volume is placed between two dielectric media, i.e. free space (40×5×40 mm^3) and the human head (40×35×40 mm^3) in order to evaluate the effectiveness of the proposed method when applied to inhomogeneous tissues. Two and four subgridding factors were used for each cell of the main FDTD. The dielectric properties for the subgridded cells located on the boundary were averaged and assigned on each field component, as shown in Figure 6.6. The variations of the unaveraged specific absorption rate (SAR) along the *y*-axis with and without subgridding are shown in the Figure 6.7.

These unaveraged SAR values without subgridding, with subgridding factor of 2 and with subgridding factor of 4 were observed at every 1.25 mm, 0.625 mm and 0.3125 mm points respectively in the ascending order along the *y*-axis. The outcomes were promising for the case of without subgridding and both with subgridding factors applied in this example due to the convergence of the results into one line.

The averaged SAR of 6.8 W/kg was computed by numerical simulation (in this case, it was averaged) over 10 gm of human tissue. The value of 6.8 W/kg (averaged over 10 gm of tissue) when compared with [15, 41] were very close. In order to prove the proposed technique is computationally viable, the situation shown in Figure 6.5 was also modeled using FDTD with coarse and fine grids, and using FDTD/MoM methods. The required computational times for each method are compared in Table 6.1.

In the coarse grid FDTD model, a 2.5 mm cell size was used while in the fine grid FDTD model, 1.25 mm and 0.625 mm cell sizes were adopted. It should be noted that all simulations were performed on an Intel Core-i7-2000 desktop with 3.4 GHz CPU and 16 GB of RAM.

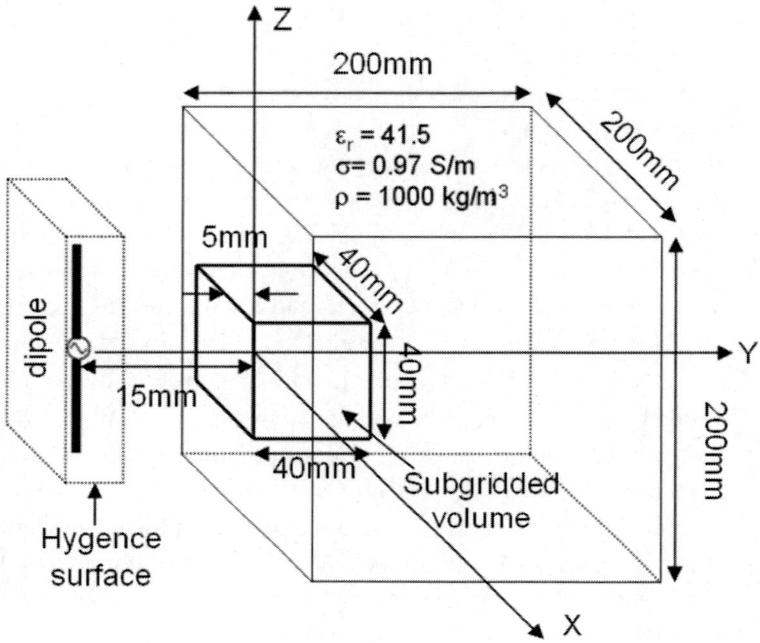

Figure 6.5. Basic geometry model of COST244.

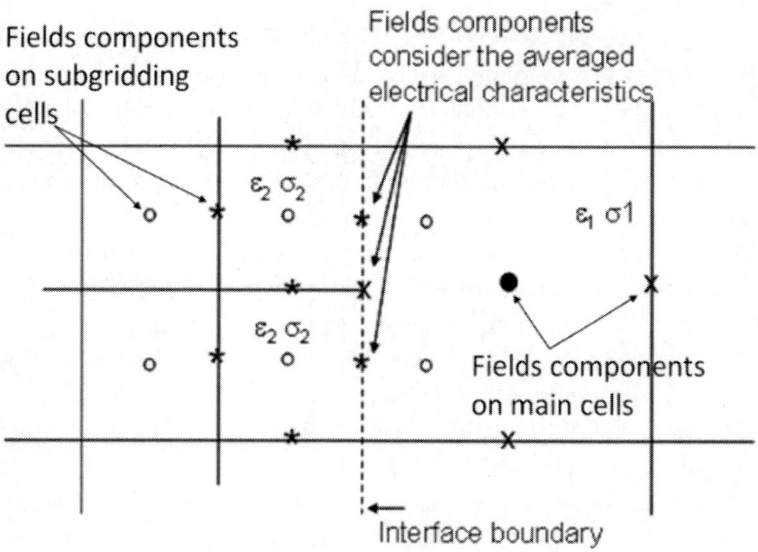

Figure 6.6. Distribution of field components on the interface boundary model of two different mediums of the proposed subgridding model.

Table 6.1. Simulation time for different numerical techniques

Method	Problem space size	Cell size (mm)	Time steps (ps)	No of steps/ number of cycles	Simulation Time (hours)
FDTD	127×127×127	2.5	3.3	8417 (25 cycles)	1.5
FDTD	254×254×254	1.2	1.041	26683 (25 cycles)	4.5
FDTD	508×508×508	0.625	0.7	39682 (25 cycles)	17
MOM/FDTD	127×127×127	2.5	3.3	8417 (25 cycles)	3.8
MOM/FDTD/SG (subgridding factor 2)	127×127×127 SG 16×16×16	2.5	3.3	8417 (25 cycles)	4.1
MOM/FDTD/SG (subgridding factor 4)	127×127×127 SG 16×16×16	2.5	3.3	8417 (25 cycles)	4.3

As can be noticed, the proposed method exhibits reasonable computational time in comparison with the standard coarse-gridded FDTD and MoM/FDTD. However, when compared with fine-gridded FDTD, the proposed algorithm requires less computational effort, yielding roughly a five-fold speed improvement.

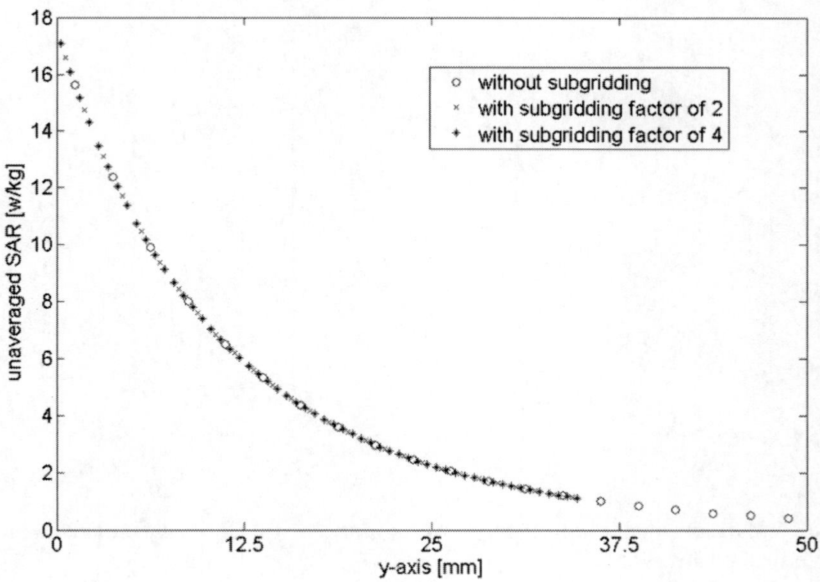

Figure 6.7. The variations of the unaveraged SAR along y-axis of the model shown in Figure 6.5 [15, 41].

6.4. Hybrid Model Technique

A Huygens surface around the proposed antenna is first introduced as illustrated in Figure 6.8 with volume of 16×4×10 and 10×4×16 FDTD cells for horizontal and vertical polarization respectively. The surface basically acts as a virtual box enclosing the radiating element such that the fields are enclosed within this region.

The equivalent electric and magnetic sources on the Huygens surface are produced at each time step from the electric and magnetic fields created by the antenna in free space using NEC code [42]. This field data is used as an input source for the FDTD code. The overall FDTD spatial volume is 118×77×327 points. The equivalence principle is carried out in 3-D by applying the hybrid electromagnetic method described in [40] with cell size $dx = dy = dz = 6.0$ mm and $dx = dy = dz = 3.0$ mm for the coarse and fine FDTD lattices respectively. The time step is set at 7.0 ps.

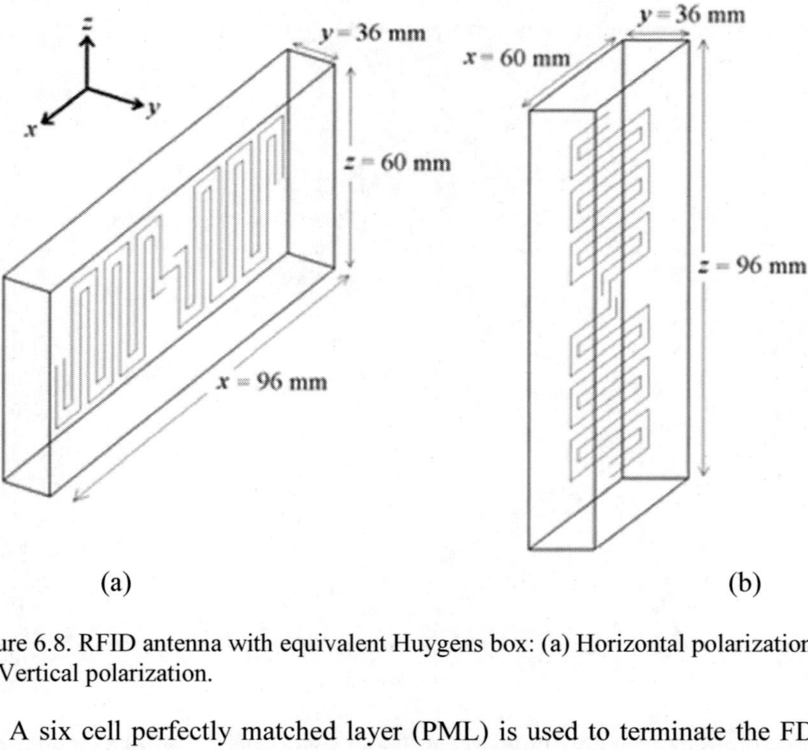

Figure 6.8. RFID antenna with equivalent Huygens box: (a) Horizontal polarization, (b) Vertical polarization.

A six cell perfectly matched layer (PML) is used to terminate the FDTD space, and the distance between the antenna Huygens surface box and the human body is 12.0 mm (or 2 cells). The human body model employed in this work was developed by Mason et al. in 2000 [43]. The near-field and far-field radiation of the antenna at 900 MHz for different locations has been analyzed in order to investigate the performance of the antenna in proximity to the human body. The number in Figure 6.9 indicates the location of the RFID antenna.

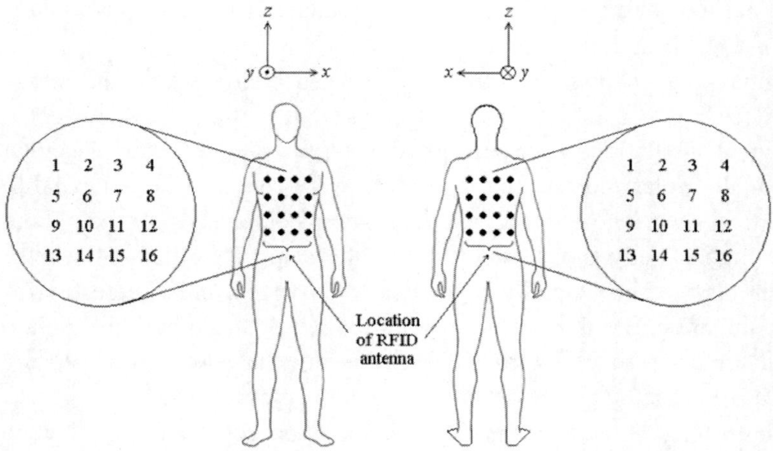

Figure 6.9. Location of the antennas, represented by black dots, in proximity to the human body.

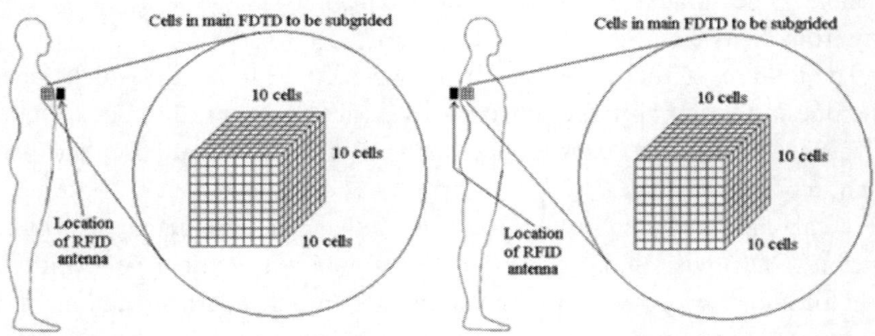

Figure 6.10. Subgrid cells of 10×10×10 FDTD cells are taken inside the human body: (a) Front location, (b) Back location (subgridded).

A total of 32 locations were investigated; 16 on the back, and 16 on the front. Each placement was investigated with horizontal and vertical polarizations, in order to obtain the clearest possible results with a restricted number of samples. Figure 6.10 shows the location of the antenna in front and at the back of human body respectively, in which subgrid cells of 10×10×10 FDTD cells were taken inside the human body for near-field analysis. The near-field and far-field of the antenna are calculated and analyzed at various locations in order to build up a realistic picture of the antenna performance while in close proximity with the human body. Figure 6.11 shows the electric field distributions using an equivalent dB scale in the immediate neighbourhood and the interior of the body model, for the vertical polarization state, in *x-y*, *x-z* and *y-z* planes.

Figure 6.12 shows the electric field distribution inside the subgridded region for *x-y*, *x-z* and *y-z* planes (represented in the equivalent dB scale). Additional simulations were performed using the same antenna placements for horizontally polarized antenna, and these were found to be comparable with the vertical state. The electric field distributions obtained in the neighbourhood of the human body model were almost the same regardless of whether the antenna tag was horizontally or vertically polarized. It is interesting to note that a similar observation was also produced within the subgridded region. The electric fields were very strong when the antenna was located close to the body, as represented by the red colouring in the plots.

Figure 6.13 illustrates the far-field radiation pattern for the antenna (in horizontal polarization state) at the front and back of the human body model. It should be noted that the underlying computation was normalized to 1 W input power. The variation in the far-field patterns implies that the field distributions are more concentrated in the direction facing the normal antenna axis, and away from the body.

The field magnitude is reduced by between 10 dB to 20 dB, which appears to be due to the tailing effect of the body. Once more, the field distributions for vertical polarization were quite similar to the horizontal case, and hence are not reproduced here.

It can be concluded from Figure 6.14 that the cumulative distribution function (CDF) has similar curves for horizontal and vertical polarization in these locations. Moreover, it is apparent that the standard deviation of the radiation efficiency when the mobile is located in the front is less compared to the back.

The antenna achieves better radiation efficiency of 43% mean percentage value for both horizontal and vertical polarization comparing front and back location.

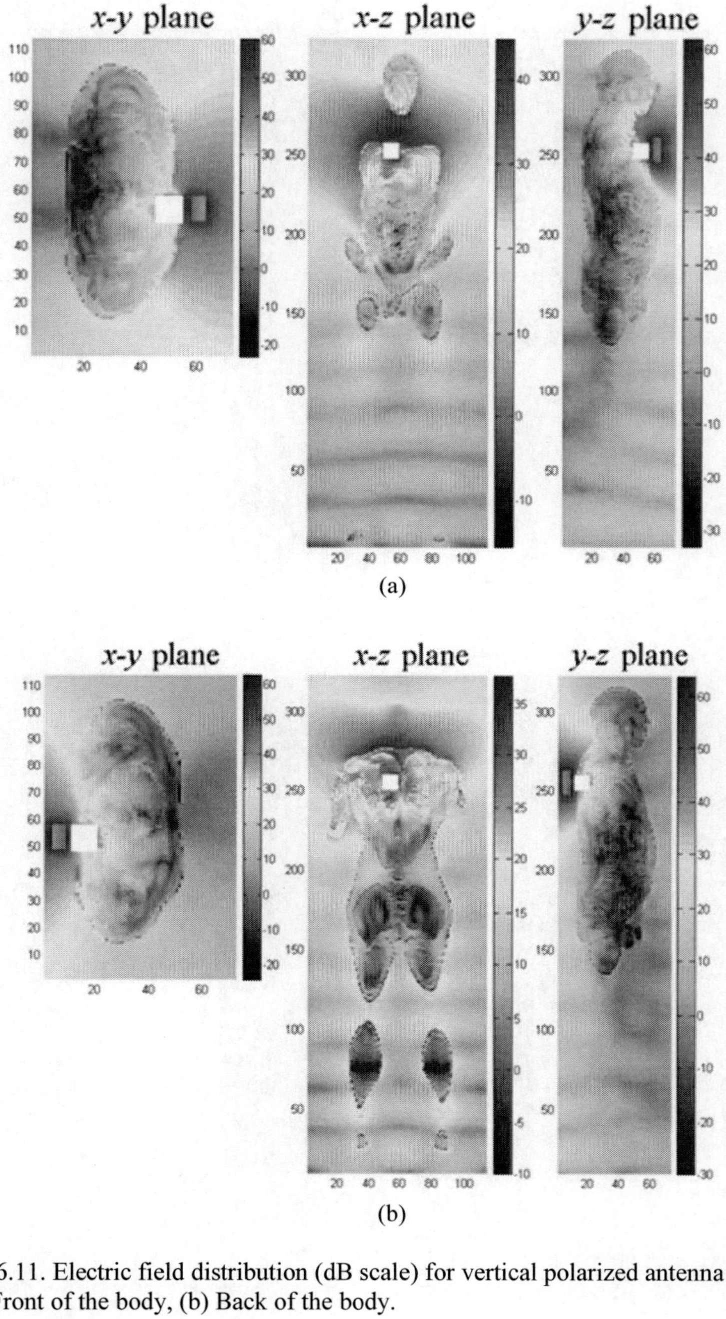

Figure 6.11. Electric field distribution (dB scale) for vertical polarized antenna placed at: (a) Front of the body, (b) Back of the body.

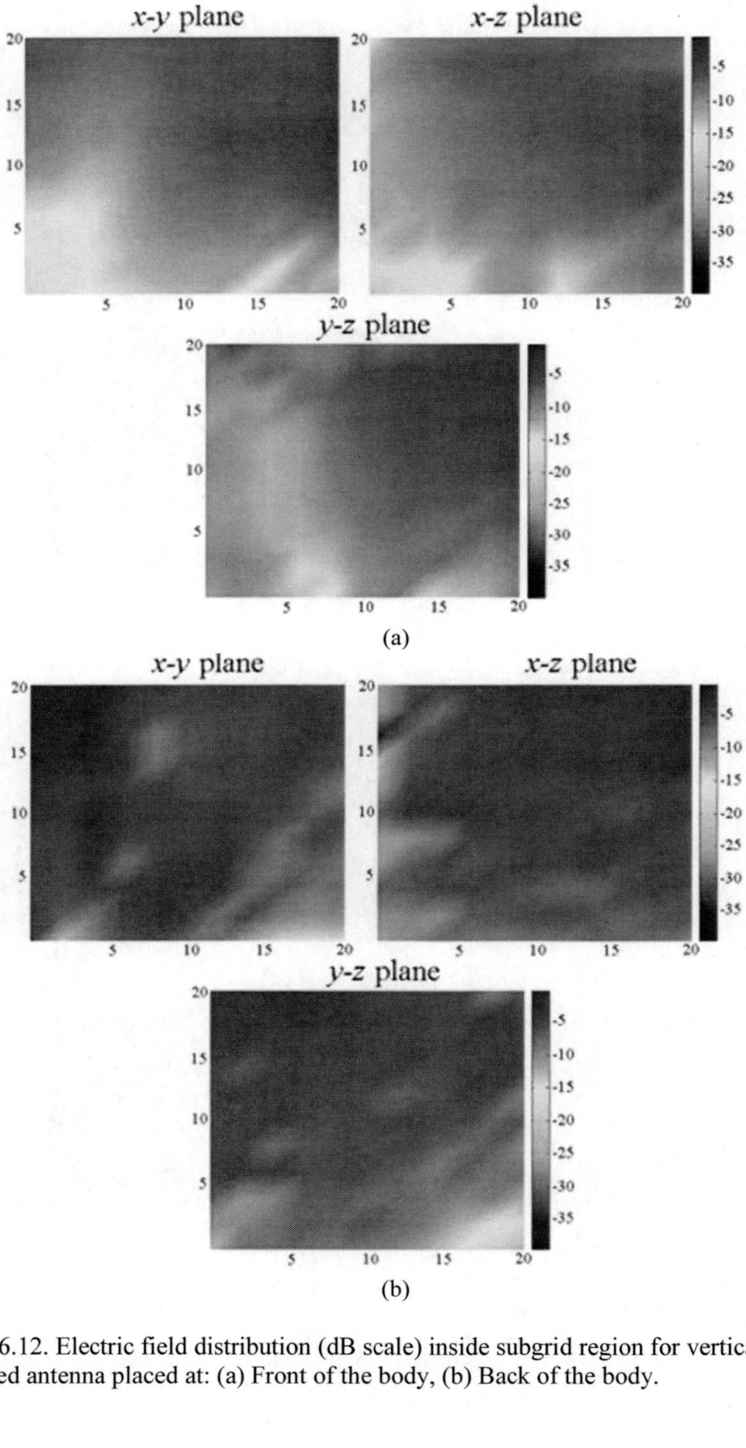

Figure 6.12. Electric field distribution (dB scale) inside subgrid region for vertically polarized antenna placed at: (a) Front of the body, (b) Back of the body.

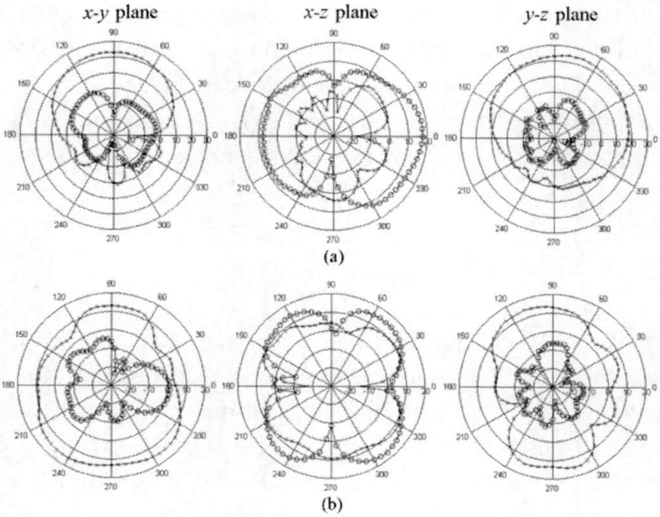

Figure 6.13. Far-field pattern for horizontally polarized antenna placed at: (a) Front of the body, (b) Back of the body; 'o-o-o': E_θ, 'x-x-x': E_ϕ.

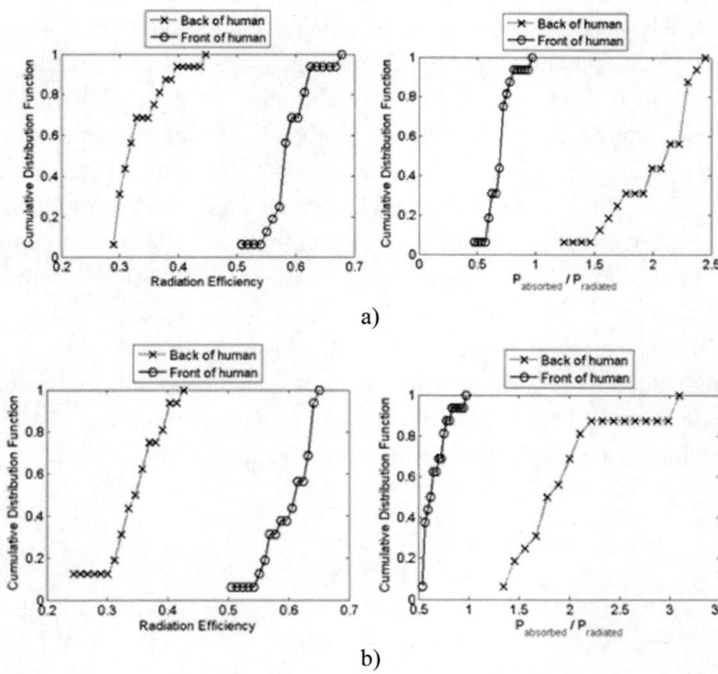

Figure 6.14. Cumulative distribution function of radiation efficiency and the ratio of $P_{absorbed}$ / $P_{radiated}$ for: (a) Horizontal polarized antenna, (b) Vertical polarized antenna.

This follows from a loss in the tissue at the front, and a greater degree of absorbed power at the back of the human body, as can be seen from the plots. Taking the previous simulation results into account, we may conclude that altering the position of the antenna on the front will not result in a large dispersion in the radiation efficiency of the antenna.

CONCLUSION

A hybrid MoM-FDTD-SGFDTD approach has been presented and adopted for modelling the interaction of the human body with a short-range RFID antenna. In order to ensure that optimal quality of results was achieved using a limited sample size of 32 points, each point was investigated using two both horizontal and vertical polarizations. The MoM technique was used to produce the electric and magnetic fields created by the antenna in free space, and the equivalent electric and magnetic sources on the Huygens surface were produced at each time step. The FDTD technique was then applied for the whole structure of the problem combined with subgridding at the object of interest, namely inside the human body near the excitation source. The human body model was designed to be inhomogeneous at close proximity to the antenna. The near-field and far-field distributions were incorporated into the study to heighten the understanding of the impact on human tissue both facing the antenna, and not directly facing the antenna. The cumulative distribution function of the radiation efficiency and the ratio of absorbed to radiated power of the antenna at these locations were also computed. The results support the conclusion that there was a clear improvement in the front of the human body model compared to the back position. The combination of hybrid MoM-FDTD-SGFDTD method approach with an arbitrary inhomogeneous human body model encourages the development of this new EM field interaction modeling approach. This study robustly supports the computing of the total power dissipated and the SAR distribution inside human tissue.

REFERENCES

[1] R. Xiong, B. Chen, Y. Yi, H. -L. Chen and H. -F. Zhao, "Two-step method for the FDTD analysis of long apertures having depth," *IEEE Antennas and Wireless Propagation Letters*, vol. 12, pp. 39-42, 2013.

[2] S. K. Kamepally, B. P. Kumar and C. S. Paidimarry, "FDTD estimation for accurate specific absorption rate in a tumor," *Annual International Conference on Emerging Research Areas and International Conference on Microelectronics, Communications and Renewable Energy (AICERA/ICMiCR)*, pp. 1-5, 2013.

[3] W. G. Whittow, C. C. Njoku and Y.C. Vardaxoglou, "Fine scale simulations of patch antennas with heterogeneous substrates," *USNC-URSI Radio Science Meeting (Joint with AP-S Symposium)*, pp. 223, 2013.

[4] C. Lesina, A. Vaccari and A. Bozzoli, "A modified RC-FDTD algorithm for plasmonics in Drude dispersive media nanostructures," *6th European Conference on Antennas and Propagation (EUCAP)*, pp. 687-690, 2012.

[5] G. M. Noetscher, Y. Xu and S. N. Makarov, "Accuracy of point source models with coincident phase centers in a cubic FDTD grid for arbitrary source orientation," *IEEE Antennas and Propagation Society International Symposium (APSURSI)*, pp. 1-2, 2012.

[6] V. Nayyeri, M. Soleimani and M. Dehmollaian, "Implementation of a PEMC boundary condition in the 2-D FDTD technique," *IEEE Antennas and Propagation Society International Symposium (APSURSI)*, pp. 1-2, 2012.

[7] D. -C. Liu and H. -C. Chang, "The second-order condition of FDTD method at sloped dielectric interfaces by averaging the permittivities along surface normal," *IEEE Transactions on Antennas and Propagation*, vol. 60, no. 11, pp. 5259-5267, 2012.

[8] M. Weiwei, S. Dong and W. Xianliang, "UPML-FDTD parallel computing on GPU," *International Conference on Microwave and Millimeter Wave Technology (ICMMT)*, vol. 4, pp. 1-4, 2012.

[9] C. Guiffaut, A. Reineix and B. Pecqueux, "New oblique thin wire formalism in the FDTD method with multiwire junctions," *IEEE Transactions on Antennas and Propagation*, vol. 60, no. 3, pp. 1458-1466, 2012.

[10] D. M. Shyroki, "Modeling of sloped interfaces on a Yee grid," *IEEE Transactions on Antennas and Propagation*, vol. 59, no. 9, pp. 3290-3295, 2011.

[11] K. S Yee, "Numerical solution of initial boundary value problems involving Maxwell's equations in isotropic media," *IEEE Transactions Antennas and Propagation*, vol. 14, no. 8, pp. 302–307, 1966.

[12] Taflove, "Application of the finite-difference time-domain method to sinusoidal steady-state electromagnetic penetration problems," *IEEE*

Transactions Electromagnetic Compatibility, vol. 22, no. 3, pp. 191-202, 1980.

[13] H. Kawai, K. Ito, "Simple Evaluation Method of Estimating Local Average SAR," IEEE *Transactions on Microwave Theory and Techniques*, Vol.52, no. 8, pp. 2021-2029, 2004.

[14] K. Fujii, M. Takahashi, and K. Ito, "Electric field distributions of wearable devices using the human body as a transmission channel," *IEEE Transactions on Antennas and Propagation*, vol. 55, no. 7, pp. 2080-2087, 2007.

[15] J. -H. Jung, S. -W. Kim, Y. -S. Kim, and S. -Y. Kim, "Electromagnetic propagation from the intestine-ingested source in the human body model," *IEEE Transactions Antennas and Propagation*, vol. 58, no. 5, pp. 1683-1688, 2010.

[16] G. Emili, A. Schiavoni, F. L. Roselli, and R. Sorrentino, "Computation of electromagnetic field inside a tissue at mobile communications frequencies," *IEEE Transactions Microwave Theory Techniques*, vol. 51, no. 1, pp. 178-186, 2003.

[17] C. H. See, R.A. Abd-Alhameed and P.S. Excell, "Computation of Electromagnetic Fields in Assemblages of Biological Cells Using a Modified Finite-Difference Time-Domain Scheme," *IEEE Transactions Microwave Theory Techniques*, vol. 55, no. 9, pp. 1986-1994, 2007.

[18] P. Monk and E. Suli, "Error estimates for Yee's method on nonuniform grids," *IEEE Transactions Magnetics*, vol. 30, no. 5, pp. 3200-3203, 1994.

[19] P. H. Aoyagi, J. F. Lee, and R. Mittra, "A hybrid Yee algorithm/scalar-wave equation approach," *IEEE Transactions on Microwave Theory and Techniques*, vol. 41, pp. 1593-1600, 1993.

[20] C. Cangellaris, M. Gribbons, and G. Sohos, "A hybrid spectral/FDTD method for the electromagnetic analysis of guided waves in periodic structures," *Microwave and Guided Wave Letters*, vol. 3, pp. 375-377, 1993.

[21] Y. Wang, S. K. Chaudhuri, and S. Safavi-Naeini, "An FDTD/ray-tracing analysis method for wave penetration through inhomogeneous walls," *IEEE Transactions on Antennas and Propagations*, vol. 50, pp. 1598-1604, 2002.

[22] M. Mrozowski, "A hybrid PEE-FDTD algorithm for accelerated time domain analysis of electromagnetic waves in shielded structures," *IEEE Microwave and Guided Wave Letters*, vol. 4, pp. 323-325, 1994.

[23] Monorchio and R. Mittra, "Time-domain (FE/FDTD) technique for solving complex electromagnetic problems," *IEEE Microwave and Guided Wave Letters*, vol. 8, pp. 93-95, 1998.

[24] X. Fierriers, J-P. Parmantier, S. Bertuol, and A.R. Ruddle, "Application of Hybrid Finite Difference/Finite Volume Method to Solve an Automotive EMC Problem," *IEEE Transactions on Electromagnetic Compatibility*, vol. 4, pp. 624-634, 2004.

[25] G. Cerri, P. Russo, A. Schiavoni, G. Tribellini, and P. Bielli, "MoM-FDTD hybrid technique for analysing scattering problems," *Electronic Letters*, vol. 34, pp. 433-440, 1998.

[26] R. Bretones, A. Monorchio, G. Manara, R. G. Martin, and R. Mittra, "Hybrid technique combining finite element, finite difference and integral equation methods in the time domain," *Electronic Letters*, vol. 36, pp. 506-508, 2000.

[27] D. Tinniswood, "Time Domain Integral Equations," PhD Dissertation, University of York, 1996.

[28] Taflove and K. R. Umashankar, "A hybrid moment method/finite difference time-domain approach to electromagnetic coupling and aperture penetration into complex geometries," *IEEE Transactions on Antennas and Propagation*, vol. AP-30, pp. 617-627, 1982.

[29] K. R. Umashankar, A. Taflove and B. Beker, "Calculation and experimental validation of induced currents on coupled wires in an arbitrary shape cavity," *IEEE Transactions on Antennas and Propagation*, vol. AP-35, pp. 1248-1257, 1987.

[30] E. L. Coffey and D. L. Kadlec, "General electromagnetic model for the analysis of complex systems (GEMACS) version 5.0," *Advanced Electromagnetic Corporation for USAF Rome Air Development Center (USA), Report No. RADC-TR-90-360*, vol. I-III, 1990.

[31] Z. Bo, H. Xiaojun, J. Bin, G. Hao and S. Xujian, "An MoM/FDTD hybrid technique for modeling HF antennas located on ground," *International Conference on Microwave and Millimeter Wave Technology (ICMMT)*, pp. 1344-1347, 2010.

[32] X. Xin, D. Wang, J. Han, Y. Feng, Q. Feng and W. Chen, "Numerical optimization of intra-operative RF coil for open vertical-field using hybrid MoM/FDTD method," *IEEE International Symposium on Biomedical Imaging: From Nano to Macro*, pp. 750-753, 2011.

[33] M. A. Mangoud, R. A. Abd-Alhameed, and P.S. Excell, "Simulation of human interaction with mobile telephones using Hybrid techniques over

coupled domains," *IEEE Transactions on Microwave Theory and Techniques*, vol. 48, pp. 2014-2021, 2000.

[34] Monorchio, A. R. Bretones, R. Mittra, G. Manara, and R. G. Martin, "A Hybrid Time-Domain Technique that Combines the Finite Element, Finite Difference and Method of Moment Techniques to Solve Complex Electromagnetic Problems," *IEEE Transactions on Antennas and Propagation*, vol. 52, pp. 2666-2674, 2004.

[35] R Alias, R A Abd-Alhameed, P S Excell and Q Gassim, "A Modified Equivalent Conducting Surface Boundary Using The Hybrid FEM-FDTD Technique," *8th IEEE International Multi-topic conference (INMIC2004)*, December 24-26, 2004 at NUCES FAST, Lahore, Pakistan, pp.698-702., 2004.

[36] Akyurtlu, D. H. Werner, V. Veremey, D. J. Steich, and K. Avdin, "Staircasing errors in FDTD at an air-dielectric interface," *IEEE Microwave and Guided Wave Letters*, vol. 9, pp. 444-446, 1999.

[37] R. A. Abd-Alhameed, P. S. Excell, and M. A. Mangoud, "Broadband antenna response using Hybrid technique combining frequency domain MoM and FDTD," *ACES Journal*, vol. 20:1, pp. 70-77, 2005.

[38] G. Alhaddad, R. A. Abd-Alhameed, D. Zhou, C.H. See, I.T.E. Elfergani and P.S. Excell, "Low Profile Dual-Band Balanced Handset Antenna with Dual-Arm Structure for WLAN Application," *IET Microwaves, Antennas and Propagation*, vol.5, pp. 1045 – 1053, 2011.

[39] K. N. Ramli, R. A. Abd-Alhameed, Y. A. S. Dama, M. S. A. Alkhambashi, M. B. Child, and P. S. Excell, "Interaction of EM fields to the human body using MoM-FDTD-SGFDTD hybrid computational method," *EMC Europe, York, UK*, 26-30 Sep., pp. 1-4, 2011.

[40] R. A. Abd-Alhameed, P. S. Excell, C. H. See, D. Zhou, and K. N. Ramli, "Accurate field distribution models for RFID applications using hybrid computational electromagnetics techniques," *Progress in Electromagnetics Research Symposium, Cambridge, USA*, pp. 436-442, 2008.

[41] S. Kajiwara, A. Yamamoto, K. Ogawa, A. Ozaki and Y. Koyanagi, "Attenuation Characteristics of the SAR in a COST244 Phantom with Different EM Source Locations and Size," *Proceeding of ISAP'04, Sendai, Japan*, pp.793-796, 2004.

[42] G. J. Burke and A. J. Poggio, "Numerical electromagnetics code (NEC): method of moments," *US Naval Ocean Systems Centre, Rep. No. TD116*, 1981.

[43] P. A. Mason, W. D. Hurt, T. J. Walters, J. A. D'Andrea, P. Gajsek, K. L. Ryan, D. A. Nelson, K. I. Smith, and J. M. Ziriax, "Effects of frequency, permittivity, and voxel size on predicted specific absorption rate values in biological tissue during electromagnetic-field exposure," *IEEE Transaction Microwave Theory and Techniques*, vol. 48, no. 11, pp. 2050-2058, 2000.

In: Development of Complex Electromagnetic … ISBN: 978-1-61122-013-1
Editors: K. N. Ramli et al.

Chapter 7

QUASI-STATIC FDTD SCHEME FOR ELECTRICALLY-SMALL REGIONS IN FREE SPACE, LOSSLESS AND LOSSY PENETRABLE MEDIA

***C. H. See*[1], *A. S. Abdullah*[2], *J. M. Noras*[3] *and R. A. Abd-Alhameed*[3]**

[1]Engineering, Sport and Sciences Academic Group,
University of Bolton, Bolton, Lancashire, UK
[2]Department of Electrical Engineering, College of Engineering,
University of Basrah, Basrah, Iraq
[3]Mobile and Satellite Communications Research Centre,
Bradford University, Bradford, UK.

A new approach to model the interaction between the electromagnetic wave and electrically small objects in free space, lossless and lossy media is discussed. The computational burden due to huge number of time steps has been reduced by implementing quasi-static approach. The accuracy of the frequency scaling method is verified with the models of homogenous and layered spheres in free space, lossless and lossy media. The numerical results are found to be in good agreement with the analytical ones. This research paves the way to the validity of using the frequency scaling FDTD to obtain the induced electric fields at low and high frequencies.

7.1. Introduction

The problem of modelling electromagnetic field interaction with structures whose dimensions are a few wavelengths has been the subject of many studies and numerical simulations for bioelectromagnetic applications. The methods used for such numerical simulations generally fall into one of three categories, method of moment (MoM) [1-12], finite-element method (FEM) [13-23] and finite-difference time-domain (FDTD) [24-36].

Method of moment does an excellent job of analyzing unbounded radiation problems and they excel at analyzing perfect electric conductor and homogeneous dielectrics. However, it is not very effective when applied to arbitrary configurations with complex geometries or inhomogeneous dielectrics. As for finite-element method, it is a very versatile technique because it allows the analysis of complex inhomogeneous structures. However, it requires unaffordable computation resources to solve extremely large problem, due to discretising the entire problem space and lack of adequate free-space boundary conditions that can be placed close to the objects modelled.

In contrast, the finite-difference time-domain, since its first introduction by Yee [37], has been used extensively over the last decade for bioelectromagnetic dosimetry – the numerical assessment of electromagnetic fields coupled to complex heterogeneous biological entities at powerline frequency and radio frequency. The use of FDTD is very attractive for this application, due mainly to its algorithmic simplicity, capable to solve complicated heterogeneous geometries and ease of modularization and parallelisation of the algorithm on massively parallel computers. Moreover, it is also capable to model electromagnetic wave interaction problems requiring the solution of considerably more than 10^8 field vector unknowns. At this level of complexity, it is possible to develop detailed, 3-D models of complete engineering systems.

The classical form of FDTD method requires extremely small time-step sizes when modelling electrically-small regions (much smaller than a wavelength). Thus, it can become impractical due to the unaffordable computation times required. This problem can be solved by implementing a quasi-static approximate version of FDTD. This approach is based on transferring the working frequency to a higher frequency, to reduce the number of time steps required. Then, the generated internal field at the higher frequency can be scaled back to the frequency of interest [37, 38-41]. It should be noted that this approach is only valid if the size of the interacting structure in the problem space is 10 times or more smaller than the wavelength and $|\sigma + j\omega\varepsilon| >> \omega\varepsilon_o$ [35, 38], where σ and ε are the conductivity and permittivity of the tissues respectively, ω is the radian frequency, and ε_o is the permittivity of free space. In order to prove the validity of the quasi-static approach, this chapter will develop a 3-D FDTD program to directly model a single homogeneous or multi-layered sphere inside a lossless or lossy problem space. The sphere was excited by a plane wave, which is replaced by an equivalent surface. The edge of the problem space was truncated by a properly modified Berenger's perfectly matched layer (PML) absorbing boundary condition (ABC), with matching impedance condition ($\sigma/\varepsilon = \sigma^*/\mu$ for lossless or lossy media, whereas $\sigma/\varepsilon_o = \sigma^*/\mu_o$ for free space) [42-45]. An optimum value of the g factor is properly chosen to provide efficient low reflection termination of the computational space for different penetrable media problem at the interface layer.

7.2. Analytical Solution

The exact solution for describing the characteristics of electromagnetic radiation scattered by a homogenous sphere when it is excited by a plane wave, was first solved by Mie [46] in 1908 and has been concisely formulated by Stratton [47] in 1941. A solution of scattering problem of a sphere with one concentric shell has been given by Aden and Kerker [48] and the scattering from an inhomogeneous spherically symmetric object has been calculated by Wyatt [49]. Since Mie series solution is simple and easy to be modified for any composition of sphere, therefore it was adapted by various authors for comparisons to investigate bioelectromagnetic problems, either to be used as an analytical solution to the problems [50-60] or as a solution to proposed numerical methods [38, 61-65] for comparisons. This section gives the solutions in a form readily amenable to machine computation for both the scattered and the interior fields. The spherical core is surrounded by arbitrary number of concentric shells. It should be noted that this exact solution will be implemented as a tool for comparison against our proposed quasi-static FDTD method throughout this chapter.

7.2.1. Sphere with Arbitrary Concentric Layers

Figure 7.1 depicts the orientation of the plane wave with respect to a rectangular and spherical coordinate system having its origin at the centre of the sphere. The plane wave is assumed to propagate in the positive z-direction and the electric field is linearly polarised in the x-direction. A stratified sphere with arbitrary thickness and electrical characteristics is shown in Figure 7.2. The basic idea of the analysis is performed by expanding the incident and secondary fields in vector spherical harmonics appropriate to each region and matching the tangential components of the fields at each boundary to determine the expansion coefficients.

The expansion of the incident plane wave is given in Stratton [47] and the same identical notation will be implemented in this work. This incident wave will induce fields in all of the regions and the general solution of the vector wave equation for the transverse fields in any shell layer can be written as a linear combination of the spherical vector wave functions, $\vec{m}$ and $\vec{n}$, that was used by Stratton [47]. In general, the radial field dependence is an arbitrary solution to any linear combination of the spherical Bessel and Hankel function of first and second kinds. Each type of Bessel and Hankel function with its associated properties will be discussed as follows:

- The spherical Bessel function of the first kind, $B_{n(1)} = J_n(r)$, indicates a standing wave.
- The spherical Bessel function of the second kind, $B_{n(2)} = Y_n(r)$, indicates a standing wave as well and is also called the spherical Neumann function or Weber function.
- The spherical Bessel function of the third kind, $B_{n(3)} = J_n(r) + iY_n(r)$, indicates an inward traveling wave and is also called the Hankel function of the first kind, $H_{n(1)}(r)$.
- The spherical Bessel function of the fourth kind, $B_{n(4)} = J_n(r) - iY_n(r)$, indicates an outward traveling wave and is also called the spherical Hankel function of second kind, $H_{n(2)}(r)$.

For the innermost layer of the sphere the spherical Bessel function of first kind, J_n, is the only admissible solution, while for the surrounding medium the fields will consist of incident waves and scattered waves in order to satisfy the radiation condition at infinity. Hence, the general solution of the electric and magnetic field vectors in p^{th} region of concentric sphere can be formulated as follows:

$$\vec{E}_p = E_o e^{i\omega t} \sum_{n=1}^{\infty} i^n \frac{2n+1}{n(n+1)} \left[a_n^p \vec{m}_{o1n}^{(1)} - i b_n^p \vec{n}_{e1n}^{(1)} + \alpha_n^p \vec{m}_{o1n}^{(3)} - i \beta_n^p \vec{n}_{e1n}^{(3)} \right] \quad (7.1)$$

$$\vec{H}_p = -\frac{k_p E_o e^{i\omega t}}{\mu_p \omega} \sum_{n=1}^{\infty} i^n \frac{2n+1}{n(n+1)} \left[b_n^p \vec{m}_{e1n}^{(1)} - i a_n^p \vec{n}_{o1n}^{(1)} + \beta_n^p \vec{m}_{e1n}^{(3)} - i \alpha_n^p \vec{n}_{o1n}^{(3)} \right] \quad (7.2)$$

where a, b, α, β are the expansion coefficients and $\vec{m}$, $\vec{n}$ are the spherical vector wave functions as given by Stratton [47].

$$\begin{aligned} \vec{m}_{\substack{o\\e}1n}^{(z)} &= \pm \frac{1}{\sin\theta} B_n^z(k_p r) P_n^1(\cos\theta) \begin{bmatrix} \cos\phi \\ \sin\phi \end{bmatrix} \hat{\theta} \\ &- B_n^z(k_p r) \frac{\partial}{\partial\theta} P_n^1(\cos\theta) \begin{bmatrix} \sin\phi \\ \cos\phi \end{bmatrix} \hat{\phi} \end{aligned} \quad (7.3)$$

$$\begin{aligned} \vec{n}_{\substack{o\\e}1n}^{(z)} &= \frac{n(n+1)}{k_p r} B_n^z(k_p r) P_n^1(\cos\theta) \begin{bmatrix} \sin\phi \\ \cos\phi \end{bmatrix} \hat{r} \\ &+ \frac{1}{k_p r} \frac{\partial}{\partial(k_p r)} \left[k_p r B_n^z(k_p r) \right] \frac{\partial}{\partial\theta} P_n^1(\cos\theta) \begin{bmatrix} \sin\phi \\ \cos\phi \end{bmatrix} \hat{\theta} \\ &\pm \frac{1}{k_p r \sin\theta} \frac{\partial}{\partial(k_p r)} \left[k_p r B_n^z(k_p r) \right] P_n^1(\cos\theta) \begin{bmatrix} \cos\phi \\ \sin\phi \end{bmatrix} \hat{\phi} \end{aligned} \quad (7.4)$$

where k_p is the complex propagation constant for the pth region:

$$\begin{matrix} \mathrm{Re} \\ \mathrm{Im} \end{matrix} (k_p) = \frac{\omega}{c} \left[\frac{\varepsilon_r^p}{2} \left(\sqrt{1 + \left(\frac{\sigma_p}{\varepsilon_o \varepsilon_r^p \omega} \right)^2} \pm 1 \right) \right]^{\frac{1}{2}} \quad (7.5)$$

ε_o is the free space permittivity, ε_r^p and σ_p are the relative permittivity and electrical conductivity of the pth layer, and the positive value of the square root is used.

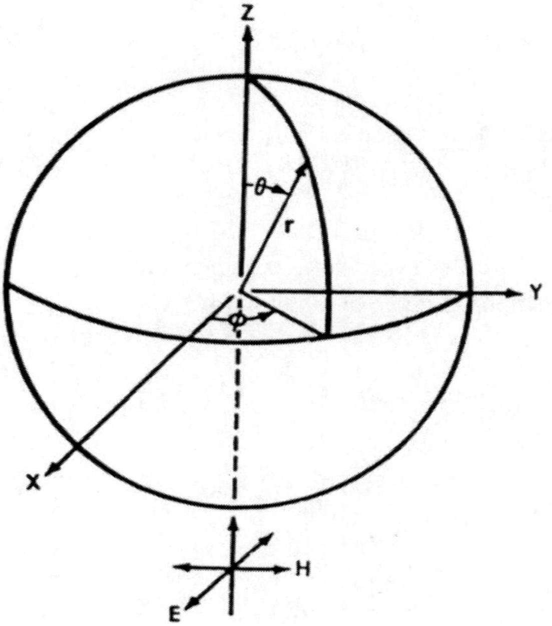

Figure 7.1. Spherical co-ordinates and incident plane wave configurations.

The n-th term of $P_n^1(\cos\theta)$ is the associated first order Legendre polynomials, properties for which can be found in literature [66] or might be computed using the follow recursive relations:

Let: $Q_n(\theta)=\dfrac{P_n^1(\cos\theta)}{\sin\theta}$ and $R_n(\theta)=\dfrac{\partial}{\partial\theta}P_n^1(\cos\theta)$ then:

$$Q_n(\theta)=\frac{(2n-1)\cos\theta Q_{n-1}(\theta)-nQ_{n-2}(\theta)}{n-1} \text{ and}$$

$$R_n(\theta)=n\cos\theta Q_n(\theta)-(n+1)Q_{n-1}(\theta)$$

which can be solved using initial value of $Q_0(\theta)=0.0, Q_1(\theta)=1.0, R_1(\theta)=\cos\theta$. The subscript "$e$" and "$o$" correspond to the even or odd character of the vector wave functions $\vec{m}$ and $\vec{n}$.

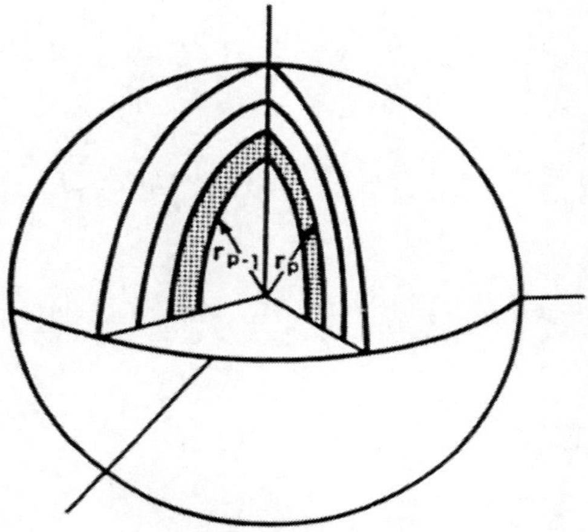

Figure 7.2. *M*-layered concentric spherical system. Shaded region represents p^{th} layer of the system with the radius of r_p and electrical parameters of ε_p, μ_p, and σ_p.

7.2.2. Solution for Expansion Coefficients

The electric and magnetic fields can be determined using equations (6.1) and (6.2) with four expansion coefficients a, b, α, β determined by the requirement that the tangential components $\vec{E}$ and $\vec{H}$ be continuous at the interfaces between all regions. This requirement yields the following four relationships at the interface to each layer:

$$\begin{aligned}(E_p)_\theta &= (E_{p+1})_\theta \\ (E_p)_\phi &= (E_{p+1})_\phi \\ (H_p)_\theta &= (H_{p+1})_\theta \\ (H_p)_\phi &= (H_{p+1})_\phi\end{aligned} \tag{7.6}$$

Substituting $\vec{m}$ and $\vec{n}$ functions into equations (7.1) and (7.2) and applying the four relationships in equation (7.6), the expansion coefficients may be written as follows [51, 67]:

$$a_n^p = \frac{1}{\Delta_n^p}\left[\left(\zeta_n^p j_n^p - \tau^p h_n^p \eta_n^{p+1}\right)a_n^{p+1} + \left(\zeta_n^p h_n^{p+1} - \tau^p h_n^p \zeta_n^{p+1}\right)\alpha_n^{p+1}\right] \tag{7.7a}$$

$$\alpha_n^p = \frac{1}{\Delta_n^p}\left[\left(\tau^p j_n^p \eta_n^{p+1} - \eta_n^p j_n^{p+1}\right)a_n^{p+1} + \left(\tau^p j_n^p \zeta_n^{p+1} - \eta_n^p h_n^{p+1}\right)\alpha_n^{p+1}\right] \quad (7.7b)$$

$$b_n^p = \frac{1}{\Delta_n^p}\left[\left(\tau_n^p \zeta_n^p j_n^{p+1} - h_n^p \eta_n^{p+1}\right)b_n^{p+1} + \left(\tau_n^p \zeta_n^p h_n^{p+1} - h_n^p \zeta_n^{p+1}\right)\beta_n^{p+1}\right] \quad (7.7c)$$

$$\beta_n^p = \frac{1}{\Delta_n^p}\left[\left(j_n^p \eta_n^{p+1} - \tau^p \eta_n^p j_n^{p+1}\right)b_n^{p+1} + \left(j_n^p \zeta_n^{p+1} - \tau^p \eta_n^p h_n^{p+1}\right)\beta_n^{p+1}\right] \quad (7.7d)$$

where:

$$j_n^p = j_n^p\left(k_p r_p\right)$$

$$h_n^p = h_n^p\left(k_p r_p\right)$$

$$\Delta_n^p = j_n^p \zeta_n^p - h_n^p \eta_n^p$$

$$\tau^p = \frac{k_{p+1}}{k_p}\frac{\mu_p}{\mu_{p+1}}$$

$$\eta_n^p = \frac{1}{\left(k_p r_p\right)}\frac{\partial}{\partial\left(k_p r_p\right)}\left(k_p r_p j_n^p\right) = \frac{1}{2n+1}\left[(n+1)j_{n-1}^p - n j_{n+1}^p\right]$$

$$\zeta_n^p = \frac{1}{\left(k_p r_p\right)}\frac{\partial}{\partial\left(k_p r_p\right)}\left(k_p r_p h_n^p\right) = \frac{1}{2n+1}\left[(n+1)h_{n-1}^p - n h_{n+1}^p\right]$$

All of the Bessel functions are now understood to be evaluated at the values of p and $p+1$ corresponding to the interface between the two regions. For the solution of coefficients a_n^p , α_n^p , b_n^p and β_n^p , equations (7.7a), (7.7b) and equations (7.7c), (7.7d) can be represented in the matrix form as follows:

$$\begin{bmatrix} a_n^p \\ \alpha_n^p \end{bmatrix} = \begin{bmatrix} Q_{11}^p & Q_{12}^p \\ Q_{21}^p & Q_{22}^p \end{bmatrix}\begin{bmatrix} a_n^{p+1} \\ \alpha_n^{p+1} \end{bmatrix} \quad (7.8)$$

$$\begin{bmatrix} b_n^p \\ \beta_n^p \end{bmatrix} = \begin{bmatrix} U_{11}^p & U_{12}^p \\ U_{21}^p & U_{22}^p \end{bmatrix} \begin{bmatrix} b_n^{p+1} \\ \beta_n^{p+1} \end{bmatrix} \tag{7.9}$$

From [43], for a *M*-layered sphere, the above equations can be stated as follows:

$$\begin{bmatrix} a_n^1 \\ \alpha_n^1 \end{bmatrix} = \begin{bmatrix} T_{11} & T_{12} \\ T_{21} & T_{22} \end{bmatrix} \begin{bmatrix} a_n^M \\ \alpha_n^M \end{bmatrix} \tag{7.10}$$

$$\begin{bmatrix} b_n^1 \\ \beta_n^1 \end{bmatrix} = \begin{bmatrix} V_{11} & V_{12} \\ V_{21} & V_{22} \end{bmatrix} \begin{bmatrix} b_n^M \\ \beta_n^M \end{bmatrix} \tag{7.11}$$

where $$T_{ij} = \prod_{p=1}^{M-1} Q_{ij}^p \text{ and } V_{ij} = \prod_{p=1}^{M-1} U_{ij}^p \tag{7.12}$$

Using the fact that $\alpha_n^1 = \beta_n^1 = 0$ (i.e., the Hankel function, in which its divergent at the origin at the innermost sphere, must be absent for this region), and $a_1^M = b_1^M = 1$. Equations (7.8) and (7.9) yield:

$$\begin{bmatrix} a_n^1 \\ 0 \end{bmatrix} = \begin{bmatrix} T_{11} & T_{12} \\ T_{21} & T_{22} \end{bmatrix} \begin{bmatrix} 1 \\ \alpha_n^M \end{bmatrix} \tag{7.13}$$

$$\begin{bmatrix} b_n^1 \\ 0 \end{bmatrix} = \begin{bmatrix} V_{11} & V_{12} \\ V_{21} & V_{22} \end{bmatrix} \begin{bmatrix} 1 \\ \beta_n^M \end{bmatrix} \tag{7.14}$$

Equations (7.13) and (7.14) represent a simple simultaneous equation form which a_n^1, α_n^M and b_n^1, β_n^M can be computed. Using the modified version of equations (7.8) and (7.9), all remaining expansion coefficients can be found as follows:

$$\begin{bmatrix} a_n^{p-1} \\ \alpha_n^{p-1} \end{bmatrix} = \begin{bmatrix} Q_{11}^{p-1} & Q_{12}^{p-1} \\ Q_{21}^{p-1} & Q_{22}^{p-1} \end{bmatrix} \begin{bmatrix} a_n^p \\ \alpha_n^p \end{bmatrix} \tag{7.15}$$

$$\begin{bmatrix} b_n^{p-1} \\ \beta_n^{p-1} \end{bmatrix} = \begin{bmatrix} U_{11}^{p-1} & U_{12}^{p-1} \\ U_{21}^{p-1} & U_{22}^{p-1} \end{bmatrix} \begin{bmatrix} b_n^p \\ \beta_n^p \end{bmatrix} \tag{7.16}$$

7.3. MODIFIED BERENGER'S PML

The primary difficulty of FDTD method when modelling open-region electromagnetic wave scattering problem, is that an artificial boundary or absorbing boundary condition (ABC) is required to be imposed at the outer grid boundaries of the computational domain, in order to truncate the size volume. Therefore, absorbing boundary conditions (ABC's) are one of the most critical elements of finite-difference time-domain analysis, which must be applied to absorb the required wave of this bounded problem. Several analytical ABC's, such as Mur [68] and Liao [69], have been invented in order to overcome this problem. The proposed ABCs provide effective reflection coefficients in order of -35 dB to -45 dB for most FDTD simulations. Recently, the capabilities of the ABC have been significantly enhanced with the introduction of Berenger's perfectly matched layer (PML) [42, 44, 45]. This is based on surrounding the FDTD problem space with highly lossy and matched non-physical absorber. It also allowed for boundary reflections below -80 dB to be realized. The PML, therefore, became the boundary of choice for the simulation problems of this chapter.

It is known that a plane wave incident on a planar boundary of a dispersionless medium from free space is perfectly transmitted without reflection (matched across a planar boundary) for all normally incident waves if the matching condition in equation (7.17) is satisfied for that medium [42].

$$\frac{\sigma}{\varepsilon_o} = \frac{\sigma^*}{\mu_o} \tag{7.17}$$

The idea of matching condition in free space for the PML will be extended to lossless and lossy media with the following modified matching condition.

$$\frac{\sigma}{\varepsilon_r \varepsilon_o} = \frac{\sigma^*}{\mu_r \mu_o} \tag{7.18}$$

where σ and σ^* are the electric conductivity and magnetic conductivity of the medium respectively. ε_o and μ_o are the free space permittivity and permeability and the relative permittivity and permeability are ε_r and μ_r. In order to reduce the reflection on the interface layer, an optimum value of the geometric grading factor g has been selected, by using an empirical expression as in [44]:

$$\sigma = -\frac{\varepsilon c}{2\Delta x} \frac{\ln g}{g^N - 1} \ln R(0) \tag{7.19}$$

where Δx is the spatial increment of FDTD mesh, $R(0)$ is the normal reflection coefficient, N is the number of the cells in the PML thickness, c is the velocity of EM waves in free space or lossy medium.

7.4. Verification Method

In order to verify the validity of quasi-static approximation (using a higher frequency f' to obtain induced E fields) in finite-difference time-domain (FDTD) method, four examples will be illustrated and discussed in this section. Due to the existence of the analytical solution for a spheres exposed to plane wave, therefore, all examples demonstrated here will be discussed subject to a homogeneous and layered sphere involving for free space, lossless and lossy penetrable media to confirm the accuracy of the solution obtained from numerical results with analytical solution.

This section is classified into two sub-sections, i.e. verification at powerline frequency and verification at mobile communication frequency, in which three examples will be given at powerline frequency and one example will be shown at mobile communication frequency. It should be noted that Examples 7.1 and 7.2 are drawn from paper [38] for proving the proposed FDTD method and compared with Mie-series solution codes that is working at powerline frequency for free space case. Then, the media of the Examples 7.1 and 7.2 are replaced with lossless case and the results are compared as shown in Example 7.3. This example gives stable results and shows the feasibility of using the quasi-static approach to biological application at mobile communication. Finally, the quasi-static concept is extended to model a biological equivalent spherical cell in lossy penetrable media at mobile communication frequency and it is clearly depicted in Example 7.4.

In FDTD computational domain, Examples 7.1 to 7.3 are modelled in a domain of 63×63×63 cm space size, with a grid resolution of 1 cm in each direction and δt of 16.67 ps. Example 7.4 uses 100×100×100 μm space size, with grid cell of 1 μm in each direction and δt of 1.3 fs. The model is excited by a plane wave of 1 V/m, which is propagating in z-direction and polarised in x-direction. Moreover, the domain was terminated by a PML (6, Gx, 40 dB) (six layers, with geometric progression profile and geometric grading factor of Gx of the media used, for which 40 dB attenuation for normal incidence is considered) on all sides. Moreover, the simulation time of order of four oscillation periods of the source field is found necessary to calculate the steady state conditions.

For the convenience of illustration of the problem, consider Figures 7.3 and 7.4, which show the 2-D and 3-D view of geometry stated in FDTD computational domain. This shows the equivalent surface (Huygens surface) that replaces the plane wave. It also shows the Berenger's perfectly matched layer (PML) that implements the absorbing boundary condition (ABC). This layer has been applied surrounding the edge of the problem space, with matching impedance condition ($\sigma/\varepsilon = \sigma^*/\mu$ for lossless or lossy media, whereas $\sigma/\varepsilon_o = \sigma^*/\mu_o$ for free space). In addition, in order to explain the results systematically, all 1-D and 2-D plots of the induced electric and magnetic fields components are standardised for ease of comparison for all test cases. For 1-D plots, it should be noted that only dominated E_x component versus y and z, it also includes E_z component versus x. Moreover, the dominated H_y vesus x and z for the central axes of the sphere are shown for discussion. For 2-D plots, they display logarithmic value of 2-D field strengths are demonstrated when eight grid cells are considered between the PML and the structure.

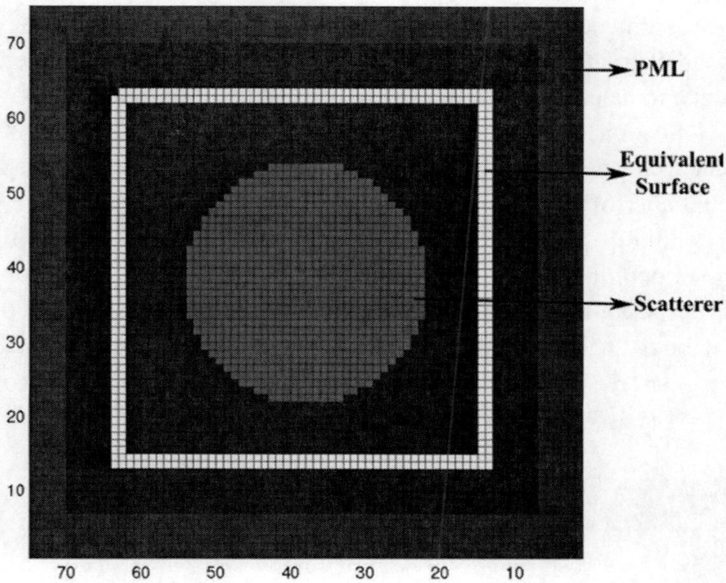

Figure 7.3. Basic structure of the problem for FDTD computational domain.

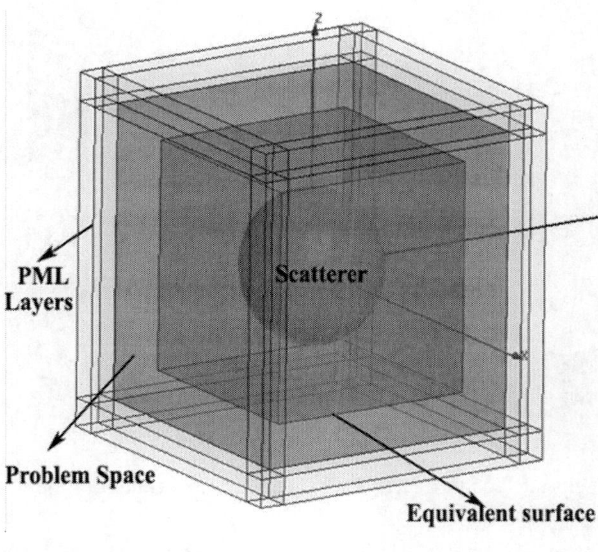

Figure 7.4. 3-D view of the problem for FDTD computational domain.

7.4.1. Verification at Low Frequency (Powerline Frequency)

This section demonstrates the feasibility of frequency scaled FDTD method to solve scattering problem at powerline frequency of 60 Hz. All the test cases in this section will initially transfer the working frequency to 20 MHz, in order to make the simulation time reasonable to be computed. Then, the obtained induced fields are scaled back to 60 Hz by using equation (4.3) in Chapter 4 Section 4.2. As can be observed, if the same problem is solved directly at frequency of 60 Hz and four oscillation-period criterion is used, then the number of the time step requires is $4\times(1/(\delta t\times 60))$. This is equivalent to about 4 billions of time steps at desired operating frequency of 60 Hz. By scaling the operating frequency to 20 MHz, the number of time steps can be reduced, i.e., from few billions to few thousands. In the other hand, due to $\sigma >> \omega\varepsilon$ at both 20 MHz and 60 Hz, the selection of the ε_r of the scatterer properties can be considered relatively insignificant for the calculations. Hence, $\varepsilon_r = 1$ is used for the scatterer.

Figure 7.5. Huygens surface in free space FDTD computational domain (logarithmic scale).

7.4.1.1. Free Space Penetrable Media

For sake of checking the correct implementation of equivalent surface (Huygens surface) that replaces the plane wave and the Berenger's PML absorbing boundary condition in FDTD computation domain, thus, the simulation is initially performed without the presence of the analysed object. Figure 7.5 shows the Huygens surface in free space of the FDTD computational domain. As can be seen, the fields inside the problem space is about 0 dB, while the numerical reflection is about -40 dB between the Huygens surface and PML region. This proves that the FDTD code is working perfectly with the Huygens surface and Berenger's PML absorbing boundary conditions.

Example 7.1: A single layer sphere of radius 16.5 cm with conductivity σ = 0.35 S/m and dielectric constant $\varepsilon_r = 1.0$. Properties of free space ($\sigma = 0$, $\varepsilon_r =$ 1.0) are assumed here in the perfectly matched layers (PML) and the problem space, where the PML (6, G5.4, 40 dB) (six layers, with geometric progression profile and geometric grading factor of 5.4, with 40 dB attenuation for normal incidence) is used on all sides. Figure 7.6 shows the intensity of E_x component in coordinate plane of the structure at 20 MHz, while Figure 7.7 depicts the electric field distribution along various central axes for a single-layer sphere in free space media at 60 Hz. The induced magnetic field inside the structure are also shown in Figure 7.8 and Figure 7.9 respectively. It can be observed from Figure 7.9 that the magnitude of magnetic field E_y component is about 1/377 A/m (which is equivalent to the incident magnetic field) constantly along x and z axis of the structure at both 20 MHz and 60 Hz, where 377 Ω is the characteristic impedance for free space medium. This result confirms that under quasi-static condition, the magnetic and electric field are decoupled. As can seen, the results of the proposed method agree well with the Mie series solution.

Example 7.2: A three-layer concentric sphere is modelled in free space here. The assumed properties are listed as follows: σ_1 = 0.52 S/m (corresponding to properties of skeletal muscle) for innermost sphere $0 < r < 8$ cm, $\sigma_2 = 0.2$ S/m (corresponding to the properties of fat) for the intermediate sphere $8 < r < 12$ cm, and $\sigma_3 = 0.04$ S/m (corresponding to the properties of bone) for the outermost sphere $12 < r < 16.5$ cm. Since $\sigma >> \omega\varepsilon_o$ for each layers at 20 MHz and 60 Hz, thus, the choice of ε_r is insignificant in calculation results. Due to the reason above, $\varepsilon_r = 1$ is assumed in all layers of the sphere. Figure 7.10 shows the E_x component in yz coordinate plane for three-layers sphere excited by plane wave of 1 V/m at 20 MHz. The calculated results of 60 Hz are well agreed with the analytical one as illustrated in Figure 7.11.

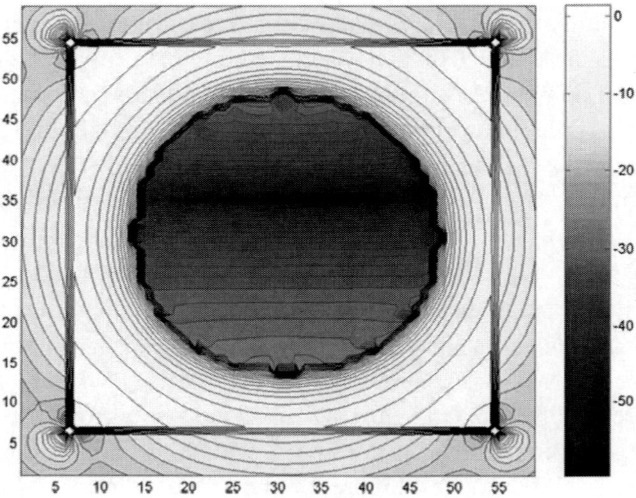

Figure 7.6. E_x component at *yz*-plane for single-layer sphere excited by plane wave of 1 V/m at 20 MHz (logarithmic scale).

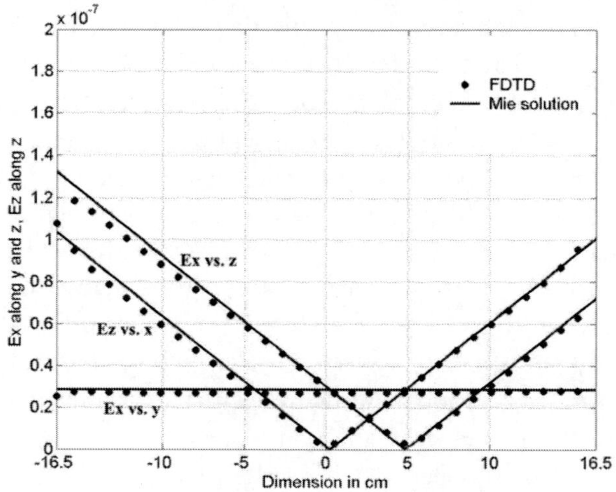

Figure 7.7. Electric fields distribution along various central axes for single-layer sphere excited by plane wave of 1 V/m at 60 Hz in free space.

Figure 7.8. H_y component at *xz*-plane for single-layer sphere excited by plane wave of 1 V/m at 20 MHz and 60 Hz (logarithmic scale).

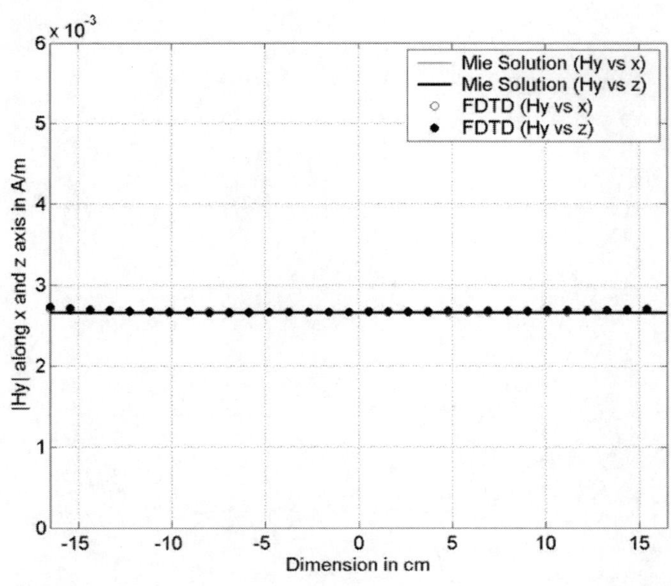

Figure 7.9. Magnetic field H_y component distribution along various central axes for single-layer sphere excited by plane wave of 1 V/m at 60 Hz and 20 MHz in free space.

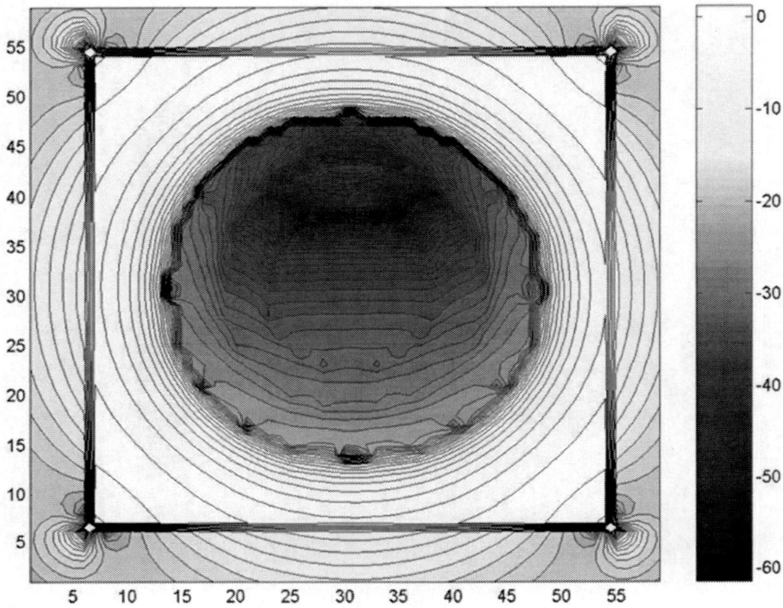

Figure 7.10. E_x component at *yz*-plane for three-layers sphere excited by plane wave of 1 V/m at 20 MHz in free space (logarithmic scale).

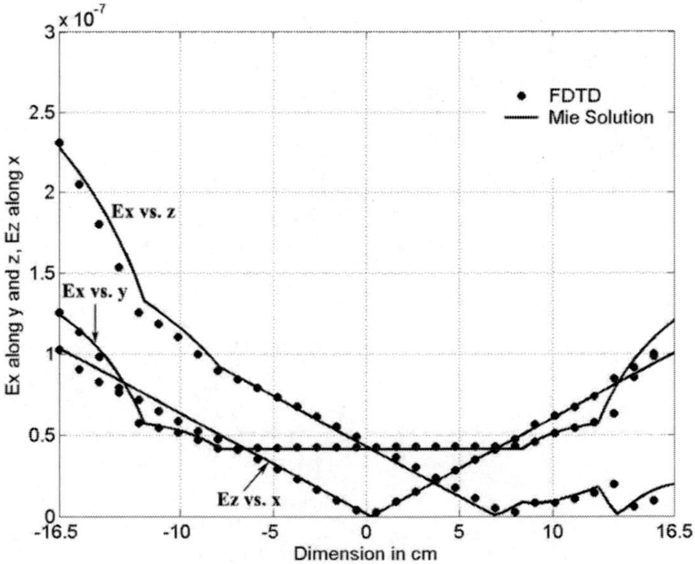

Figure 7.11. Electric fields distribution along various central axes for three-layer sphere excited by plane wave of 1 V/m at 60 Hz in free space.

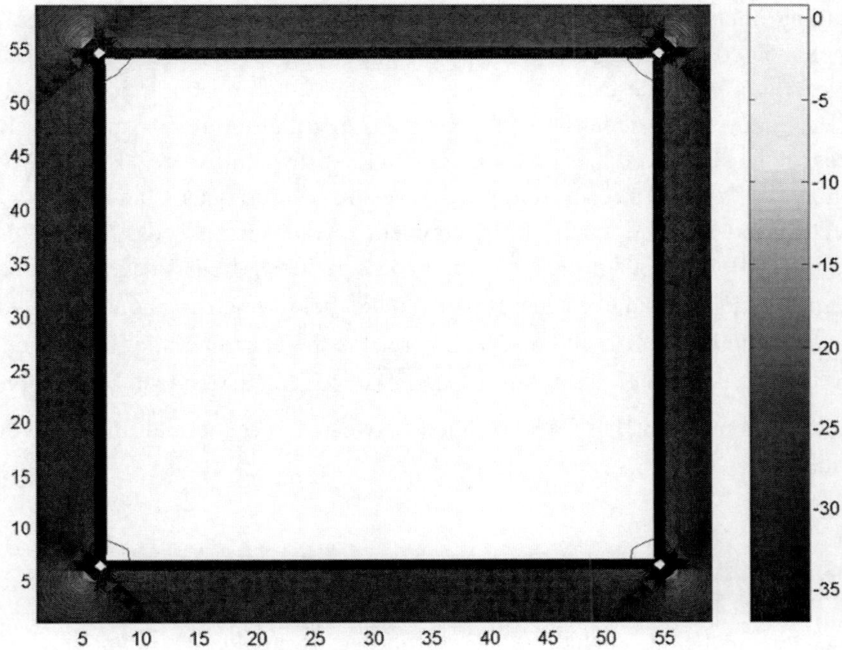

Figure 7.12. Huygens surface in lossless media FDTD computational domain (logarithmic scale).

7.4.1.2. Lossless Penetrable Media

By using equations (6.18) and (6.19), the Berenger's PML absorbing boundary conditions in FDTD computational domain is modified to permit the modelling of lossless media. To validate the correctness of PML absorbing boundary conditions, the simulation is first running without the presence of the scatterer. Figure 7.12 shows Huygens surface in lossless media of the FDTD computational domain. The results are well agreed with the predicted results, in which the fields inside the problem space is about 0 dB and numerical reflection is found about -40 dB between the Huygens surface and PML region.

Example 7.3: A homogenous single layer sphere with the same conductivity and permittivity of previous example, are being used in this example, except the properties of the surrounding media and perfectly matched layers (PML) are assumed to be ($\sigma = 13\times10^{-9} \approx 0$, $\varepsilon_r = 50$) in order to consider the sphere in a lossless media. It should be noted that the PML (6, G9.5, 40 dB) (six layers, with geometric progression profile and geometric grading factor of 9.5, with 40 dB attenuation for normal incidence) is implemented on all sides, in order to reduce the reflection of the interface layer. Figure 7.13 describes the intensity of the E_x component in *yz*-plane at 20 MHz. Figure 7.14 depicts the electric field distribution along various central axes in lossless media. The magnetic field distributions are also shown in Figure 7.15 and 7.16 respectively. It proves that within quasi-static regime, the electric and magnetic field are decoupled, in the other words, the scattered magnetic field is insignificant compared to incident field since the induced magnetic field is equal to incident magnetic field which is about $1/\eta \cong 18.8$ mA/m at both 20 MHz and 60 Hz, where η is the characteristic impedance of lossless media which is about $\sqrt{\mu_o/\varepsilon_o\varepsilon_r} \cong 53.27\ \Omega$. As can be observed, excellent agreement has been reached between the numerical and analytical result.

7.4.2. Verification at Mobile Communication Frequency

The frequency scaled FDTD technique is applied to biological cell modelling at mobile communication frequency of 2450 MHz. Since the cell size of this problem is considered about micrometer, therefore $\delta t \leq \Delta s/c\sqrt{3} \cong 10^{-15}$ s where Δs is the cell size and c is the speed of light in lossy penetrable media. Since four oscillations period of the incident wave are required for the completion of the simulation to reach the steady state. This would involve some millions of iteration, hence, apply the operating frequency to 2450 MHz to the problem will increase the number of iterations required. Consequently, frequency of 30 GHz is selected as transferable working frequency at initial state, to support the simulation time at affordable level to be computed. Then, the obtained induced fields are scaled back to 2450 MHz.

7.4.2.1. Lossy Penetrable Media

The stability and the correctness of the implementation of modified lossy Berenger's PML absorbing boundary conditions are initially performed, by running the simulation without the presence of the analysed scatterer. Figure 7.17 shows Huygens surface in lossy FDTD computation. According to Figure 7.17, the electric field inside the problem space is 0 dB, while numerical reflection between the Huygens surface and PML region is nearly -40 dB. As can be seen, the results are well agreed with the predicted one.

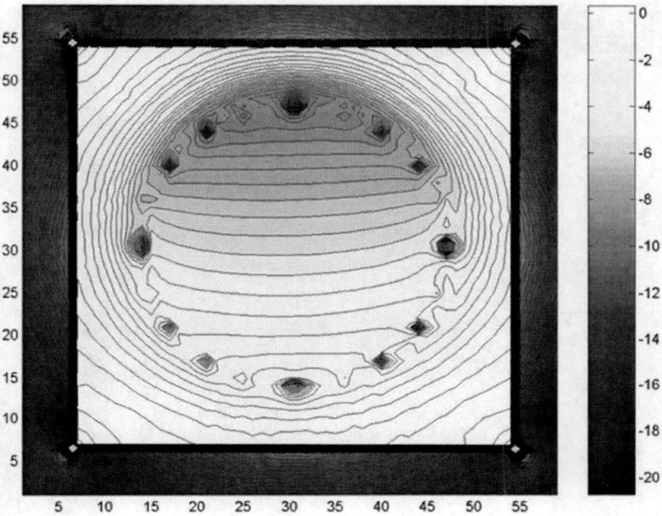

Figure 7.13. E_x component at yz-plane for single-layer sphere excited by plane wave of 1 V/m at 20 MHz in lossless media (logarithmic scale).

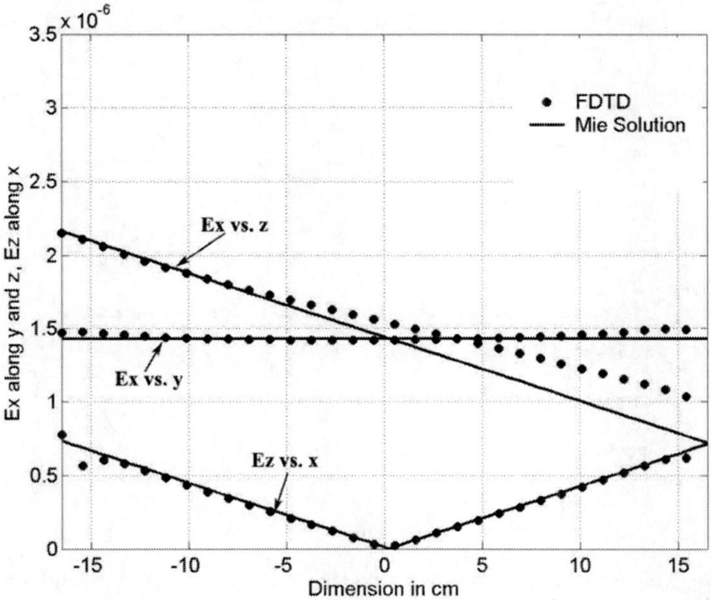

Figure 7.14. Electric fields distribution along various central axes for single-layer sphere excited by plane wave of 1 V/m at 60 Hz in lossless medium (logarithmic scale).

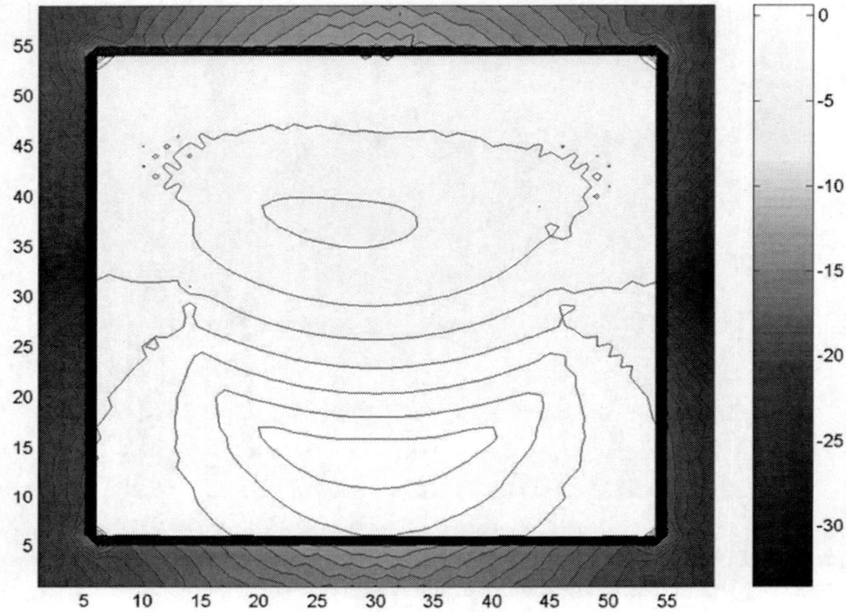

Figure 7.15. H_y component at *xz*-plane for single-layer sphere excited by plane wave of 1 V/m at 60 Hz and 20 MHz in lossless media (logarithmic scale).

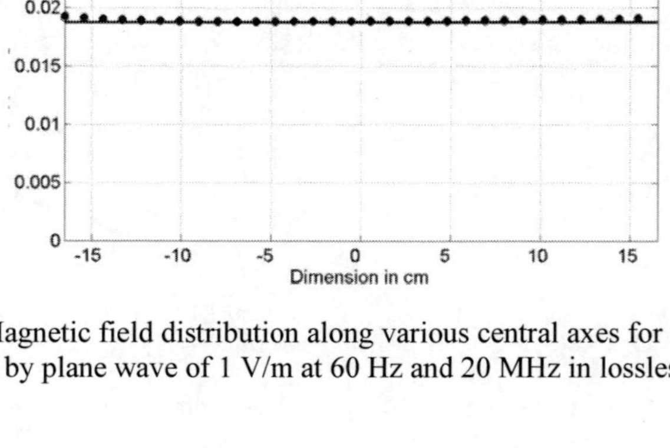

Figure 7.16. Magnetic field distribution along various central axes for single-layer sphere excited by plane wave of 1 V/m at 60 Hz and 20 MHz in lossless medium.

Example 7.4: A two-layer sphere simulating a biological cell inside a lossy medium, for which the assumed properties are as follows: cytoplasm (internal) $\varepsilon_r = 48.699$, $\sigma = 1.412$; membrane $\varepsilon_r = 11.3$, $\sigma = 0.0$; and lossy medium (external) $\varepsilon_r = 70.87$, $\sigma = 2.781$. The internal radius (internal region) is 25 μm and the membrane thickness is set to 2 μm. The operating frequency is 2.45 GHz whereas the interim transformed frequency used in this example is 30 GHz. From equation (6.19), the optimum grading factor g is calculated as 6.07 for which the FDTD cell size is 1 μm. The electric field distributions inside the simulated biological equivalent spherical cell are shown in Figures 7.18-7.20, while magnetic field distributions are depicted in Figures 7.21 and 7.22 at the operating frequency of 2450 MHz. From the inspection of Figure 7.22, the induced magnetic field is constant over the sphere at both 2.45 GHz and 30 GHz. The value of these fields, is about $1/\eta_{lossy}$ A/m, in which the η_{lossy} is the characteristic impedance of the lossy media. The value of η_{lossy} is computed and approximately is found to be 44.77 Ω. In the light of all the obtained simulated results, as can be noticed that the numerical and analytical results are indistinguishable.

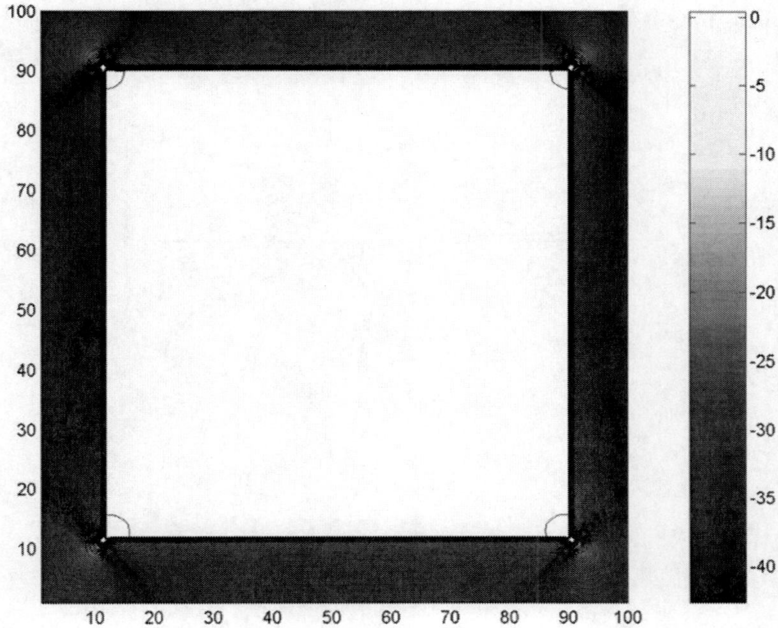

Figure 7.17. Huygens surface in lossy medium FDTD computational domain (logarithmic scale).

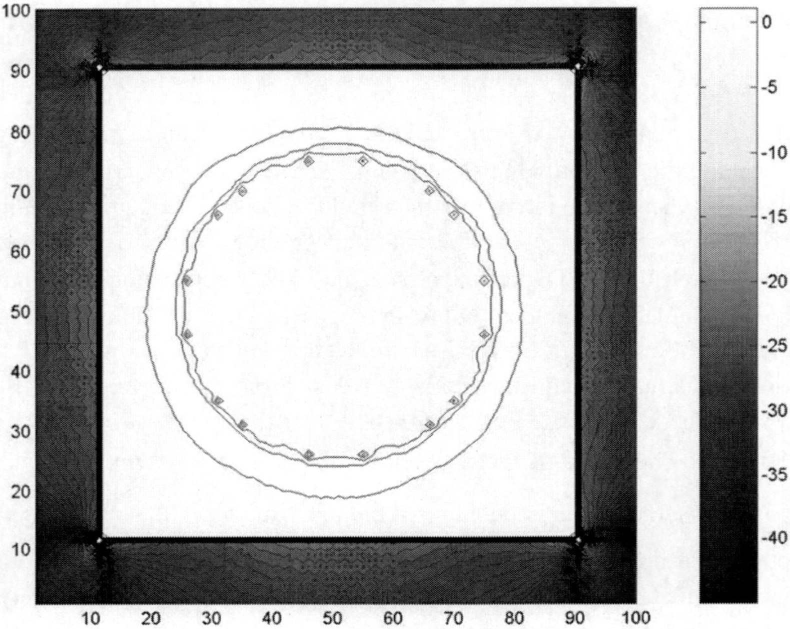

Figure 7.18. E_x component at *yz*-plane for double-layers sphere excited by plane wave of 1 V/m at 30 GHz in lossy medium (logarithmic scale).

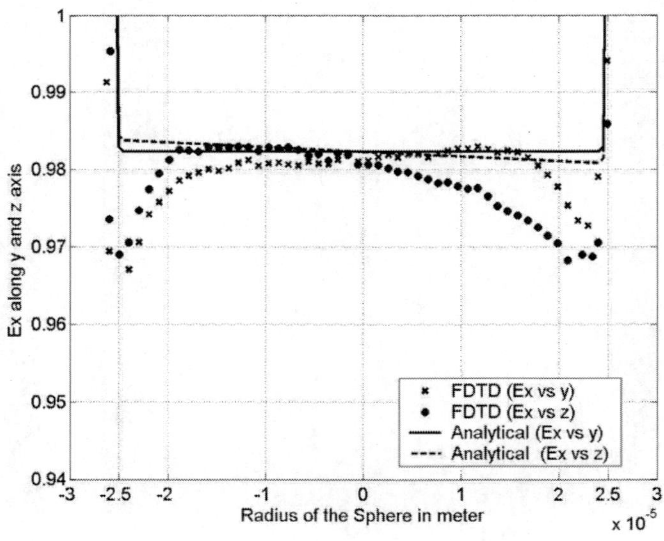

Figure 7.19. Electric field (E_x) distribution along various central axes for double-layers sphere excited by plane wave of 1 V/m at 2450 MHz in lossy medium.

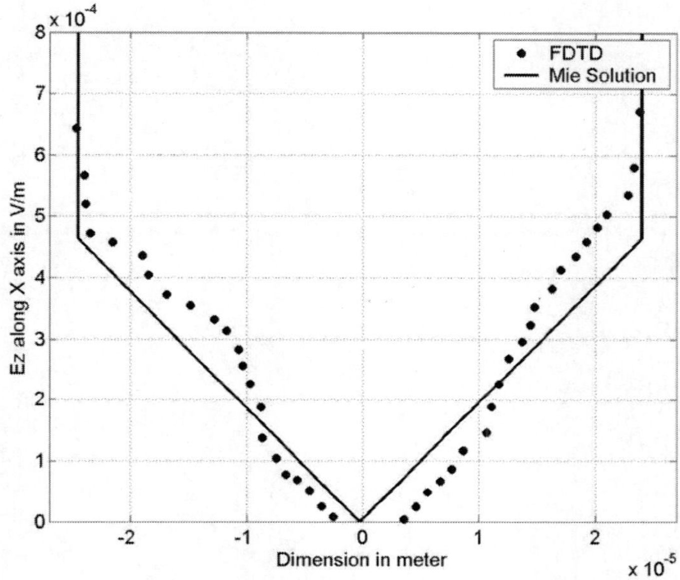

Figure 7.20. Electric field (E_z) distribution along x-axes for double-layer sphere excited by plane wave of 1 V/m at 2450 MHz in lossy medium.

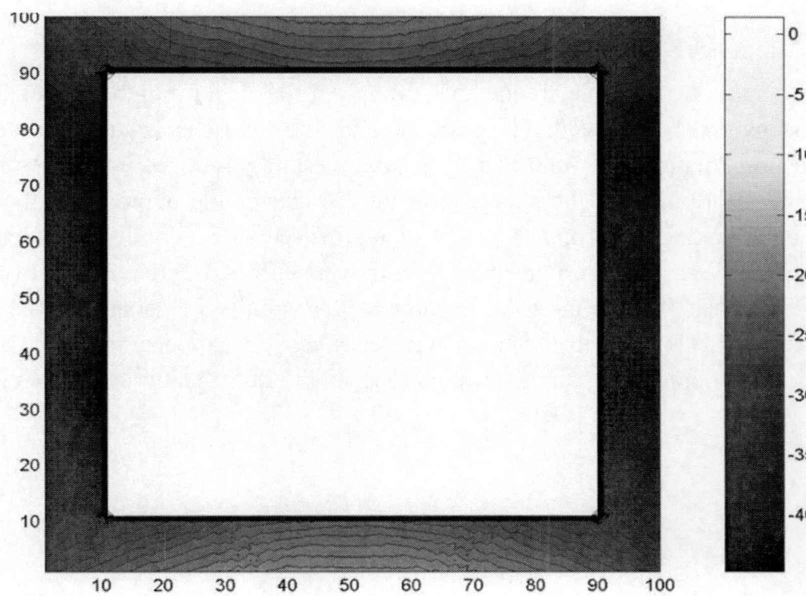

Figure 7.21. H_y component at xz-plane for double-layers sphere excited by plane wave of 1 V/m at 2450 MHz and 30 GHz in lossy medium (logarithmic scale).

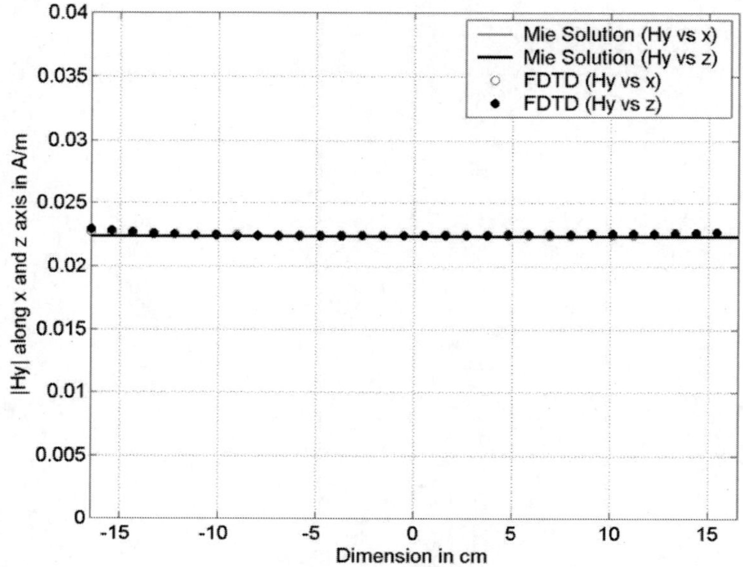

Figure 7.22. Magnetic field distribution along various central axes for double-layer sphere excited by plane wave of 1 V/m at 2450 MHz and 30 GHz in lossy medium.

CONCLUSION

By implementing the frequency scaling approach, the number of FDTD time steps can be reduced. The reflection on the interface layers inside the FDTD computation domain, has also been successfully reduced in lossless and lossy penetrate media. The accuracy of the FDTD scaling approach with the models of homogenous and layered spheres in free space, lossless and lossy media, was verified. The numerical results were in good agreement with the analytical ones. This lending its support to the validity of using the scaled-frequency FDTD method to obtain induced electric fields at power-line and mobile communication frequencies, as long as the quasi-static conditions are valid.

REFERENCES

[1] Zang SR; Bergmann, JR. "Analysis of omnidirectional dual-reflector antenna and feeding horn using method of moments," *IEEE*

Transactions on Antennas and Propagation, vol. 62, no. 3, 1534-1538, 2014.

[2] Pan, X–M; Sheng, X-Q. "Improved algebraic preconditioning for MoM solutions of large-scale electromagnetic problems," *IEEE Antennas and Wireless Propagation Letters*, vol. 13, 106-109, 2014.

[3] Echeverri Bautista, MA; Francavilla, MA; Vipiana, F; Vecchi, G. "A hierarchical fast solver for EFIE-MoM analysis of multiscale structures at very low frequencies," *IEEE Transactions on Antennas and Propagation*, vol. 62, no. 3, 1523-1528, 2014.

[4] Amir, M; Bedra, S; Benkouda, S; Fortaki, T. "Bacterial foraging optimisation and method of moments for modelling and optimisation of microstrip antennas," *IET Microwaves, Antennas and Propagation*, vol. 8, no. 4, 295-300, 2014.

[5] Shafieipour, M; Jeffrey, I; Aronsson, J; Okhmatovski, VI. "On the equivalence of RWG method of moments and the locally corrected nyström method for solving the electric field integral equation," *IEEE Transactions on Antennas and Propagation*, vol. 62, no. 2, 772-782, 2014.

[6] Schroder, A; Rasek, GA; Bruns, H-D; Reznicek, Z; Kucera, J; Loos, SE; Schuster, C. "Analysis of high intensity radiated field coupling into aircraft using the method of moments," *IEEE Transactions on Electromagnetic Compatibility*, vol. 56, no. 1, 113-122, 2014.

[7] Hendijani, N; Cheng, J; Adams, RJ. "Combined field integral equation using a constraint-based Helmholtz decomposition," *IEEE Transactions on Antennas and Propagation*, vol. 62, no. 3, 1500-1503, 2014.

[8] Alvarez, J; Angulo, LD; Cabello, MR; Rubio Bretones, A; Garcia, SG. "An analysis of the leap-frog discontinuous Galerkin method for Maxwell's equations," *IEEE Transactions on Microwave Theory and Techniques*, vol. 62, no. 2, 197-207, 2014.

[9] Vorobev, YU. Method of moments in applied mathematics (translated from Russian by Seckler): New York: Gordon & Breach, 1965.

[10] Harrington, RF. Field computation by moment methods: The Macmillan Co., New York, 1968.

[11] Strait, BJ. "Applications of the method of moments to electromagnetics," St. Cloud, FL: SCEEE Press, 1980.

[12] Ney, MM. "Method of moments as applied to electromagnetic problems," *IEEE Transactions on Microwave Theory and Techniques*, vol. 33, 972-980, 1985.

[13] Tuncer, O; Shanker, B; Kempel, LC. "Discontinuous Galerkin inspired framework for vector generalized finite-element methods," *IEEE Transactions on Antennas and Propagation*, vol. 62, no. 3, 1339-1347, 2014.

[14] Wagner, N; Schwing, M; Scheuermann, A. "Numerical 3-D FEM and experimental analysis of the open-ended coaxial line technique for microwave dielectric spectroscopy on soil," *IEEE Transactions on Geoscience and Remote Sensing*, vol. 52, no. 2, 880-893, 2014.

[15] Amparo Callejon, M; Reina-Tosina, J; Naranjo-Hernandez, D; Roa, LM. "Galvanic coupling transmission in intrabody communication: A finite element approach," *IEEE Transactions on Biomedical Engineering*, vol. 61, no. 3, 775-783, 2014.

[16] Lindberg, J; Aldegunde, M; Nagy, D; Dettmer, WG; Kalna, K; Garcia-Loureiro, A. J. Peric, D. "Quantum corrections based on the 2-D Schrödinger equation for 3-D finite element Monte Carlo simulations of nanoscaled FinFETs," *IEEE Transactions on Electron Devices*, vol. 61, no. 2, 423-429, 2014.

[17] Thomas, P; Le Menach, Y. "An arbitrary thick shell finite element for eddy-current dual vector-scalar potential formulations," *IEEE Transactions on Magnetics*, vol. 49, no. 5, 1725-1728, 2013.

[18] Bender, VC; Iaronka, O; Marchesan, TB. "Study on the thermal performance of LED luminaire using finite element method," *39th Annual Conference of the IEEE Industrial Electronics Society (IECON)*, 6099-6104, 2013.

[19] Dean, J; Kohn, A; Kovacs, A; Allwood DA; Suess, D; Schrefl, T. "Influence of thermal energy on exchange-bias studied by finite-element simulations," *Applied Physics Letters*, vol. 103, no. 4, 042410 - 042410-5, 2013.

[20] Webb, JP. "Developments in a finite element method for three-dimensional electromagnetic problems," PhD Dissertation, University of Cambridge, 1981.

[21] Maile, GL. "Three-dimensional analysis of electromagnetic problems by finite element methods," PhD Dissertation, University of Cambridge, 1979.

[22] Jin, J. The finite element method in electromagnetics, 2nd ed., New York: John Wiley & Sons Inc., 2002.

[23] Sadiku, MNO. "A simple introduction to finite element analysis of electromagnetic problems," *IEEE Transactions on Education*, vol. 32, 85-93, 1989.

[24] Xiong, R; Chen, B; Yi, Y; Chen, H-L; Zhao, H-F. "Two-step method for the FDTD analysis of long apertures having depth," *IEEE Antennas and Wireless Propagation Letters*, vol. 12, 39-42, 2013.

[25] Kamepally, SK; Kumar, BP; Paidimarry, CS. "FDTD estimation for accurate specific absorption rate in a tumor," *Annual International Conference on Emerging Research Areas and International Conference on Microelectronics, Communications and Renewable Energy (AICERA/ICMiCR)*, 1-5, 2013.

[26] Whittow, WG; Njoku, CC; Vardaxoglou, YC. "Fine scale simulations of patch antennas with heterogeneous substrates," *USNC-URSI Radio Science Meeting (Joint with AP-S Symposium)*, 223, 2013.

[27] Lesina, AC; Vaccari, A; Bozzoli, A. "A modified RC-FDTD algorithm for plasmonics in Drude dispersive media nanostructures," *6th European Conference on Antennas and Propagation (EUCAP)*, 687-690, 2012.

[28] Noetscher, GM; Xu, Y; Makarov, SN. "Accuracy of point source models with coincident phase centers in a cubic FDTD grid for arbitrary source orientation," *IEEE Antennas and Propagation Society International Symposium (APSURSI)*, 1-2, 2012.

[29] Nayyeri, V; Soleimani, M; Dehmollaian, M. "Implementation of a PEMC boundary condition in the 2-D FDTD technique," *IEEE Antennas and Propagation Society International Symposium (APSURSI)*, 1-2, 2012.

[30] Liu, D-C; Chang, H-C. "The second-order condition of FDTD method at sloped dielectric interfaces by averaging the permittivities along surface normal," *IEEE Transactions on Antennas and Propagation*, vol. 60, no. 11, 5259-5267, 2012.

[31] Weiwei, M; Dong, S; Xianliang, W. "UPML-FDTD parallel computing on GPU," *International Conference on Microwave and Millimeter Wave Technology (ICMMT)*, vol. 4, 1-4, 2012.

[32] Guiffaut, C; Reineix, A; Pecqueux, B. "New oblique thin wire formalism in the FDTD method with multiwire junctions," *IEEE Transactions on Antennas and Propagation*, vol. 60, no. 3, 1458-1466, 2012.

[33] Shyroki, DM. "Modeling of sloped interfaces on a Yee grid," *IEEE Transactions on Antennas and Propagation*, vol. 59, no. 9, 3290-3295, 2011.

[34] Kunz, KS; Lee, KM. "A three-dimensional finite-difference solution of the external response of an aircraft to a complex transient EM environment I: The method, and its implementation," *IEEE Transaction on Electromagnetic Compatibility*, vol. 20, 328-333, 1978.

[35] Taflove, A. Advances in computational electrodynamics: The finite-difference time-domain method, Artech House, 627-632, 1998.

[36] Taflove, A; Hagness, SC. Computational electromagnetics: The finite-different time-domain method, 3rd ed: Artech House, 2005.

[37] Yee, KS. "Numerical solution of initial boundary value problems involving Maxwell's equations," *IEEE Transaction on Antennas and Propagation*, vol. 3, 302-307, 1966.

[38] Gandhi, OP; Chen, J. "Numerical dosimetry at powerline frequencies using anatomically based models," *Bioelectromagnetics*, vol. 1, 43-60, 1992.

[39] Moerloose, JD; Dawson, TW; Stuchly, MA. "Application of the finite difference time domain algorithm to quasi-static field analysis," *Radio Science*, vol. 32, 329-341, 1997.

[40] Potter, ME; Okoniewski, M; Stuchly, MA. "Low frequency finite difference time domain (FDTD) for modelling of induced fields in humans close to line sources," *Journal of Computational Physics*, vol. 162, 82-103, 2000.

[41] Emili, G; Schiavoni, A; Roselli, FL; Sorrentino, R. "Computation of electromagnetic field inside a tissue at mobile communications frequencies," *IEEE Transaction on Microwave Theory and Techniques*, vol. 51, 178-186, 2003.

[42] Berenger, J. "A perfectly matched layer for absorption of electromagnetic waves," *Journal of Computational Physics*, vol. 114, 185-200, 1994.

[43] Katz, DS; Thiele, ET; Taflove, A. "Validation, and extension to three dimensions of the Berenger PML absorbing boundary condition for FDTD meshes," *IEEE Microwave and Guided Wave Letters*, vol. 4, 268-270, 1994.

[44] Berenger, J. "Perfectly matched layer for the FDTD solution of wave-structure interaction problems," *IEEE Transaction on Antennas and Propagation*, vol. 44, 110-117, 1996.

[45] Berenger, J. "Improved PML for the FDTD solution of wave-structure interaction problem," *IEEE Transaction on Antennas and Propagation*, vol. 45, 466-473, 1997.

[46] Mie, G. "Contributions to the optics of diffusing media," *Ann. Physic*, vol. 25, 377, 1908.

[47] Stratton, J. Electromagnetic theory, McGraw-Hill, New York, 563-573, 1941.

[48] Aden, AL; Kerker, M. "Scattering of electromagnetic waves from two concentric spheres," *Journal of Applied Physics*, vol. 22, 1242-1246, 1951.

[49] Wyatt, P. "Scattering of electromagnetic plane waves from inhomogeneous spherically symmetric objects," *Phys. Rev.*, vol. 5, 1837-1842, 1962.

[50] Anne, A; Satio, M; Salati, OM; Schwan, HP. "Relative microwave absorption cross section of biological significance," *Biological effect of microwave radiation*, vol. 1, 153-176, 1960.

[51] Shapiro, AR; Lutomirski, RF; Yura, HT. "Induced fields, and heating within a cranial structure irradiated by an electromagnetic plane wave," *IEEE Transactions on Microwave Theory and Techniques*, vol. 19, 187-196, 1971.

[52] Kritikos, HN; Schwan, HP. "Hot spots generated in conducting spheres by electromagnetic waves, and biological implications," *IEEE Transactions on Biomedical Engineering*, vol. 19, 53-58, 1972.

[53] Joines, WT; Spiegel, RJ. "Resonance absorption of microwaves by the human skull," *IEEE Transactions on Biomedical Engineering*, vol. 21, 46-48, 1974.

[54] Weil, CM. "Absorption characteristics of multi-layered sphere models exposed to UHF/microwave radiation," *IEEE Transactions on Biomedical Engineering*, vol. 22, 468-476, 1975.

[55] Kritikos, HN; Schwan, HP. "Formation of hot spots in multi-layer spheres," *IEEE Transactions on Biomedical Engineering*, vol. 23, 168-172, 1976.

[56] Hizal, A; Baykal, YK. "Heat potential distribution in an inhomogeneous spherical model of a cranial structure exposed to microwaves due to loop or dipole antennas," *IEEE Transactions on Microwave Theory and Techniques*, vol. 26, 607-612, 1978.

[57] Li, J; Lin, W. "Inverse scattering for dielectric spheres," *Antennas and Propagation Society International Symposium, Merging Technologies for the 90's Digest*, vol. 2, 1048-1051, 1990.

[58] Liu, LM; Cleary, SF. "Absorbed energy distribution from radiofrequency electromagnetic radiation in a mammalian cell model: effect of membrane-bound water," *Bioelectromagnetics*, vol. 16, 160-171, 1995.

[59] Smith, DD; Fuller, KA. "Mie scattering by concentric multilayers," *Quantum Electronics and Laser Science Conference*, 119-120, 2002.

[60] Apollonio, F; Liberti, M; Inzeo, GD; Tarricone, L. "Integrated models for analysis of biological effects of EM fields used for mobile communications," *IEEE Transactions on Microwave Theory and Techniques*, vol. 48, 2082-2093, 2000.

[61] Sullivan, DM; Borup, DT; Gandhi, OP. "Use of the finite-difference time-domain method in calculating EM absorption in human tissue," *IEEE Transactions on Biomedical Engineering*, vol. 34, 148-157, 1987.

[62] Yuan, X. "Three-dimensional electromagnetic scattering from inhomogeneous objects by the hybrid moment, and finite element method," *IEEE Transactions on Microwave Theory and Techniques*, vol. 38, 1053 - 1058, 1990.

[63] El-Babli, I; Sebak, A; Simons, N. "Application of the TLM method to the interaction of EM fields with dispersive dielectric bodies," *Microwaves, Antennas and Propagation, IEE Proceedings*, vol. 147, 211-217, 2000.

[64] Potter, ME; Okaniewski, M; Stuchly, MA. "Low frequency finite difference time domain (FDTD) for modeling of induced fields in humans close to line sources," *Journal of Computational Physics*, vol. 162, 82-103, 2000.

[65] Schechter, RS; Kragalott, M; Kluskens, MS; Pala, WP. "Splitting of material cells, and averaging properties to improve accurary of the FDTD method at interfaces," *ACES Journal*, vol. 17, 198-208, 2002.

[66] Abramowitz, M; Stegun, IA. "Handbook of mathematical functions," NBS Math Series, Dover Publications, New York., 1970.

[67] Ruck, GT; Barrick, DE; Stuart, WD; Krichbaum, CK. Radar cross section handbook, vol. 1., Plenum Press, New York, 1970.

[68] Mur, G. "Absorbing boundary conditions for the finite-difference approximation of the time-domain electromagnetic-field equations," *IEEE Transaction on Electromagnetic Compatibility*, vol. EMC-23, 377-382, 1981.

[69] Liao, ZP; Wong, HL; Yang, BP; Yuan, YF. "A transmitting boundary for transient wave analysis," *Scientia Sinica (series A)*, 1063-1076, 1984.

In: Development of Complex Electromagnetic … ISBN: 978-1-61122-013-1
Editors: K. N. Ramli et al.

Chapter 8

COMPUTATION OF ELECTROMAGNETIC FIELD INSIDE A TISSUE USING QUASI-STATIC AND LUMPED-ELEMENT FDTD SCHEME

***C. H. See*[1], *R. A. Abd-Alhameed*[2], *M. J. Ngala*[3] *and P. S. Excell*[3]**

[1]Engineering, Sport and Sciences Academic Group,
University of Bolton, Bolton, Lancashire, UK
[2]Mobile and Satellite Communications Research Centre,
Bradford University, Bradford, UK
[3]Heaton Education, Glydegate, Bradford, UK
[4]Institute for Arts, Science and Technology,
Glyndwr University, Wrexham, UK

An approach to model the biological cells incorporating a generic lumped-element membrane model is presented and analysed. Linear assemblages of cells are investigated and then Floquet periodic boundary conditions are imposed to imitate the effect of periodic replication of the assemblages. The analysis of a large structure of cells is made more computationally efficient than the modeling of the entire structure. The total fields of the simulated structures are shown to give reasonable and stable results at 900, 1800, and 2450 MHz. This method will facilitate deeper investigation of the phenomena in the interaction between electromagnetic fields and biological systems.

8.1. INTRODUCTION

Research into possible mechanisms of interaction of electromagnetic (EM) fields with biological tissues and cells in culture has motivated a growing need for accurate models describing the EM behaviour of cells exposed to these fields. Therefore, several numerical models have been set up in order to study the interaction between EM fields and biological entities, at tissue level, cell level and ionic level. In this area, the most frequently used technique for computing the EM field is the finite-difference time-domain (FDTD) method [1-12], due to its independence from the material parameters.

The standard FDTD method requires extremely small time-step when modelling electrically-small regions (much smaller than a wavelength): this is especially the case when modelling biological cells, since they have maximum dimensions of a few tens of micrometers. Thus, it can become impractical due to the unaffordable computation times required. This problem can be solved by implementing a quasi-static approximate version of FDTD. This approach is based on transferring the working frequency to a higher frequency, to reduce the number of time steps required. Then, the generated internal field at the higher frequency can be scaled back to the frequency of interest [13-16].

Cells are surrounded by thin membranes, typically about a few nanometres thick [17]. They are the major barrier in the cell, separating the inside of the cell from the exterior medium. This structure will allow cells to selectively interact with their environment. Therefore, the cell membrane has been identified as the primary target of the action of the EM field on biological structures. Since the thickness of the membrane is about 1000 times smaller than the biological cell, and if the standard FDTD procedure were to be blindly applied to model detail in the membrane within a complete cell model, then this will take some millions of iterations to complete one cycle of simulation. This again will cause excessive computation time. To overcome this problem in standard FDTD, the lumped element finite-different time-domain (LE-FDTD) method [18-22] was implemented in such a way to model the behaviour of the membrane, based on the Hodgkin-Huxley (HH) model [23-27] on the surface of the biological cell.

This chapter presents a new approach to model and analyse of the HH membrane model [23-27]. The HH model is represented as an electrical circuit on the surface of the biological equivalent spherical cells. For the sake of simplicity, the analysed structure has been represented with spherical cells and then Floquet periodic boundary conditions [28-31] have been applied to the border of the analysed structure in order to mimic the presence of the surrounding cells. Despite the cellular tissues are not perfectly periodic and the living cells are not precisely spheres, this approximation allows the modelling of biological tissue with only a small part of the structure and alleviates the problem of huge requirement of computer resources for the simulation of a complete tissue. Since the external medium of the biological tissue is lossy fluid, the modified Berenger's perfectly matched layer (PML) absorbing boundary condition (ABC) is used to truncate the computation grid [32-35], in order to reduce the reflections on the interface layers. It should be noted that Berenger's PML ABC is more accurate than the Mur ABC [36], used in other recent work [16].

A further difficulty is the limited extent of studies on the dielectric properties of cell tissues [37]; thus, the complex permittivity of each cell tissue is not clearly established for radio frequencies. However, in this study, an analytical method for estimating the electrical properties of cell tissues in the RF band [38] will be adopted throughout the analysis. Earlier work only considered two media (water and membrane) [16], but the procedure adopted here enables the tissue model to consist of three media (lossy medium, membrane and cytoplasm). In addition, a mass of connected biological tissue is simulated by creating an equivalent stack of compact cell structures (such as spherical and cylindrical with interstices, and cubical). The total electric fields along the central axes of rows of these spherical, cubical and cylindrical cell structures will be investigated.

8.2. Floquet Periodic Boundary Condition (PBC)

Many structures of electromagnetic interest are extremely large and periodic in one or more dimensions. In order to perform the EM analysis of these types of structure with reasonable computational time, the structures are assumed to be an infinite grid and subsequently reducing the problem into a unit cell analysis by using Floquet boundary condition. This will enable the numerical solution to simulate the effect of the periodic replication. In this area, photonic bandgap (PBG) structure, frequency selective surface (FSS) and antenna array are the typical candidates for EM analysis that use the same boundary condition. These structures may contain complex arbitrary shape of scatterers comprised of dielectrics and conductors, thus, numerical methods that can easily handle complex and inhomogeneous geometries are indispensable in this scope of research.

For the purpose of numerical modelling, the FDTD technique appears to be an attractive choice for this problem due to its simplicity and comprehensibility. In fact, the Floquet theorem has been successfully implemented inside the FDTD to analyse the case of normal [28, 39] and oblique incidence [40, 41] for two- and three-dimensional problems. The techniques used to perform FDTD with the Floquet periodic boundary condition can be classified into two categories, i.e., direct field methods and field transformation methods. Direct field methods comprise of normal incidence, sine-cosine, multiple unit cell and angled update method, whereas the field transformation methods consist of multi-spatial grid and split-field method [12]. In order to understand basic theory and implementation of Floquet periodic boundary into FDTD, the case of normal incidence method for 2-D and 3-D problems is demonstrated in the following context.

8.2.1. Three-Dimensional Case

Figure 8.1 shows a simple geometry for the elucidation of three-dimensional periodic boundary implementation. As can be observed, the periodic boundaries are imposed on the x- and y-sides of the structures, while ABCs are applied to truncate the space lattice along z-axis. The coordinate point (i_0, j_0, k_0) and (i_N, j_M, k_p) denote a space point in a uniform rectangular lattice, where i_0, j_0, k_0 are the smallest lattice grid number in x, y, z direction respectively and i_N, j_M, k_p are the largest lattice gird number in x, y, z direction respectively.

Consider the problem space is filled with the lossless medium ($\sigma = 0$, $\varepsilon = \varepsilon_0\varepsilon_r$ and $\mu = \mu_0$) and the normal incident plane wave is propagating along z-axis. The tangential electric fields distribution on plane i_0, i_N, j_0 and j_M are illustrated in Figure 8.2. As can be seen, the red arrows are representing the tangential electric field components which are located on edge of the surface plane, while the black arrows are indicated as the least of the tangential electric field components which are located on the surface plane. It should be noted that the explanation of the implementation method of the periodic boundary condition into FDTD computational domain in the following context, are based on the normal incidence methods [12], therefore, it is only applicable when the normal incidence plane wave is used. For the sake of simplicity and consistency of explanation of updating equations for the periodic boundary condition on the surfaces of the geometry shown in Figure 8.1, the updating equations of the tangential electric components (E_x, E_z and

E_y) which are not on the edge of the surface will be firstly to be discussed, and subsequently, the updating equations of the edged tangential electric component (E_z) are demonstrated. Consider the four surfaces planes at $i = i_0$, i_N and $j = j_0$, j_M, in which the Floquet periodic boundary condition should be applied as shown in Figure 8.1.

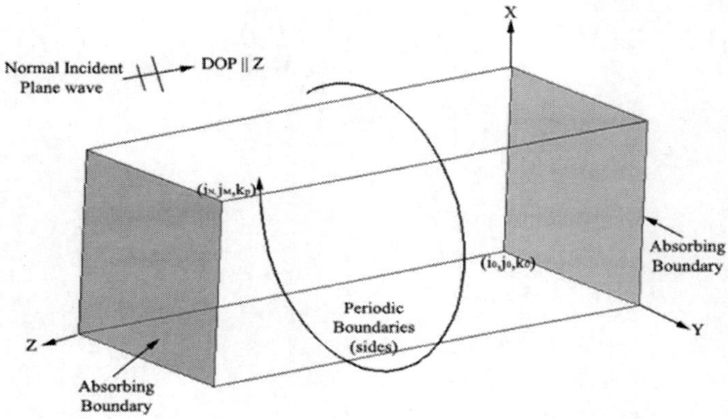

Figure 8.1. Geometry used in the analysis of 3-D infinite periodic structure illuminated by a normal incident plane wave.

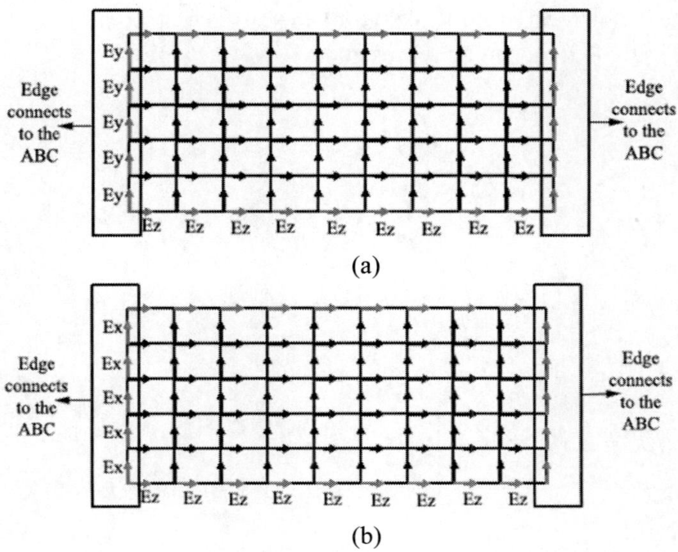

Figure 8.2. (a) Location of $E_z(\rightarrow)$ and $E_y(\uparrow)$ components in plane $i = i_0$ and i_N, (b) Location of $E_z(\rightarrow)$ and $E_x(\uparrow)$ components in plane $j = j_0$ and j_M.

From Figure 8.2(a), the 3-D updating equations for the tangential electric components (E_y and E_z) which are not located on the edge of surface plane $i = i_0$ and i_N, can be written as:

$$E_y^{n+1}(i,j,k) = E_y^n(i,j,k) + \frac{\Delta t}{\varepsilon\Delta}\begin{pmatrix} H_x^{n+\frac{1}{2}}(i,j,k+1/2) - H_x^{n+\frac{1}{2}}(i,j,k_p - 1/2) \\ + H_z^{n+\frac{1}{2}}(i-1/2,j,k) - H_z^{n+\frac{1}{2}}(i+1/2,j,k) \end{pmatrix} \quad (7.1a)$$

$$E_y^{n+1}(i_N, j, k) = E_y^{n+1}(i,j,k) \quad (8.1b)$$

$$E_z^{n+1}(i,j,k) = E_z^n(i,j,k) + \frac{\Delta t}{\varepsilon\Delta}\begin{pmatrix} H_y^{n+\frac{1}{2}}(i+1/2,j,k) - H_y^{n+1/2}(i_N - 1/2,j,k) \\ + H_x^{n+\frac{1}{2}}(i,j-1/2,k) - H_x^{n+\frac{1}{2}}(i,j+1/2,k) \end{pmatrix} \quad (8.2a)$$

$$E_z^{n+1}(i_N, j, k) = E_z^{n+1}(i,j,k) \quad (8.2b)$$

where Δt is the time increment and Δ is the space lattice increment.

As for the 3-D updating equations for the tangential components (E_x and E_z) which are not located on the edge of the surface at plane $j = j_0$ and j_M as shown in Figure 8.2(b), can be represented by:

$$E_x^{n+1}(i,j,k) = E_x^n(i,j,k) + \frac{\Delta t}{\varepsilon\Delta}\begin{pmatrix} H_z^{n+\frac{1}{2}}(i,j+1/2,k) - H_z^{n+\frac{1}{2}}(i,j_M - 1/2,k) \\ + H_y^{n+\frac{1}{2}}(i,j,k-1/2) - H_y^{n+\frac{1}{2}}(i,j,k+1/2) \end{pmatrix} \quad (8.3a)$$

$$E_x^{n+1}(i, j_M, k) = E_x^{n+1}(i,j,k) \quad (8.3b)$$

$$E_z^{n+1}(i,j,k) = E_z^n(i,j,k) + \frac{\Delta t}{\varepsilon\Delta}\begin{pmatrix} H_y^{n+\frac{1}{2}}(i+1/2,j,k) - H_y^{n+1/2}(i-1/2,j,k) \\ + H_x^{n+\frac{1}{2}}(i,j_M-1/2,k) - H_x^{n+\frac{1}{2}}(i,j+1/2,k) \end{pmatrix} \quad (8.4a)$$

$$E_z^{n+1}(i, j_M, k) = E_z^{n+1}(i,j,k) \quad (8.4b)$$

As for the case of the edged tangential electric components (E_x, E_y, E_z) on plane $i = i_0$, i_N and $j = j_0$, j_M and due to the E_x and E_y components are on the edge of the absorbing boundary conditions as seen in Figure 8.2(a) and (b), therefore, they are assumed to be updated by the ABC updating equations. It should be noted that the only edged E_z tangential components will be considered here for the periodic boundary condition. According to Figure 8.3, due to the normal incidence plane wave, hence, four equations of E_z tangential components can be updated simultaneously:

$$E_z^{n+1}(i,j,k) = E_z^{n}(i,j,k) + \frac{\Delta t}{\varepsilon\Delta}\begin{pmatrix} H_y^{n+\frac{1}{2}}(i+1/2,j,k) - H_y^{n+1/2}(i_N-1/2,j,k) \\ +H_x^{n+\frac{1}{2}}(i,j_M-1/2,k) - H_x^{n+\frac{1}{2}}(i,j+1/2,k) \end{pmatrix} \quad (8.5a)$$

$$E_z^{n+1}(i_N,j_M,k) = E_z^{n+1}(i,j,k) \quad (8.5b)$$

$$E_z^{n+1}(i_N,j_0,k) = E_z^{n+1}(i,j,k) \quad (8.5c)$$

$$E_z^{n+1}(i_0,j_M,k) = E_z^{n+1}(i,j,k) \quad (8.5d)$$

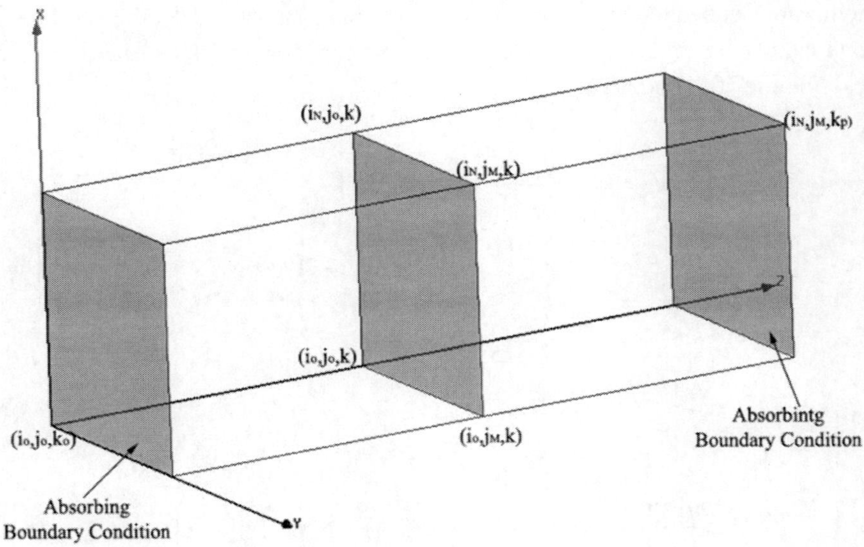

Figure 8.3. Location of the edged tangential component (E_z).

8.2.2. Hodgkin and Huxley (HH) Membrane Model

Cells are surrounded by thin membranes. It is the major barrier in the cell, separating cell contents from outside. Since the cell is separated from its environment, the membrane must be able to accommodate the required cell functions. Therefore, the membrane has to act as a selective barrier, allowing nutrients to pass in but keeping out many harmful substances to the cell, and as a dynamic barrier medium, constantly adapting to changing environmental conditions (different concentrations of ions).

The dimensions of a biological cell are around a few tens of micrometres while the thickness of the membranes is in the scale of a few nanometres, strongly depending on the type of the tissue. Depending on the type of the cell, voltages in the range of 20-200 mV can be obtained across the membrane. When the cell is in a resting state, the current across the membrane averages zero, but more generally it depends on the variation on the membrane voltage [23, 24].

Hodgkin and Huxley (HH) gave a general description of the time course of the current which flows through the membrane of a squid giant axon when the potential difference across the membrane is suddenly changed from its steady state [23, 24]. The results in [23], suggest that the behaviour of a membrane may be represented by the electrical circuit shown in Figure 8.4. Current can be carried through the membrane either by charging the membrane capacitance or by movement of ions through the nonlinear conductance in parallel with the membrane capacitance. The equations that describe the HH model are [23]:

$$\begin{aligned} I = C_m \frac{dV_m}{dt} &+ g_k n^4[V,t](V_m - E_k) \\ &+ g_{Na} m^3[V,t]h[V,t](V_m - E_{Na}) \\ &+ g_l(V_m - E_L) \end{aligned} \tag{8.6}$$

$$\frac{dn[V,t]}{dt} = a_n(V)(1 - n[V,t]) - b_n(V)n[V,t] \tag{8.7}$$

$$\frac{dm[V,t]}{dt} = a_m(V)(1 - m[V,t]) - b_m(V)m[V,t] \tag{8.8}$$

$$\frac{dh[V,t]}{dt} = a_h(V)(1 - h[V,t]) - b_h(V)h[V,t] \tag{8.9}$$

where:

$$V = V_m - V_r \tag{8.10}$$

The element definitions of the HH model of the above equations are given as follows:

- g_l is the leakage conductance of chloride or other ionic channels.
- *m*, *n*, *h* are the probability parameters determining the percentage of open channels (where an ion passes through the membrane). All variables change between 0 and 1 as functions of time and voltage.
- C_m is the membrane capacity per unit area.
- V_m is the voltage across the membrane.
- E_k, E_{na}, and E_l are the resting potentials of the various ionic channels. They contribute to the steady-state voltage on the membrane for each ionic channel.
- g_k and g_{na} are the nonlinear conductances of potassium and sodium channels.
- V_r is the resting potential of the membrane.
- V is the potential difference between the resting potential and the potential of the membrane, expressed in mV.
- I is the total ionic current across the membrane due to flux of ions. It is equal to zero if the membrane voltage equals the resting potential.

Table 8.1 quotes the constant value of all the components in HH model, as stated in [23]. The percentage of the open potassium channels is n^4 and the percentage of the open sodium channels is m^3h. It also needs to be highlighted that the quantities appearing in equations (8.7) to (8.9) are defined as [23]:

Table 8.1. Constant values of the parameters in equation (8.6)

C_m	E_k	g_k	E_{Na}	g_{Na}	E_L	g_L
0.01 F/m^2	72 mV	360 S/m^2	-55 mV	1200 S/m^2	50 mV	3 S/m^2

$$a_n(V) = \frac{0.01(V + 10)}{e^{((V+10)/10)} - 1} \tag{8.11}$$

$$a_m(V) = \frac{0.1(V + 25)}{e^{((V+25)/10)} - 1} \tag{8.12}$$

$$a_h(V) = 0.07e^{(V/20)} \tag{8.13}$$

$$b_n(V) = 0.125e^{(V/80)} \tag{8.14}$$

$$b_m(V) = 4e^{(V/18)} \tag{8.15}$$

$$b_h(V) = \frac{1}{e^{((V+30)/10)} + 1} \tag{8.16}$$

Equations from (8.11) to (8.16) can be redefined and rearranged in more general form as follows:

$$\frac{d\psi}{dt} = -(a_\psi + b_\psi)\psi + \alpha_\psi \tag{8.17}$$

where:

$$\Psi \in \{n, m, h\}$$

After solving the first order partial differential equation of (8.17), the expression for ψ can be obtained:

$$\psi(t) = \frac{a_\psi}{a_\psi + b_\psi}\left(1 - \frac{e^{-(a_\psi + b_\psi)t}}{K}\right) \tag{8.18}$$

where K is the constant that defines the initial condition. As can be observed, if the time t is much greater than the time constant $\tau = 1/(a_\psi+b_\psi)$, the parameter ψ can be considered and replaced to the following constant:

$$\psi(t)_{t>>\tau} \cong \frac{a_\psi}{a_\psi + b_\psi} \tag{8.19}$$

Consider a biological cell excited by a plane wave of 1 V/m. Since the dimension of the membrane is just a few nanometres, the potential difference on the membrane can be estimated as a few nV. This leads to the conclusion that the variables *m*, *n*, *h* in equations (8.11)-(8.16) can be assumed as constant values at steady state, and the HH model can be treated as a linear model. Figure 8.5 shows the percentage variations of a_ψ and b_ψ versus V in the range of ±1 μV. Table 8.2 quotes the constant values of *n*, *m* and *h* at steady state.

8.3. Implementation and Validation

8.3.1. Implementation of Floquet Boundary Condition

This section demonstrates the implementation of Floquet boundary conditions, quasi-static FDTD and the present modified PML for a lossy medium excited by a 100 V/m plane wave at an operating frequency of 900 MHz. The lossy medium properties were $\varepsilon_r = 1.0$, $\sigma = 25$ S/m. The problem space and cell sizes were 21×21×121 and 10 μm respectively. The Floquet boundary conditions were imposed on four sides of the lossy medium. The remaining two sides were each terminated by a PML of 6 cells. The analyses were performed at 10, 15 and 20 GHz and then transferred back to the desired operating frequency of 900 MHz. As can be observed in Figure 8.6, the analytical and computed results are in good agreement.

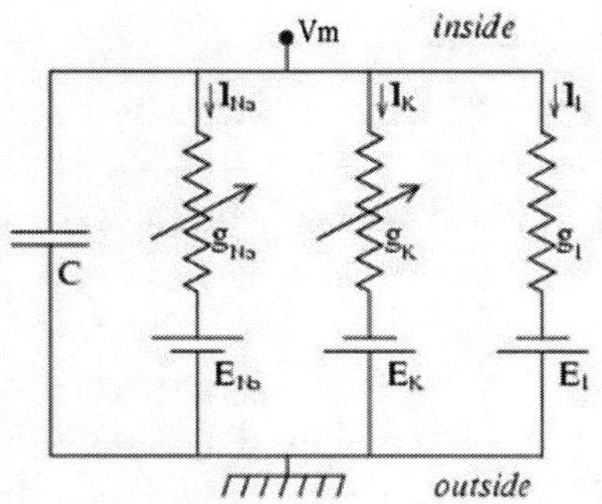

Figure 8.4. Equivalent electrical circuit for the cell's membrane.

Table 8.2. Steady-state values

n	m	h
0.3177	0.0529	0.5961

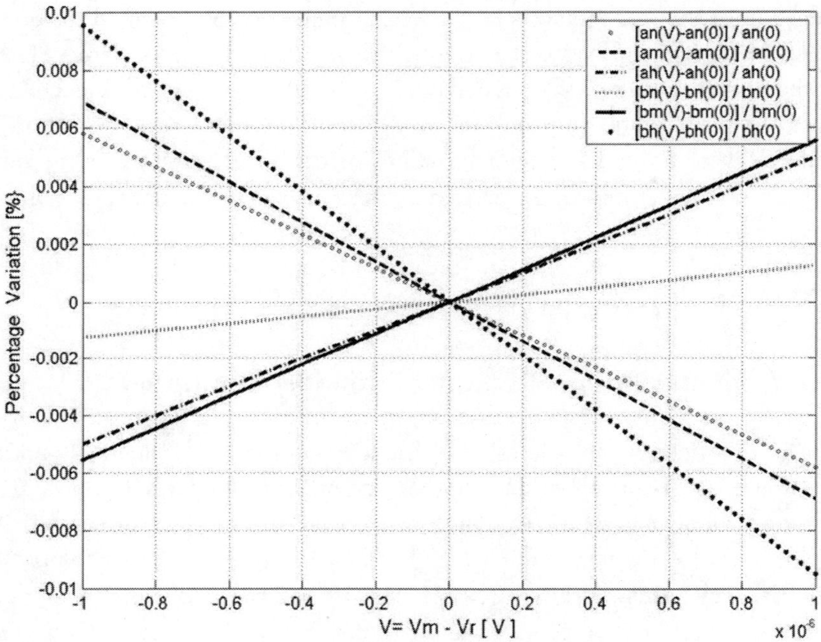

Figure 8.5. Percentage variation of a_ψ and b_ψ ($\psi = n, m, h$) versus voltage.

8.3.2. Implementation of HH Model

To verify the correctness of the implementation of the HH model within the FDTD framework, the results of the analytically computed solution have been used for comparison. In the first test case, the HH model will be included in one FDTD grid cell which is aligned with the grid. In order to observe the transient phenomena and the steady-state with a reasonable period of computation time, the cell size of the FDTD framework is assumed to be 10 cm. Since the entire system is linear, these assumptions do not affect the validity of this test case. No external excitation will be given in this test case until the steady state is reached. It should be noted that the direction of the cell's membrane is normal to the x-axis. Therefore, the following equations are exploited for this analysis in FDTD computational domain.

$$E_x\Big|^{n+1/2}_{i,j+1/2,k+1/2} = C_{a,E_x|i,j+1/2,k+1/2}\, E_x\Big|^{n-1/2}_{i,j+1/2,k+1/2} + C_{b,E_x|i,j+1/2,k+1/2} \cdot \begin{pmatrix} H_z\Big|^{n}_{i,j+1,k+1/2} - H_z\Big|^{n}_{i,j,k+1/2} \\ + H_y\Big|^{n}_{i,j+1/2,k} - H_y\Big|^{n}_{i,j+1/2,k+1} \end{pmatrix} \tag{8.20}$$
$$+ C_{a,E_x|i,j+1/2,k+1/2} \cdot \left(g_k n^4 V_k\Big|^n + g_{Nd} m^3 h V_{Na}\Big|^n + g_l V_l\Big|^n \right)$$

$$C_{a_{m,n,p}}\Big|_{i,j,k} = \frac{\left(1 - \dfrac{\sigma_{i,j,k}\Delta t}{2\varepsilon_{i,j,k}} + \dfrac{C\Delta_m}{\Delta_n\Delta_p\varepsilon_{i,j,k}} - \dfrac{g_k n^4 \Delta_m \Delta t}{2\Delta_n\Delta_p\varepsilon_{i,j,k}} - \dfrac{g_{Nd} m^3 h \Delta_m \Delta t}{2\Delta_n\Delta_p\varepsilon_{i,j,k}} - \dfrac{g_l \Delta_m \Delta t}{2\Delta_n\Delta_p\varepsilon_{i,j,k}}\right)}{\left(1 + \dfrac{\sigma_{i,j,k}\Delta t}{2\varepsilon_{i,j,k}} + \dfrac{C\Delta_m}{\Delta_n\Delta_p\varepsilon_{i,j,k}} + \dfrac{g_k n^4 \Delta_m \Delta t}{2\Delta_n\Delta_p\varepsilon_{i,j,k}} + \dfrac{g_{Nd} m^3 h \Delta_m \Delta t}{2\Delta_n\Delta_p\varepsilon_{i,j,k}} + \dfrac{g_l \Delta_m \Delta t}{2\Delta_n\Delta_p\varepsilon_{i,j,k}}\right)} \tag{8.21}$$

$$C_{b_{m,n,p}}\Big|_{i,j,k} = \frac{\left(\dfrac{\Delta t}{\varepsilon_{i,j,k}\Delta_m}\right)}{\left(1 + \dfrac{\sigma_{i,j,k}\Delta t}{2\varepsilon_{i,j,k}} + \dfrac{C\Delta_m}{\Delta_n\Delta_p\varepsilon_{i,j,k}} + \dfrac{g_k n^4 \Delta_m \Delta t}{2\Delta_n\Delta_p\varepsilon_{i,j,k}} + \dfrac{g_{Nd} m^3 h \Delta_m \Delta t}{2\Delta_n\Delta_p\varepsilon_{i,j,k}} + \dfrac{g_l \Delta_m \Delta t}{2\Delta_n\Delta_p\varepsilon_{i,j,k}}\right)} \tag{8.22}$$

$$C_{c_{m,n,p}}\Big|_{i,j,k} = \frac{\left(\dfrac{1}{\Delta_n\Delta_p}\right)}{\left(1 + \dfrac{\sigma_{i,j,k}\Delta t}{2\varepsilon_{i,j,k}} + \dfrac{C\Delta_m}{\Delta_n\Delta_p\varepsilon_{i,j,k}} + \dfrac{g_k n^4 \Delta_m \Delta t}{2\Delta_n\Delta_p\varepsilon_{i,j,k}} + \dfrac{g_{Nd} m^3 h \Delta_m \Delta t}{2\Delta_n\Delta_p\varepsilon_{i,j,k}} + \dfrac{g_l \Delta_m \Delta t}{2\Delta_n\Delta_p\varepsilon_{i,j,k}}\right)} \tag{8.23}$$

where V_k = 72 mV, V_{Na} = -55 mV, V_L = 50 mV

$$g_k = 360\Delta_n\Delta_p,\; g_{Na} = 1200\Delta_n\Delta_p,\; g_k = 3\Delta_n\Delta_p,\; C = 0.01\Delta_n\Delta_p$$

Now, by setting current $I = 0$ in equation (8.6), the steady state condition for the cell can be attained, which means the average rate of the ions crossing the membrane is zero. Under this steady state conditions, the resting voltage is equal to the membrane's voltage ($V_r = V_m$), then voltage (V_r) can be computed as follows:

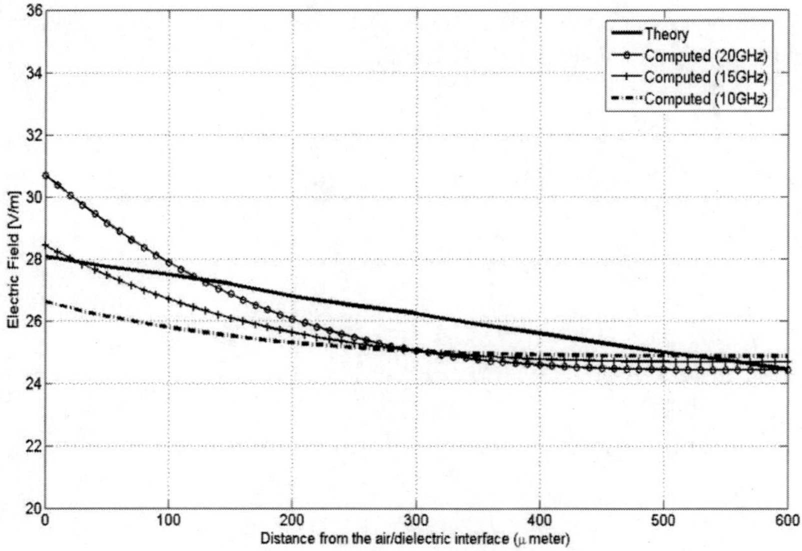

Figure 8.6. Electric field along the centre of the lossy medium.

$$V_r = \frac{G_K n^4 E_k + G_{Na} m^3 h E_{Na} + G_L E_L}{G_K n^4 + G_{Na} m^3 h + G_L} = 60.27 \text{ mV}$$

Figure 8.7 shows the comparison between the FDTD model and the analytical computed membrane voltage versus time. As can be seen, the analytical and FDTD results are indistinguishable in the steady state. Since the HH model contains a large capacitor, it causes the transient of the HH circuit to require hundreds of thousands of time steps to reach the steady state when simulation considers the dimensions of a realistic biological cell. To speed up the computation time, the actual FDTD simulation is performed in two steps. In the first step (known as DC simulation), simulation is performed with the capacitor in the circuit and then removed and its effects on the system is energized by the step voltage (DC) sources. Once the steady state is reached, the electric and magnetic fields on each FDTD discretisation cell are used as the initial values for the second step. During the transient simulation, the capacitor and the sinusoidal source are activated. In short, to speed up the computation time, the value obtained in the steady state without excitation can be used as the initial value of the membrane voltage. This procedure is similar to the approach used by the SPICE computer program [24] for the transient and AC analysis of electronic circuits.

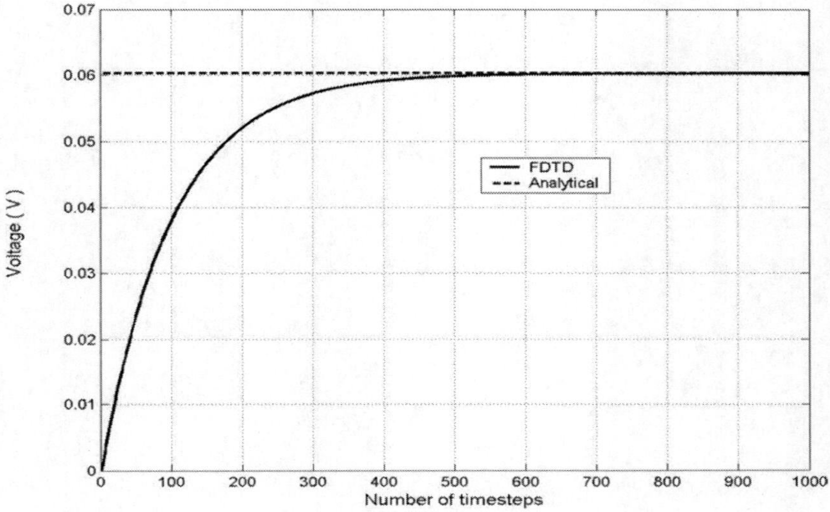

Figure 8.7. Steady state membrane's voltage verse number of time steps (the normal is directed along *X*).

In the second test case, it was shown that the method above gives the same results as when complete transient simulation is performed. The HH model was again inserted in one of the FDTD grid cells, with dimension of 10 cm. Then, it was excited by a plane wave at a frequency of 10 MHz. Two initial conditions were assumed in the simulation; firstly the membrane voltage was set at zero, and then at 60.27 mV. Figure 8.8 depicts the comparison of the two initial conditions, leading to the conclusion that the computation time can be speeded up by assigning the resting voltage on the cell's membrane at the initial state of the simulation. In a third test case, all the parameters in equation (8.11) were set to zero, except for n = 1 and g_k = 2,000,000 S/m^2, so the HH model behaves like a resistor whose value is:

$$R = \frac{1}{g_k \Delta s^2} = 50\ \Omega$$

In this case, the grid cell's size is 0.1 mm, and a plane wave excitation was used propagating in the *z*-direction and polarised in the *x*-direction at 40 GHz. Figure 8.9 shows the distribution of voltage and current on the membrane. The ratio between the peak values of voltage and current on the membrane was 50 Ω.

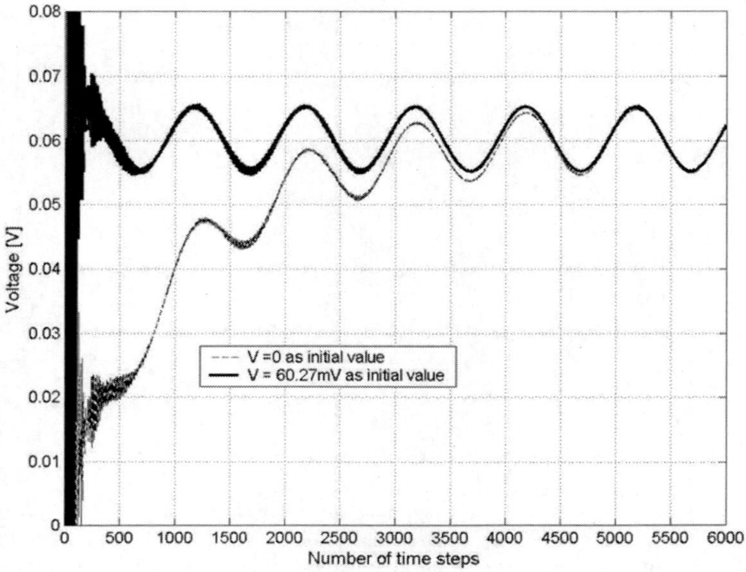

Figure 8.8. Comparison between membrane voltage with different initial conditions.

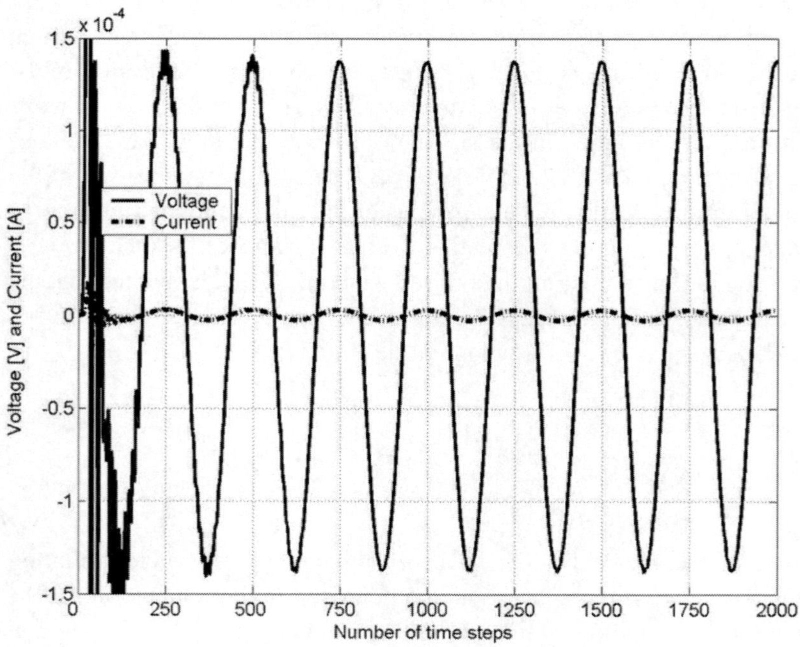

Figure 8.9. Voltage and current on the membrane.

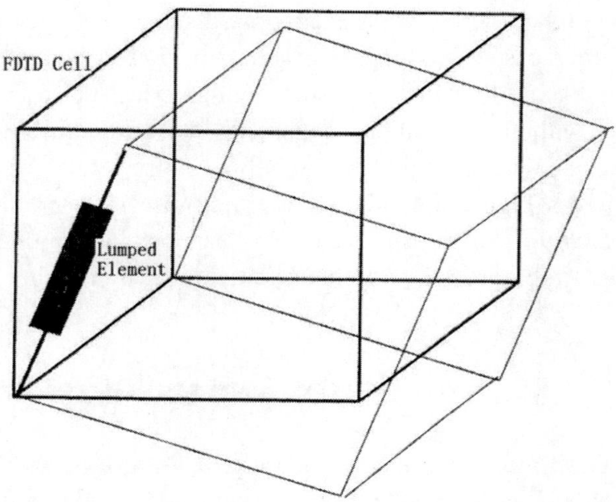

Figure 8.10. Modified LE-FDTD cell on normal FDTD cell grid.

8.3.3. Single Cell Validation

In this section, the HH model was modelled on a spherical structure with diameter 50 μm and discretised with 1 μm steps, in order to check the polarization voltage of 60.27 mV on the membrane. The HH model is included on the surface of the cell. The inner and external mediums of the cell geometry, have been considered as cytoplasm (ε_r = 48.699, σ = 1.412 S/m) and lossy medium (ε_r = 70.87, σ = 2.781 S/m) respectively. It should be noted that the LE-FDTD method has been successfully modified in order to accommodate the case when the position of the lumped element inside the membrane's cell is not aligned with the FDTD grid (see Figure 8.10), so that it can be excited by the electric field component normal to its surface. The rotation was done subject to the fields in spherical coordinates and then transferred to the actual grid of the FDTD at each time step. A matrix transformation was implemented on each FDTD cell surface.

Figure 8.11 depicts the polarization voltage value of 60.27 mV on the membrane of the spherical structure without any external excitation. From Figure 8.12, it can be observed that the field in the centre of the spherical structure is not zero. This effect can be described to a type of numerical noise which is created by the geometry of the structure. In order to verify whether the noise is random or deterministic, two simple simulations have been carried out. In one, the structure was irradiated with a plane wave of 100 kV/m at 100

GHz and in the other there was no excitation. It should be noted that the plane wave in the first case is propagating along the z-axis and electric field is parallel to the x-axis. Figure 8.13 shows the numerical noise in the centre of the structure without excitation. Figure 8.14 shows that the dominant component (E_x) is corrupted by the numerical noise along the z-axis of the structure, whereas Figure 8.15 shows that the noise is deterministic and the harmonic excitation can be retrieved. This can be done by subtracting the numerical noise in Figure 8.13 from the excitation field in Figure 8.14.

8.4. Simulation and Results

By incorporating the aforementioned and validated theories, i.e., Floquet periodic boundary conditions (FPBCs), modified PML, quasi-static and lumped-element FDTD approach, this section will model biological tissue with different structures that consider a cluster of spherical, cubical and cylindrical cells and examine the electromagnetic field behaviour inside the tissue when it is exposed by a plane wave.

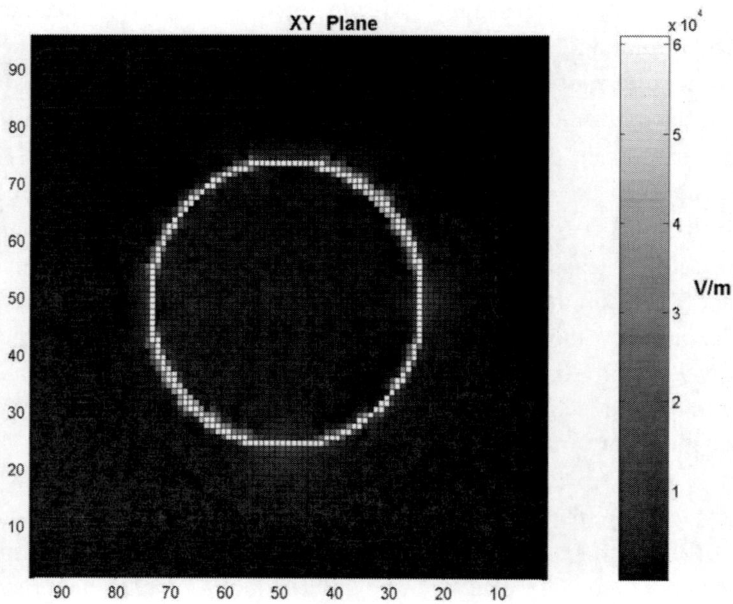

Figure 8.11. Field distribution on coordinate plane.

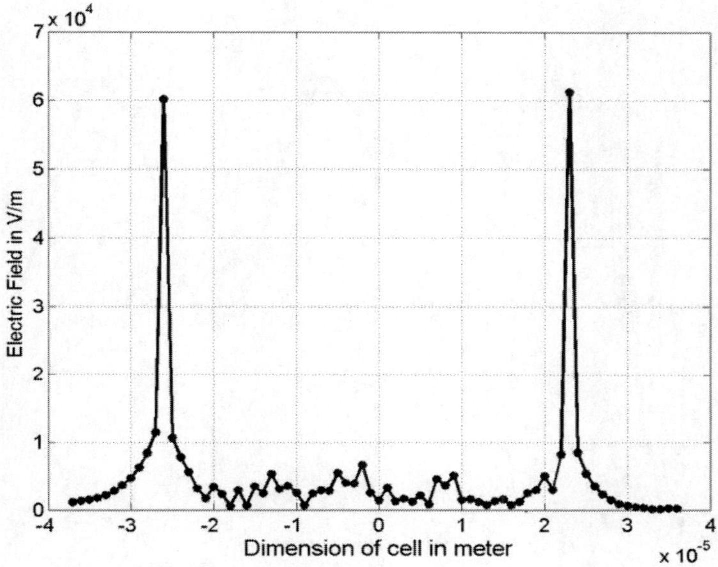

Figure 8.12. Total electric field along central axis of the structure.

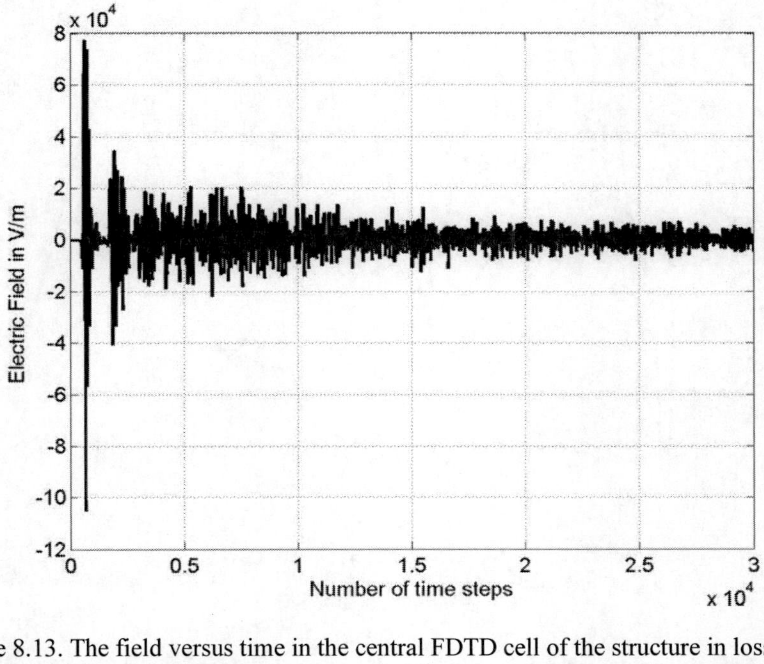

Figure 8.13. The field versus time in the central FDTD cell of the structure in lossy medium without excitation.

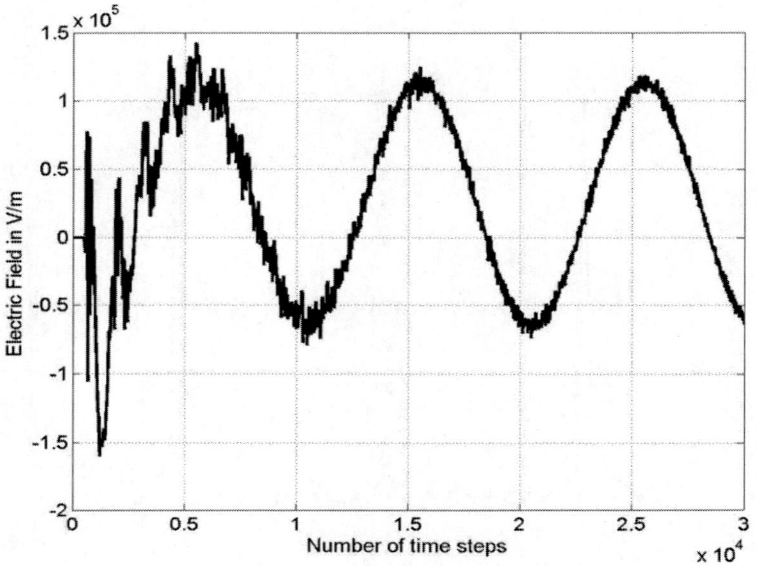

Figure 8.14. The corrupted excitation field in the centre of the FDTD cell of the structure.

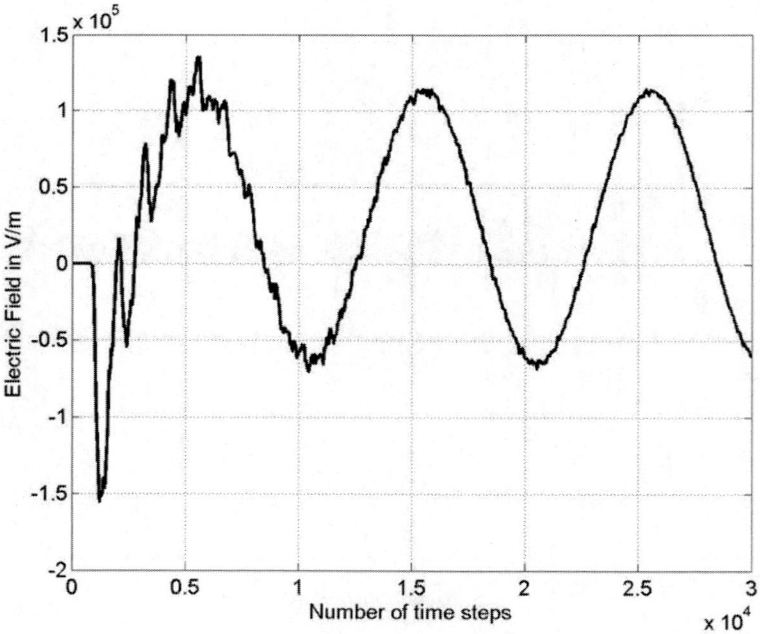

Figure 8.15. Excitation field removed from the numerical noise.

In general, the radius of the each biological cell in a tissue was 10 μm. The biological cell consists of three layers, which are cytoplasm, membrane and extracellular medium and the dielectric properties of these were obtained from [38]. The equations of the effective conductivity $\sigma(\omega)$ and effective dielectric permittivity $\varepsilon(\omega)$ of the material of biological cells are given in the following two forms:

$$\sigma(\omega) = \sigma(0) + \omega \sum_{k=1}^{n} \frac{\Delta\varepsilon_k \omega \tau_k}{1+\omega^2 \tau_k^2} \tag{8.24}$$

$$\varepsilon(\omega) = \varepsilon(0) - \sum_{k=1}^{n} \frac{\Delta\varepsilon_k \omega^2 \tau_k^2}{1+\omega^2 \tau_k^2} \tag{8.25}$$

where $\omega = 2\pi f$, f is the frequency of the field, n is the number of steps of dielectric relaxation of the material, $\Delta\varepsilon_k$ and τ_k are the magnitude of the k-th relaxation step and the time constant of the k-th step respectively, while $\sigma(0)$ and $\varepsilon(0)$ are the conductivity and dielectric permittivity of the material and it is measured at $\omega << 1/\tau_1$. It should be noted that all the fixed parameters that govern the equations are well defined in Tables 8.3 and 8.4. Once all the parameters are known, the material of the biological cell, as function of field frequency, can be established easily as shown in Figures 8.16 and 8.17 which are the ones in [38]. From the inspection of Figures 8.16 and 8.17, the electrical properties of cytoplasm, membrane and extracellular medium are found and all the desired parameters of this analysis at various frequency of interest, i.e. 900 MHz, 1800 MHz, 2000 MHz, 2450 MHz and 10 GHz, are tabulated in Table 8.5 respectively.

Table 8.3. Electrical properties of biological cell model

Parameter	Symbol	Value
Permittivity of cytoplasm	$\varepsilon_i(0)$	6.4×10^{-10} As/Vm
Conductivity of cytoplasm	$\sigma_i(0)$	0.3 S/m
Permittivity of membrane	$\varepsilon_m(0)$	6.4×10^{-10} As/Vm
Conductivity of membrane	$\sigma_m(0)$	3.0×10^{-7} S/m
Permittivity of extracellular medium	$\varepsilon_e(0)$	4.4×10^{-11} As/Vm
Conductivity of extracellular medium	$\sigma_e(0)$	1.2 S/m

Table 8.4. Parameters of dielectric relaxation of cytoplasm, membrane and extracellular medium

Parameter	Symbol	Value
Dielectric relaxation of cytoplasm and extracellular medium:		
First relaxation time	τ_{re}/τ_{ri}	6.2×10^{-12} s
First relaxation step	$\Delta\varepsilon_e/\Delta\varepsilon_i$	5.9×10^{-10} As/Vm
Dielectric relaxation of membrane:		
First relaxation time	τ_{rm1}	3.0×10^{-9} s
First relaxation step	$\Delta\varepsilon_{rm1}$	2.3×10^{-11} As/Vm
Second relaxation time	τ_{rm2}	4.6×10^{-10} s
Second relaxation step	$\Delta\varepsilon_{rm2}$	7.4×10^{-12} As/Vm

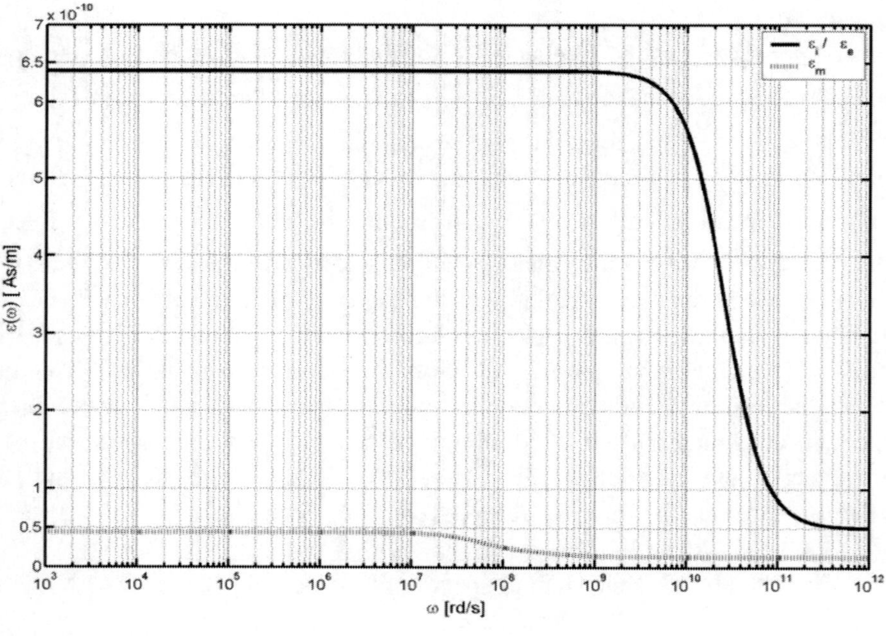

Figure 8.16. Effective permittivity of cytoplasm, membrane and extracellular medium of the biological cells as a function of frequency.

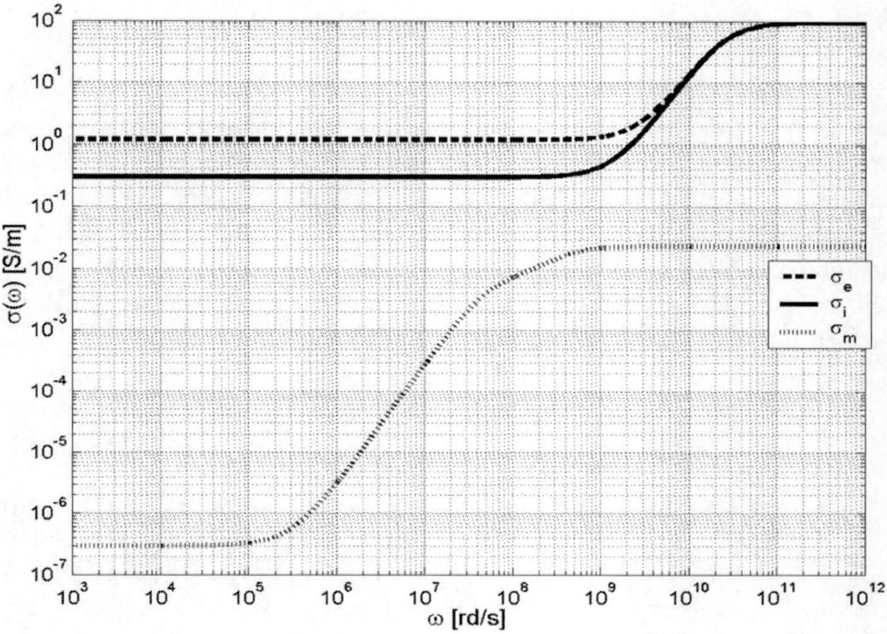

Figure 8.17. Effective conductivities of cytoplasm, membrane and extracellular medium of the biological cells as a function of frequency.

Table 8.5. Electrical property of the simulated structure at mobile communication frequencies

900 MHz	1800 MHz	2000 MHz	2450 MHz	10 GHz
$\varepsilon_i = 72.2003$	$\varepsilon_i = 71.956$	$\varepsilon_i = 71.88$	$\varepsilon_i = 71.6806$	$\varepsilon_i = 63.5023$
$\sigma_i = 0.4168$	$\sigma_i = 0.7656$	$\sigma_i = 0.8742$	$\sigma_i = 1.1590$	$\sigma_i = 12.8384$
$\varepsilon_m = 1.6526$	$\varepsilon_m = 1.5680$	$\varepsilon_m = 1.5621$	$\varepsilon_m = 1.5536$	$\varepsilon_m = 1.5371$
$\sigma_m = 0.0217$	$\sigma_m = 0.0232$	$\sigma_m = 0.0233$	$\sigma_m = 0.0234$	$\sigma_m = 0.0237$
$\varepsilon_e = 72.2003$	$\varepsilon_e = 71.956$	$\varepsilon_e = 71.88$	$\varepsilon_e = 71.6806$	$\varepsilon_e = 63.5023$
$\sigma_e = 1.3168$	$\sigma_e = 1.6656$	$\sigma_e = 1.7742$	$\sigma_e = 2.0590$	$\sigma_e = 13.7384$

8.4.1. Connected Tissue Model Using Spherical Cells

A stack of ten spherical cells was investigated, as shown in Figures 8.18 and 8.19. The radius of the each cell was 10 μm. A plane wave of 100 V/m, propagating in the z-direction and polarized in the x-direction was used as the excitation. Note that the incident plane wave excitation was applied on a plane lying between the PML region and the outer limit of the FDTD grid. In addition, in order to reduce high-frequency transients [42, 43] and DC offsets [44, 45], associated with unramped sine wave excitations, the following ramped sinusoidal source in equation (8.26) was adopted [44]:

$$f(t) = \begin{cases} 0 & t < 0 \\ 0.5[1-\cos(\omega_r t)]\sin(\omega_r t) & 0 \le t \le \frac{T_r}{2} \\ \sin(\omega_r t) & t > \frac{T_r}{2} \end{cases} \tag{8.26}$$

where T_r is the period of the ramped cosine which is about 3 source cycles.

The PML, shown in Figure 8.19, was six FDTD elements wide, the grading factor g was 10.1383 and the grid structure was effectively extended to infinity in the x- and y-directions, by imposing the Floquet boundary condition along the x- and y-axes. The Floquet periodic boundary condition plays an important role to mimic the presence of an extended 3-D structure of biological cells, simulating connected tissue. This can be easily imagined in two dimensions, as shown in Figure 8.18. The FDTD problem space was 220×20×20 FDTD elements of size 1 μm while a discretization time step δt of 1.3 fs was chosen to drive the FDTD computation, to meet the requirements of the Courant stability criterion.

Before implementing the Hodgkin-Huxley model into the simulated structure, the effect of moving the Floquet boundaries gradually away from the simulated structure was studied. Figures 8.20 and 8.21 depict the field distribution through the centre of the simulated structure at 10 GHz with varying locations of the Floquet boundaries, where Ncell is the number of FDTD elements between the Floquet boundaries and the boundaries of the biological cells, in the x and y directions. Figure 8.22 shows the field distribution on the xz-plane of the simulated structure for the case of Ncell = 10. When the Floquet boundaries are exactly adjacent to the simulated structure (Ncell = 0), the strongest coupling effect between cells can be obtained. The highest induced field on the membrane and lowest induced field in the cytoplasm of the cell can be observed. Conversely, when the Floquet boundaries are far away from the simulated structure (Ncell = 10), the lowest induced field on the membrane and highest induced field in the cytoplasm of the cell are observed. It should be noted that all the following analysis will be based on Ncell = 0, which is assumed to be the most appropriate model for the real living biological tissues or cells in this microdosimetry study. The simulations were performed at the transformed intermediate frequency of 10 GHz and the overall model was then transformed to the intended lower frequencies. Table 8.6 reports the transformation factors at 900 MHz, 1800 MHz, 2000 MHz and 2450 MHz used in the analysis.

Figure 8.23 illustrates the 10 GHz field distribution on the *xz*-plane of the simulated structure. The distributions of the electric field through the centre of the simulated structure, along the incident wave propagation direction, at 900 MHz, 1800 MHz, 2000 MHz and 2450 MHz are given in Figures 8.24 and 8.25, where Figure 8.25 is an enlarged version of Figure 8.24. The magnified version of Figure 8.25 is also plotted in Figure 8.26 and 8.27 as well. From inspection of Figures 8.25, 8.26 and 8.27, the field inside the cells is not constant and the induced field intensity is directly proportional to the frequency. In other words, the higher the operating frequency that is used to excite the model, the higher the electric field intensity that will be induced within the analyzed structures.

To complete the simulation, the Hodgkin-Huxley models were embedded in the surface of the spherical cells, in a direction normal to the surface, to represent the membrane effect of the tissue model. Versions including this were studied at frequencies of 900 MHz and 2450 MHz. As can be seen in Figures 8.28 and 8.29, there is a difference of approximately 15% in the field strength due to the contribution of the membrane effect from the Hodgkin-Huxley model. These variations were in well agreement with expectations [16, 23, 38,46,47,48].

8.4.2. Connected Tissue Model Using Cubical Cells

Since living cells, when compacted into connected tissue, are not perfectly spheres, a cluster of cubical cells was also chosen for study on the foundation of the previous spherical cells analysis. Figure 8.30 depicts the proposed cluster of cubical cells in a three dimensional view of the FDTD computational domain. In order to compare the results obtained from the previous model with this analysis, an FDTD simulation was executed, keeping the same parameter values unchanged. The 2-D view of the electric field inside the cubical cell tissue is shown in Figure 8.31. The field distributions along the propagation direction of the incident wave, through the centre of the simulated structure at various frequencies, are illustrated in Figures 8.32 and 8.33, while the magnified version of Figure 8.33 are depicted in Figures 8.34 and 8.35. The contribution of the Hodgkin-Huxley model to the cubical tissue model has also been investigated, as shown in Figures 8.36 and 8.37. The effects of adding the Hodgkin-Huxley model are about 15% difference in field magnitude, as can be easily observed in the figures.

The peak field values on the membrane of the cubical structure are observed to be about three times higher than in the cytoplasm, which agrees well with the results from the structure based on spherical cells. However, the absolute field strength is approximately doubled in the spherical cell case, presumably because of the curvature at the points studied. It is to be expected that much higher fields would be observed at the corners of the cubical cells, but it might be argued that, as a localised matter, these points do not correspond well with biological reality.

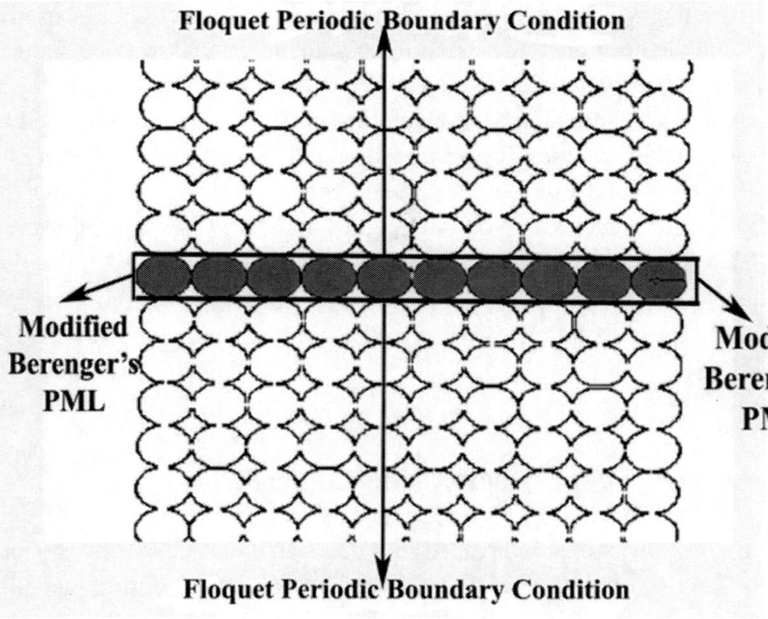

Figure 8.18. The 2-D view of the simulated structures in FDTD computational domain.

8.4.3. Connected Tissue Model Using Cylindrical Cells

The shape of the living cells can be so diverse. In order to have better understanding on the subject of EM field interaction with different geometry of biological tissue, a cluster of cylindrical cells model of the tissue is proposed, as depicted in Figure 8.38. This analysis is performed in which the material properties unchanged as in the case of spherical and cubical cell structures. Figure 8.39 describes the 2-D view of electric field distribution of the proposed cylindrical cell tissue at 10 GHz, while the electric field distribution along the centre of the analysed structure is shown in Figure 8.40 to 8.43, where the Figures 8.41-8.43 are the amplified version of the Figure 8.40.

The loading effect of the HH model into the cylindrical cell tissue has also been studied. Figure 8.44 and 8.45 demonstrate the difference of 15% in field magnitude with and without the presence of HH model in the proposed simulated structure. The results show consistent difference with the previous spherical and cubical cells tissue simulated structures.

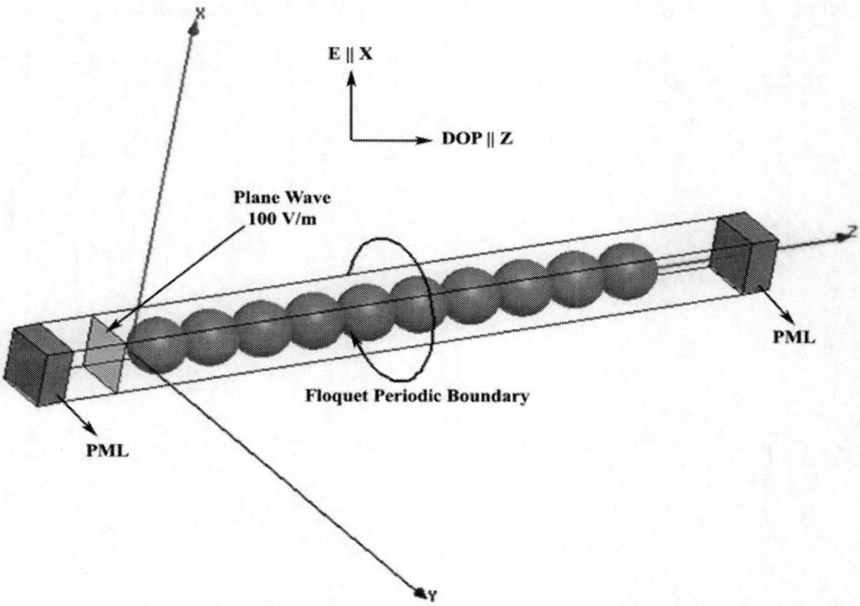

Figure 8.19. The 3-D view of the simulated spherical structures in FDTD computational domain.

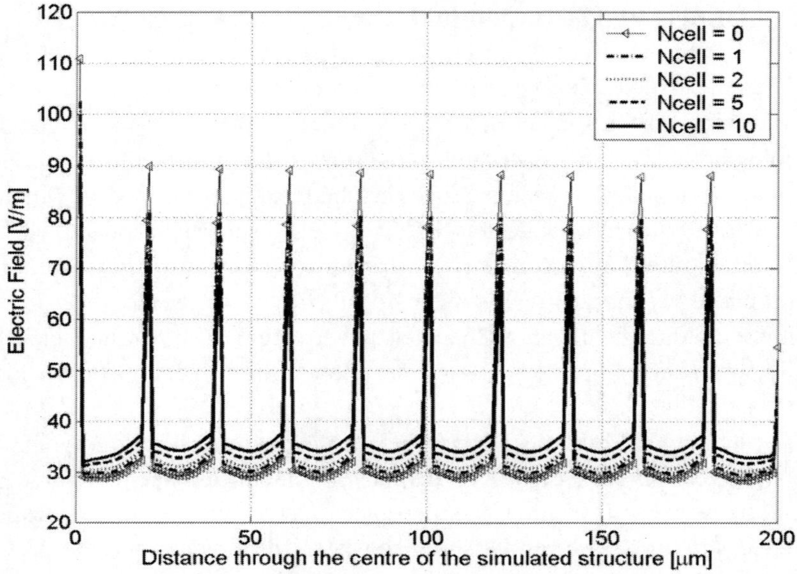

Figure 8.20. Penetration of electric field along z-axis, through the centre of the simulated structure with different location of the Floquet boundary condition.

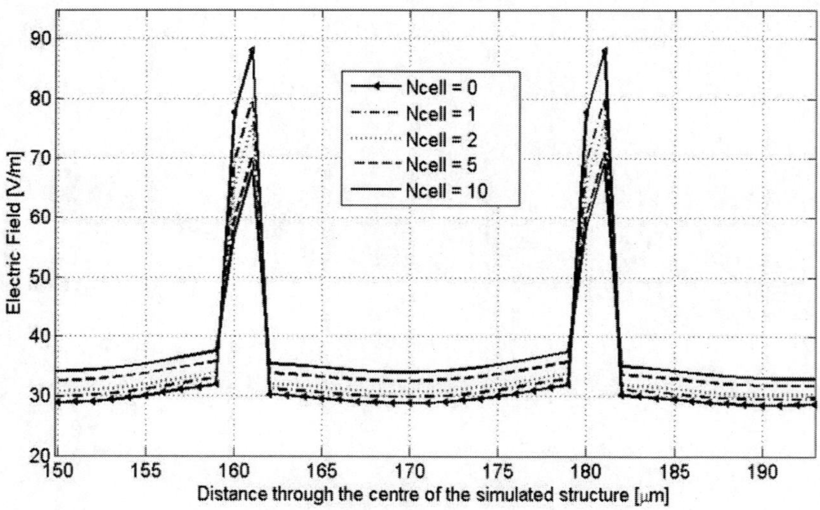

Figure 8.21. Penetration of electric field (enlargement of Figure 8.20).

The comparison of the field distribution of spherical, cubical and cylindrical cells tissue model, through the centre of the analysed structure, is elucidated in Figure 8.46. The peak field on the membrane of the cylindrical structure is found to be about 1.7 times higher than in the cytoplasm, which is distinct with the previous two models. As can be noticed, the peak field value of this cylindrical model is higher than the cubical model and lower than the spherical model, whereas the peak field value on the cytoplasm is the about the same as in the spherical model and double the value found in the cubical structure.

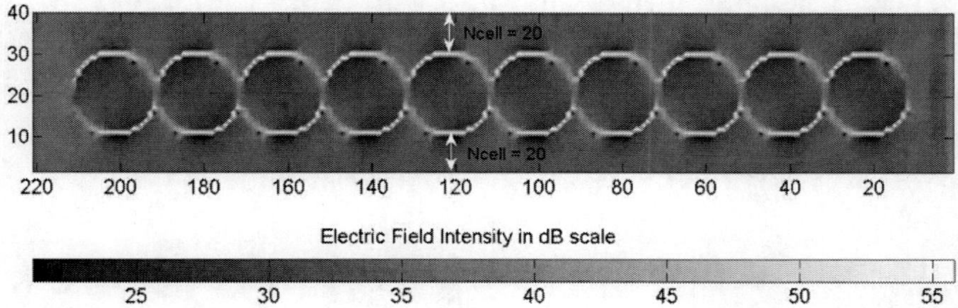

Figure 8.22. Ncell = 20, modulus of the electric field on *xz*-plane at intermediate frequency 10 GHz (logarithmic scale).

Table 8.6. Frequency scaling transformation factor from 10 GHz to the following mobile communication frequencies

Parameter	900 MHz	1800 MHz	2000 MHz	2450 MHz
Cytoplasm	0.9296	0.9337	0.9344	0.9360
Membrane	0.9000	0.9700	0.9756	0.9838
Extracellular	0.8867	0.9226	0.9254	0.9301

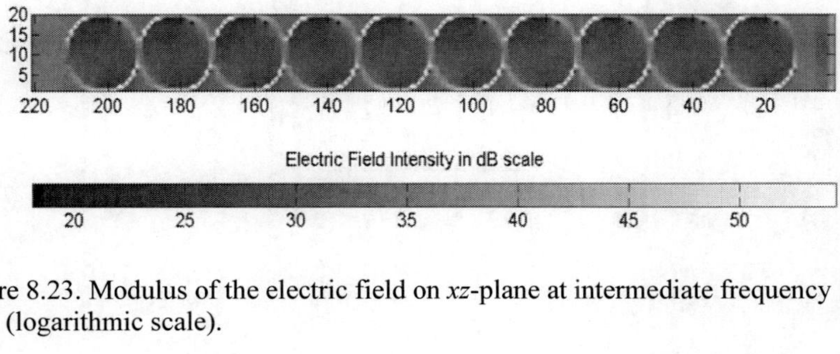

Figure 8.23. Modulus of the electric field on *xz*-plane at intermediate frequency 10 GHz (logarithmic scale).

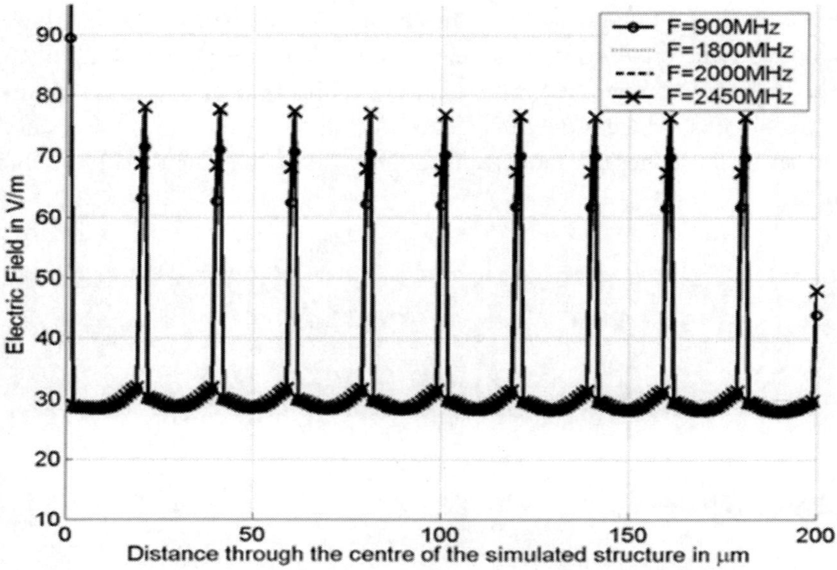

Figure 8.24. Penetration of electric field along *z*-axis, through the centre of the simulated structure.

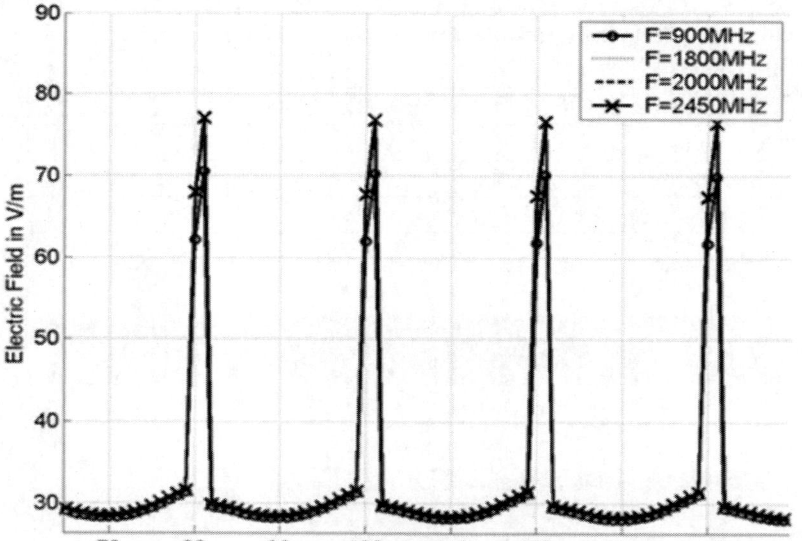

Figure 8.25. Penetration of electric field (enlargement of Figure 8.24).

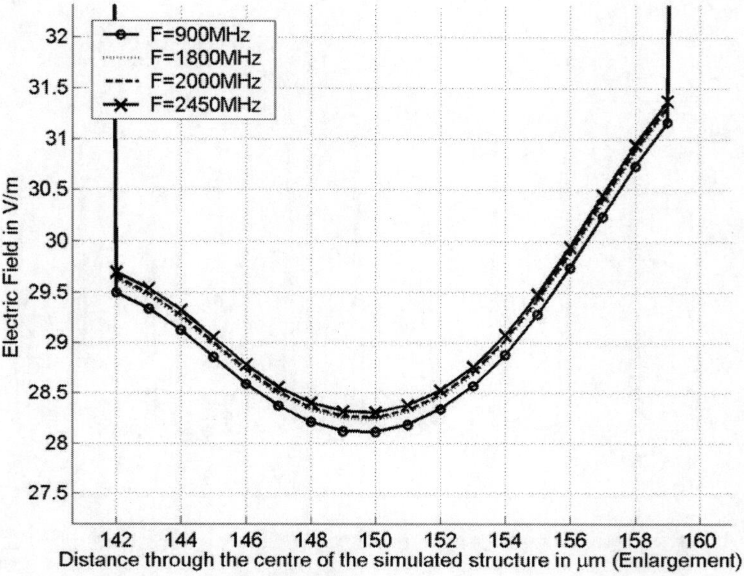

Figure 8.26. Penetration of electric field on the cytoplasm (enlargement of Figure 8.25).

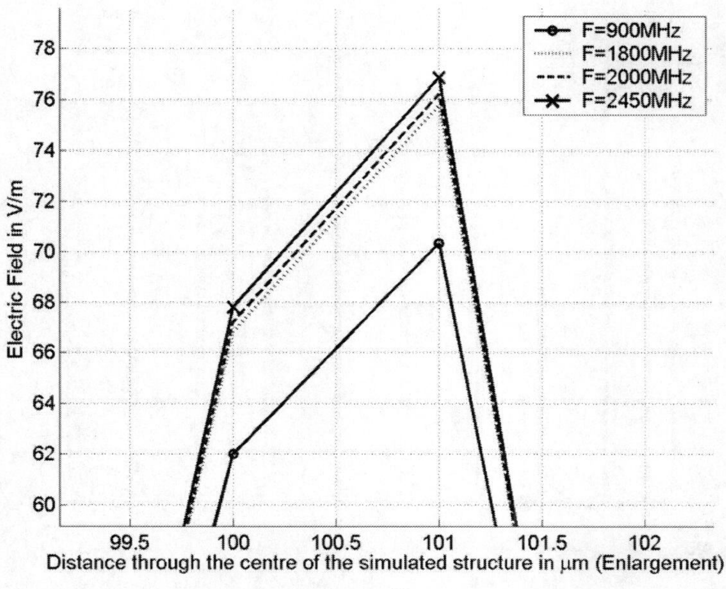

Figure 8.27. Penetration of electric field on the membrane (enlargement of Figure 8.25).

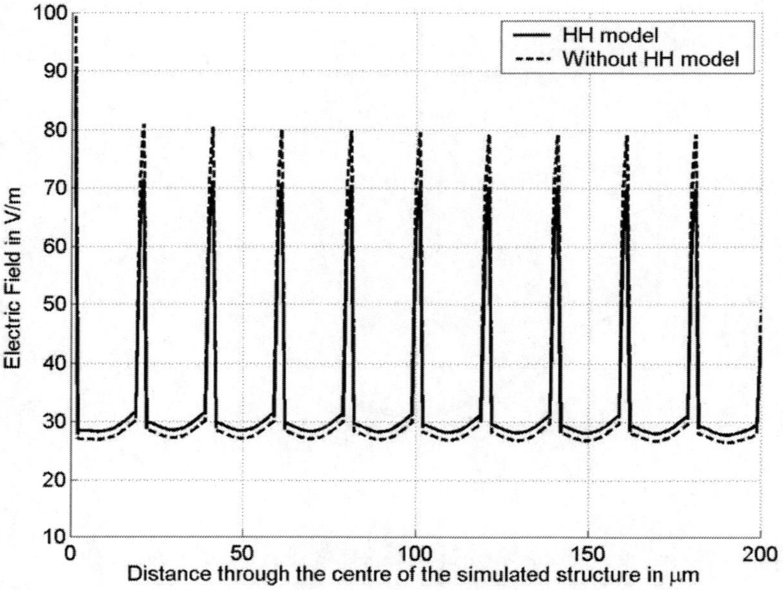

Figure 8.28. Electric field distribution along *z*-axis, through the centre of the simulated spherical structure in Figure 8.19, incorporating Hodgkin-Huxley model and driven at 900 MHz.

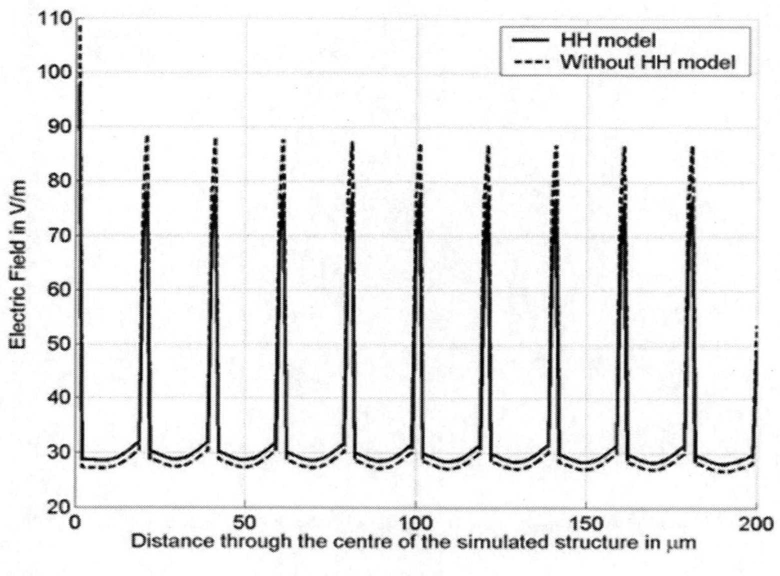

Figure 8.29. As Figure 8.28, driven at 2450 MHz.

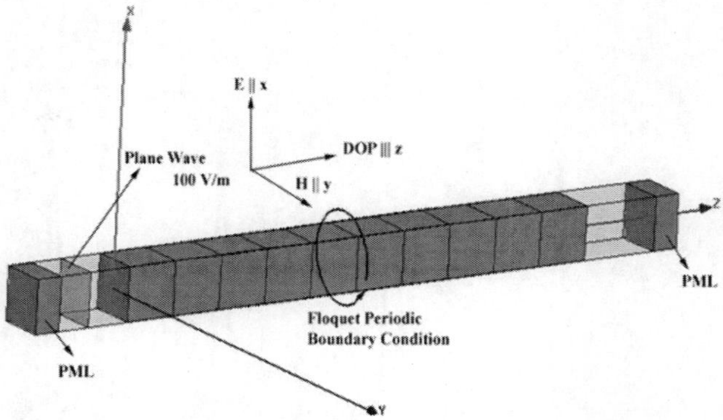

Figure 8.30. The 3-D view of the simulated cubical structures in FDTD computational domain.

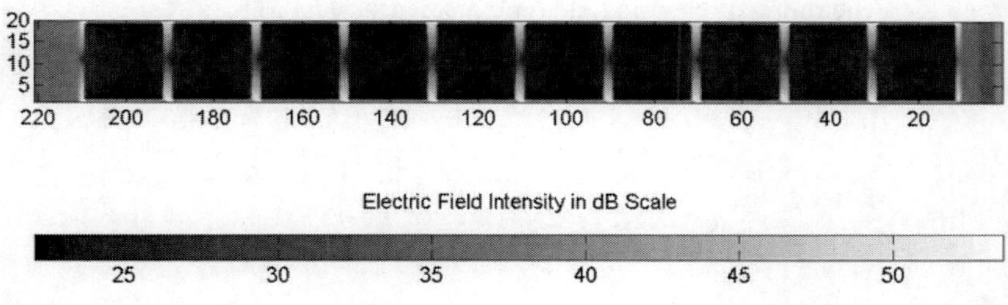

Figure 8.31. Modulus of the electric field on *xz*-plane at intermediate frequency 10 GHz (logarithmic scale).

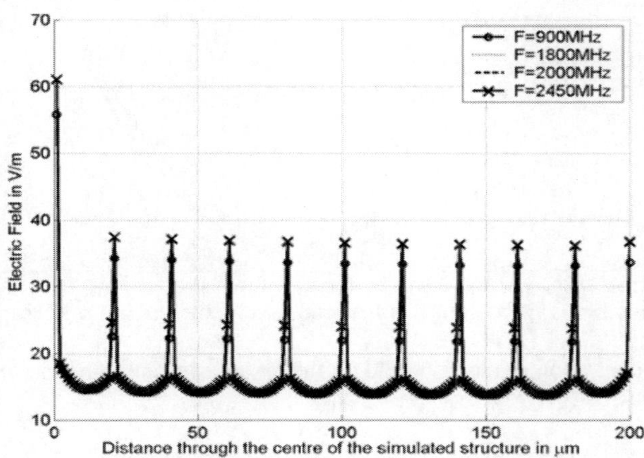

Figure 8.32. Penetration of electric field along *z*-axis, through the centre of the simulated structure.

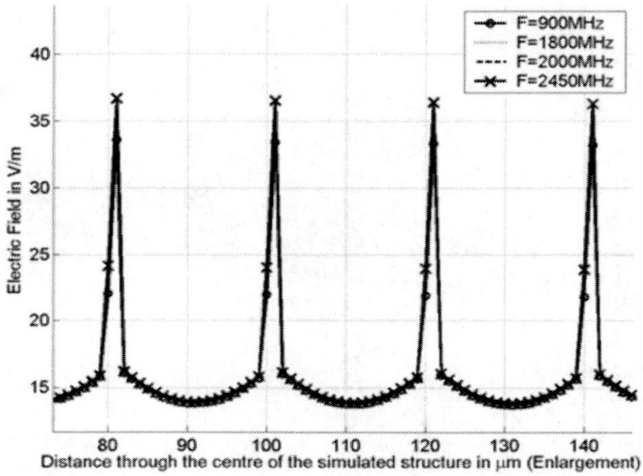

Figure 8.33. Penetration of electric field (enlargement of Figure 8.32).

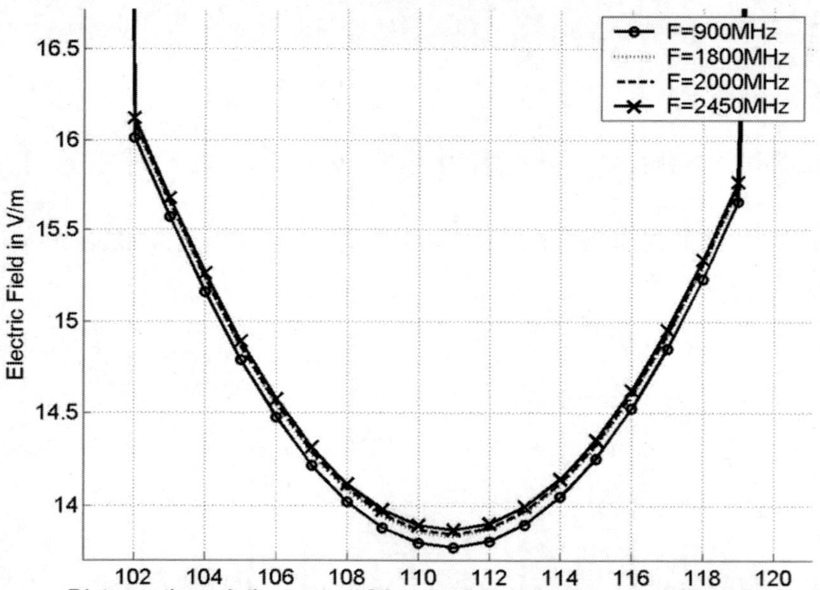

Figure 8.34. Penetration of electric field on the cytoplasm (enlargement of Figure 8.33).

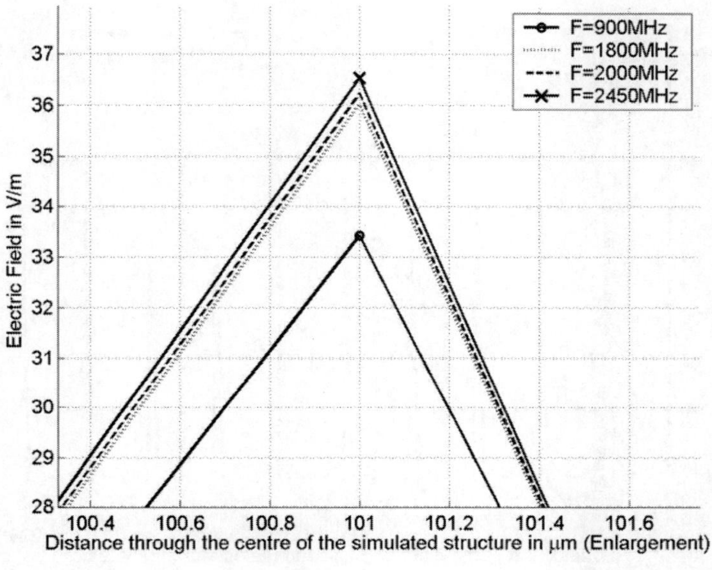

Figure 8.35. Penetration of electric field on the membrane (enlargement of Figure 8.33).

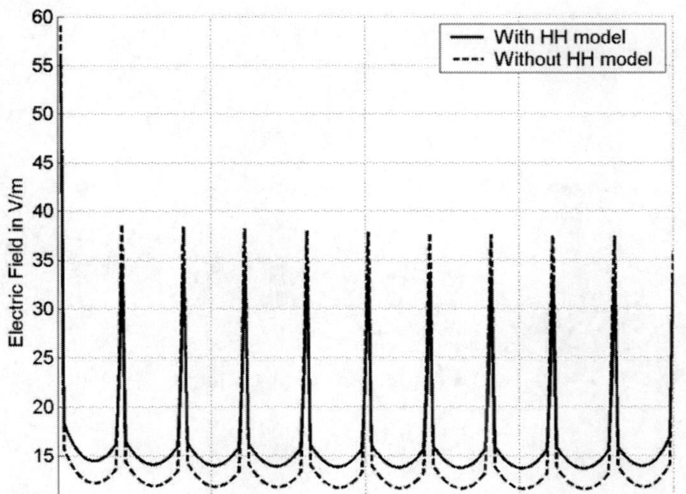

Figure 8.36. Electric field distribution along z-axis, through the centre of the simulated cubical cell structure in Figure 8.30, incorporating Hodgkin-Huxley model and driven at 900 MHz.

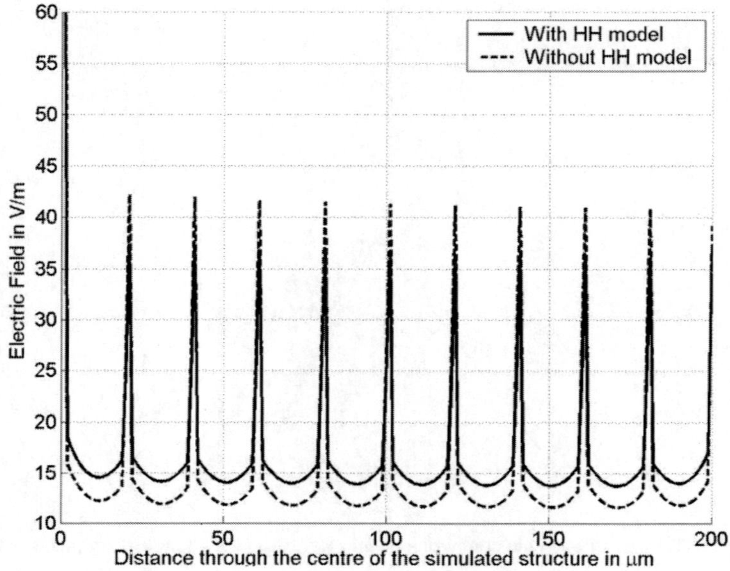

Figure 8.37. As Figure 8.36, driven at 2450 MHz.

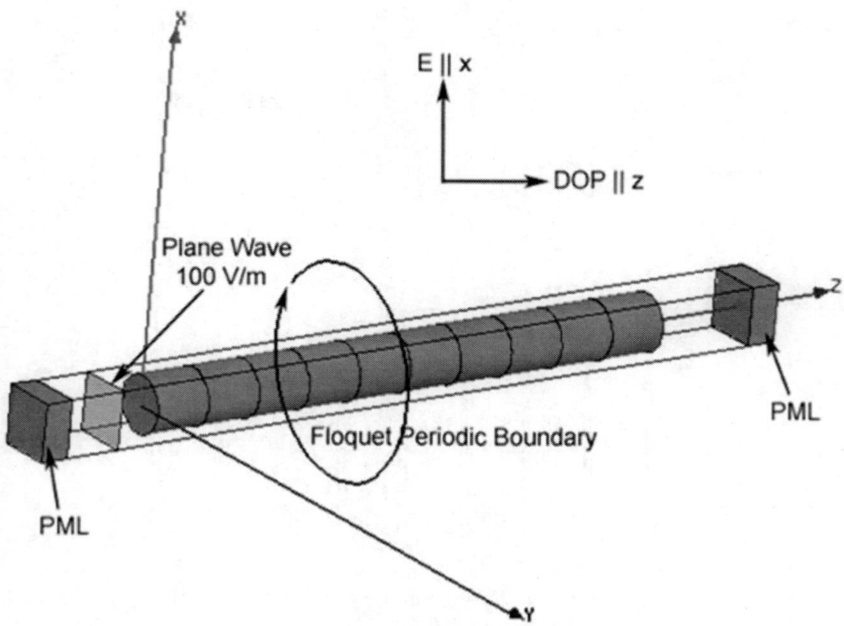

Figure 8.38. The 3-D view of the simulated cylindrical structures in FDTD computational domain.

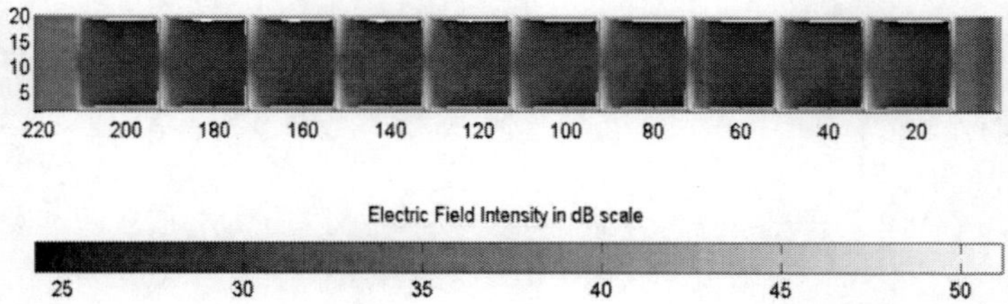

Figure 8.39. Modulus of the electric field on *xz*-plane at intermediate frequency 10 GHz (logarithmic scale).

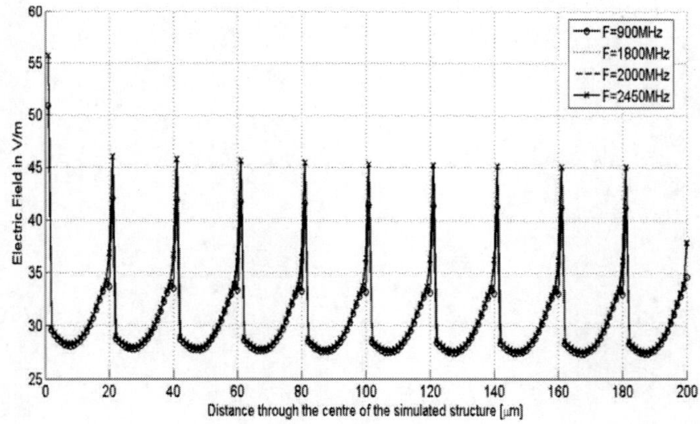

Figure 8.40. Penetration of electric field along *z*-axis, through the centre of the simulated structure.

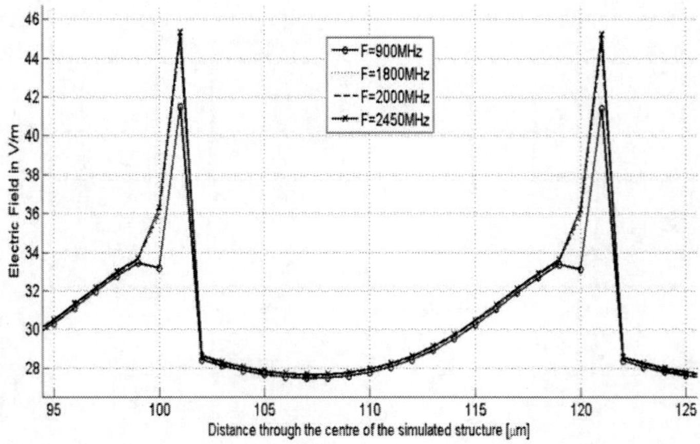

Figure 8.41. Penetration of electric field (enlargement of Figure 8.40).

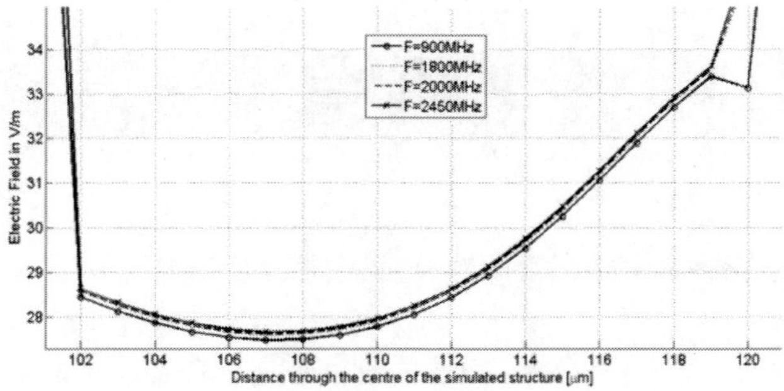

Figure 8.42. Penetration of electric field on the cytoplasm (enlargement of Figure 8.41).

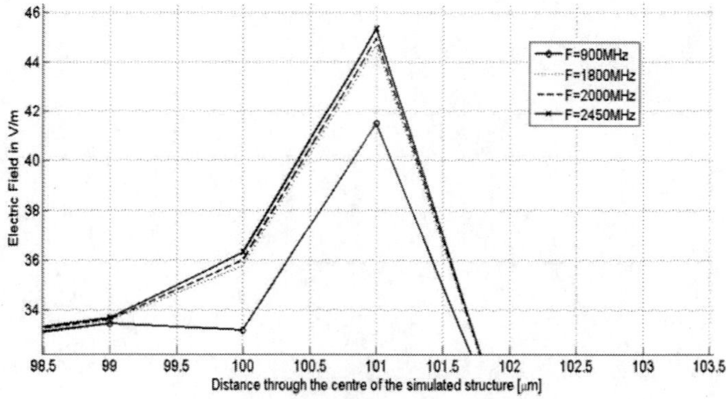

Figure 8.43. Penetration of electric field on the membrane (enlargement of Figure 8.41).

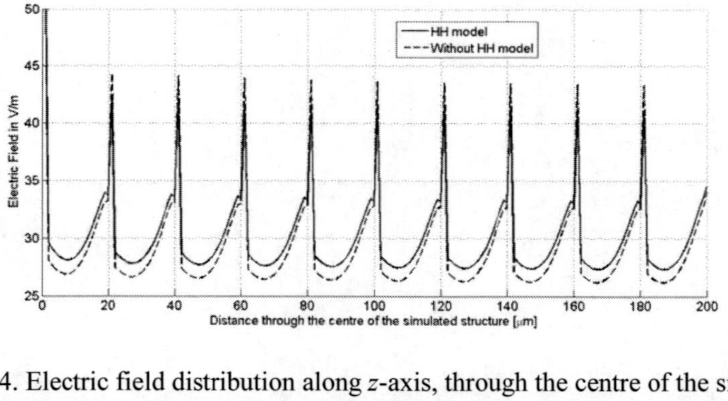

Figure 8.44. Electric field distribution along z-axis, through the centre of the simulated cubical cell structure in Figure 8.38, incorporating Hodgkin-Huxley model and driven at 900 MHz.

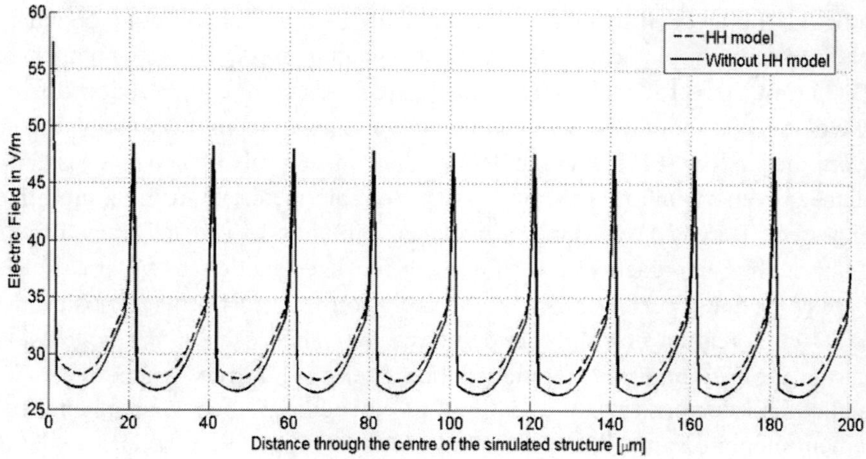

Figure 8.45. As Figure 8.44, driven at 2450 MHz.

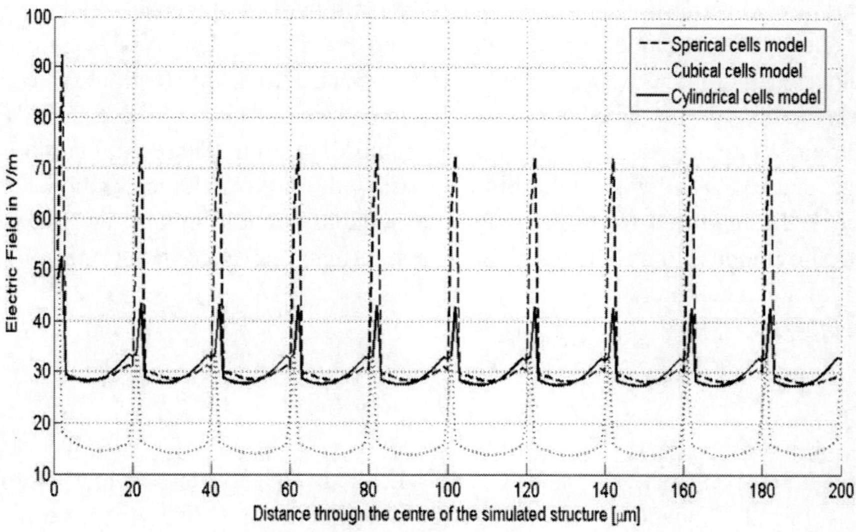

Figure 8.46. Comparison of three different simulated structures at 900 MHz.

Conclusion

An approach to microdosimetric modeling of bioelectromagnetic interactions at the cellular level has been presented. This uses the FDTD method, combined with an arbitrarily-oriented implementation of the

Hodgkin-Huxley cell-membrane model and the Floquet periodic boundary condition. By implementing a frequency-scaling approach, the number of FDTD time steps for such an electrically-small structure can be reduced from several millions to a few tens of thousands. The reflection on the interface layers inside the FDTD computation domain has also been successfully reduced, even though it is within lossy penetrable media, by using a modified version of Berenger's absorbing boundary condition. The accuracy of the FDTD scaling approach was verified with idealized models of spherical cells in lossy media. The feasibility of the inclusion of the HH model inside the FDTD computation domain was demonstrated. This leads to the conclusion that the application of the HH model allows cells of arbitrary geometries to be handled and demonstrates the viability of embedding other types of lumped-element model for membrane behavior. It can be argued that the HH model is imperfect for microwave frequencies, but it is reasonable to use it as a working hypothesis (as have others [6]) to develop modeling techniques while operational versions of improved models are still in development.

Use of the Floquet boundary condition enables a non-trivial region of connected biological tissue to be simulated. Such a tool will facilitate deeper investigation of the phenomena in the interaction between EM fields and biological systems at various levels of spatial definition. The combination of quasi-static FDTD with an arbitrarily-oriented lumped element membrane model, the modified Berenger absorbing boundary condition and the Floquet periodic boundary condition represents a significant advance in verisimilitude of biological cell modeling.

REFERENCES

[1] Xiong, R; Chen, B; Yi, Y; Chen, H-L; Zhao, H-F. "Two-step method for the FDTD analysis of long apertures having depth," *IEEE Antennas and Wireless Propagation Letters*, vol. 12, 39-42, 2013.

[2] Kamepally, SK; Kumar, BP; Paidimarry, CS. "FDTD estimation for accurate specific absorption rate in a tumor," *Annual International Conference on Emerging Research Areas and International Conference on Microelectronics, Communications and Renewable Energy (AICERA/ICMiCR)*, 1-5, 2013.

[3] Whittow, WG; Njoku, CC; Vardaxoglou, YC. "Fine scale simulations of patch antennas with heterogeneous substrates," *USNC-URSI Radio Science Meeting (Joint with AP-S Symposium)*, 223, 2013.

[4] Lesina, AC; Vaccari, A; Bozzoli, A. "A modified RC-FDTD algorithm for plasmonics in Drude dispersive media nanostructures," *6th European Conference on Antennas and Propagation (EUCAP)*, 687-690, 2012.

[5] Noetscher, GM; Xu, Y; Makarov, SN. "Accuracy of point source models with coincident phase centers in a cubic FDTD grid for arbitrary source orientation," *IEEE Antennas and Propagation Society International Symposium (APSURSI)*, 1-2, 2012.

[6] Nayyeri, V; Soleimani, M; Dehmollaian, M. "Implementation of a PEMC boundary condition in the 2-D FDTD technique," *IEEE Antennas and Propagation Society International Symposium (APSURSI)*, 1-2, 2012.

[7] Liu, D-C; Chang, H-C. "The second-order condition of FDTD method at sloped dielectric interfaces by averaging the permittivities along surface normal," *IEEE Transactions on Antennas and Propagation*, vol. 60, no. 11, 5259-5267, 2012.

[8] Weiwei, M; Dong, S; Xianliang, W. "UPML-FDTD parallel computing on GPU," *International Conference on Microwave and Millimeter Wave Technology (ICMMT)*, vol. 4, 1-4, 2012.

[9] Guiffaut, C; Reineix, A; Pecqueux, B. "New oblique thin wire formalism in the FDTD method with multiwire junctions," *IEEE Transactions on Antennas and Propagation*, vol. 60, no. 3, 1458-1466, 2012.

[10] Shyroki, DM. "Modeling of sloped interfaces on a Yee grid," *IEEE Transactions on Antennas and Propagation*, vol. 59, no. 9, 3290-3295, 2011.

[11] Yee, KS. "Numerical solution of initial boundary value problems involving Maxwell's equation in isotropic media," *IEEE Transaction on Antennas Propagation*, vol. AP-14, 302-307, 1966.

[12] Taflove, A; Hagness, SC. Computational electrodynamics: The finite-difference time-domain, 3rd ed., Artech House, Boston, 2005.

[13] Gandhi, OP; Chen, J. "Numerical dosimetry at power-line frequencies using anatomically based models," *Bioelectromagnetics*, vol. 1, 43-60, 1992.

[14] Gandhi, OP. "FDTD in bioelectromagnetics: Safety assessment and medical applications," in Advances in computational electrodynamics: The finite-difference time-domain method, A. Taflove, 1st ed., Artech House, 627-632, 1998.

[15] Potter, ME; Okoniewski, M; Stuchly, MA. "Low frequency finite difference time domain (FDTD) for modelling induced fields in humans

close to line sources," *Journal of Computational Physics*, vol. 162, 82-103, 2000.

[16] Emili, G; Schiavoni, A; Roselli, FL; Sorrentino, R. "Computation of electromagnetic field inside a tissue at mobile communications frequencies," *IEEE Transaction on Microwave Theory and Techniques*, vol. 51, 178-186, 2003.

[17] Liu, LM; Cleary, SF. "Absorbed energy distribution from radio frequency electromagnetic radiation in a mammalian cell model: Effect of membrane-bound water," *Bioelectromagnetics*, vol. 16, 160-171, 1995.

[18] Sui, W; Christensen, DA; Durne, CH. "Extending the two-dimensional FDTD method to hybrid electromagnetic systems with active and passive lumped elements," *IEEE Transaction on Microwave Theory and Techniques*, vol. MTT-40, 724-730, 1992.

[19] Piket-May, M; Taflove, A; Baron, J. "FDTD modeling of digital signal propagation in 3D circuits with passive and active loads," *IEEE Transaction on Microwave Theory and Techniques*, vol. MTT-42, 1514-1523, 1994.

[20] Ciampolini, P; Mezzanotte, P; Roselli, L; Sereni, D; Torti, P; Sorrentino, R. "Simulation of HF circuits with FDTD technique including nonideal lumped elements," *IEEE MTT-S Int. Microwave Symposium Dig. Orlando, Florida*, 361-364, 1995.

[21] Ciampolini, P; Mezzanotte, P; Roselli, L; Sorrentino, R. "Accurate and efficient circuit simulation with lumped-element FDTD technique," *IEEE Transaction on Microwave Theory and Techniques*, vol. 44, 2207-2214, 1996.

[22] Kuo, CN; Houshmand, B; Itoh, T. "Full-wave analysis of packaged microwave circuits with active and nonlinear devices: An FDTD approach," *IEEE Transactions on Microwave Theory and Techniques*, vol. 45, 819-826, 1997.

[23] Hodgkin, AL; Huxley, AF. "A quantitative description of membrane current and its application to conduction and excitation in nerve," *J. Physiol.*, vol. 117, 500-544, 1952.

[24] Hodgkin, AL; Huxley, AF. "Current carried by sodium and potassium ions through the membrane of the giant axon of Loligo," *J. Physiol.*, vol. 116, 449-472, 1952.

[25] Hodgkin, AL; Huxley, AF; Katz, B. "Measurement of current-voltage relations in the membrane of the giant axon of Loligo," *J. Physiol.*, vol. 116, 424-448, 1952.

[26] Hodgkin, AL; Huxley, AF. "The components of membrane conductance in the giant axon of Loligo," *J. Physiol.*, vol. 116, 473-496, 1952.

[27] Hodgkin, AL; Huxley, AF. "The dual effect of membrane potential on sodium conductance in the giant axon of Loligo," *J. Physiol.*, vol. 116, 497-506, 1952.

[28] Tsay, W-J; Pozar, DM. "Application of the FDTD technique to periodic problems in scattering and radiation," *IEEE Microwave and Guided Wave Letters*, vol. 3, 250-252, 1993.

[29] Ren, J; Gandhi, OP; Walker, LR; Fraschilla, J; Boerman, CR. "Floquet-based FDTD analysis of two-dimensional phased array antennas," *IEEE Microwave and Guided Wave Letters*, vol. 4, 109-111, 1994.

[30] Alexanian, A; Kolias, NJ; Compton, RC; York, RA. "Three-dimensional FDTD analysis of quasi-optical arrays using Floquet boundary conditions and Berenger's PML," *IEEE Microwave and Guided Wave Letters*, vol. 6, 138-140, 1996.

[31] Maloney, JG; Kesler, MP. "Analysis of periodic structures," in Computational electrodynamics: The finite-difference time-domain method, A. Taflove, 2nd ed., Artech House, London, 569-625, 2000.

[32] Berenger, J. "A perfectly matched layer for absorption of electromagnetic waves," *Journal of Computational Physics*, vol. 114, 185-200, 1994.

[33] Katz, DS; Thiele, ET; Taflove, A. "Validation and extension to three dimensions of the Berenger PML absorbing boundary condition for FD-TD Meshes," *IEEE Microwave and Guided Wave Letters*, vol. 4, 268-270, 1994.

[34] Berenger, J. "Perfectly Matched layer for the FDTD Solution of wave-structure interaction problems," *IEEE Transaction on Antennas and Propagation*, vol. 44, 110-117, 1996.

[35] Berenger, J. "Improved PML for the FDTD solution of wave-structure interaction problem," *IEEE Transaction on Antennas and Propagation*, vol. 45, 466-473, 1997.

[36] Mur, G. "Absorbing boundary conditions for the finite-difference approximation of the time-domain electromagnetic-field equations," *IEEE Transaction on Electromagnetic Compatibility*, vol. EMC-23, 377-382, 1981.

[37] Ebara, H; Tani, K; Onishi, T; Uebayashi, S; Hashimoto, O. "Method for estimating complex permittivity based on measuring effective permittivity of dielectric mixtures in radio frequency band," *IEICE Transactions on Communication*, vol. E88-B, 3269-3274, 2005.

[38] Kotnik, T; Miklavcic, D. "Theoretical evaluation of the distributed power dissipation in biological cells exposed to electric fields," *Bioelectromagnetics*, vol. 21, 385-394, 2000.

[39] Navarro, EA; Gimeno, B; Cruz, JL. "Modelling of periodic structures using the finite-difference time-domain method combined with the Floquet theorem," *Electronic Letters*, vol. 29, 446-447, 1993.

[40] Prescott, DT; Shuley, NV. "Extensions to the FDTD method for the analysis of infinitely periodic arrays," *IEEE Microwave and Guided Wave Letters*, vol. 4, 352-354, 1994.

[41] Veysoglu, ME; Shin, RT; Kong, JA. "A finite-difference time-domain analysis of wave scattering from periodic surfaces: Oblique incidence case," *J. Electron. Waves Appl.*, vol. 7, 1595-1607, 1993.

[42] Furse, CM; Chen, JY; Gandhi, OP. "The use of the frequency-dependent finite-difference time-domain method for induced current and SAR calculations for a heterogeneous model of the human body," *IEEE Transactions on Electromagnetic Compatibility*, vol. 36, 128-133, 1994.

[43] Roper, DH; Baird, JM. "Analysis of overmoded waveguides using the finite-difference time-domain method," *IEEE MTT-S International Microwave Symposium Digest*, 401-404, 1992.

[44] Furse, CM; Mathur, SP; Gandhi, OP. "Improvements to the finite-difference time-domain method for calculating the radar cross section of a perfectly conducting target," *IEEE Transactions on Microwave Theory and Techniques*, vol. 38, 919-927, 1990.

[45] Furse, CM; Roper, DH; Buechler, DN; Christensen, DA; Durney, CH. "The problem of treatment of DC offsets in FDTD Simulations," *IEEE Transactions on Antennas and Propagations*, vol. 48, 1198-1201, 2000.

[46] See, CH; Abd-Alhameed, RA; Excell, PS. "Computation of Electromagnetic Fields in Assemblages of Biological Cells Using a Modified Finite Difference Time Domain Scheme," *IEEE Transactions on Microwave Theory and Techniques*, vol. 55, 1986-1994, 2007.

[47] See, CH; Abd-Alhameed, RA; Excell, PS; Zhou, D. "Microscopic Biological Cell Level Model Using Modified Finite-difference Time-domain at Mobile Radio Frequencies," *PIERS Online*, vol. 4, 6-10, 2008.

In: Development of Complex Electromagnetic … ISBN: 978-1-61122-013-1
Editors: K. N. Ramli et al.

Chapter 9

SIMULATION OF ANTENNAS COUPLED TO LOSSY DIELECTRIC VOLUMES

***M. A. Mangoud*[1], *R. A. Abd-Alhameed*[1] *and P. S. Excell*[2]**
[1]Mobile and Satellite Communications Research Centre,
Bradford University, Bradford, UK
[2]Institute for Arts, Science and Technology,
Glyndwr University, Wrexham, UK

ABSTRACT

A heterogeneous hybrid computational electromagnetic method is presented in this chapter. Different parts of an antenna simulation problem are treated by using different methods enabling the most appropriate method to be used for each part. The method employs the method of moments and finite-difference time-domain techniques to compute the fields in two regions. The two regions are interfaced by surfaces on which effective sources are defined by the application of the equivalence principle. An extension to this permits the conduction currents to cross the boundary between the different computational domains. Several examples are studied and the results are compared and analysed. The method paves the way for accurate simulation of the behaviour of an antenna that is closely coupled with lossy dielectric volumes.

9.1. INTRODUCTION

Electromagnetic computation algorithms are widely used for the prediction of the behaviour of antennas of all types. In general, these are broadly classified into two types, integral equation methods, often known as the method of moments (MoM) [1-10] and differential equation methods (such as finite-difference time-domain) [11-20], which have significantly different strengths and weaknesses. Integral equation methods are widely used for simulation of broadcasting antennas since they can simulate the behaviour of arbitrarily shaped metallic structures with relative ease and a good level of fidelity. However, problems arise if the antenna is associated with substantial dielectric volumes, particularly lossy dielectrics such as earth, sea water and the human body. The integral equation methods inherently compute the coupling between each discretization element in the model and every other element. This does not cause a problem for metal because it is considered to be effectively impenetrable and hence only the surface has to be discretized; further, wires and thin rods can be represented as one-dimensional elements. Penetrable dielectrics, on the other hand, have to be represented with a 3-D assembly of volumetric elements. Leaving aside the mathematical complexity of modelling such elements, the key difficulty that arises is that, in being a 3-D structure, the number of elements required in the model rapidly becomes very large and the computational task can become unmanageable. An alternative approach for penetrable dielectrics that is computationally much more tractable is to use a differential equation method, such as the FDTD method. This is a well-known and highly developed algorithm that directly implements a discretized form of Maxwell's equations. It is very efficient and relatively easy to implement, but it has one major disadvantage in that it requires the problem space to be discretized into three-dimensional rectangular elements. This results in a significant deficiency in the modelling of structures with surfaces that are not parallel to the principal axes, since the surfaces then have a stepped shape (known as 'staircasing') as a result of conformance to the rectangular quantization grid. This poses particular problems when attempts are made to model linear (wire-type) antennas with components that are not parallel to the axes. These two methods have unique advantages in their own domain of application but it is not possible to apply either to the other domain without significant disadvantages (e.g., using integral equation methods to model the ground or using FDTD to model wires). A method was thus developed to enable both of them to be used for a single problem, combining them via a hybridization algorithm in which a bounding surface was established on which fictitious equivalent sources were considered to exist. The first stage of hybridization between these methods was achieved by analysing each part of a problem with the most appropriate method [21, 22], the coupling between these methods being computed by using the equivalence principle theorem. In this theorem, the objects should not be physically connected, but separated by a distance of at least a moderate fraction of a wavelength to place the equivalence-principle surface in isolation between them [23].

9.2. Conduction Current Crossing Domain Boundaries in a Heterogeneous Hybrid MoM-FDTD Formulation

If conduction currents are required to cross the boundary between the two hybridised domains, a special treatment is required to ensure current continuity across the boundary surface. To include this case, the previous work [23-26] can be extended for the case where the source region is subdivided, one part replaced by equivalent surface currents using the equivalence principle (computed by MoM) and the other part handled with the use of direct impressed currents (suitable to couple with FDTD). An industry-standard frequency-domain MoM program [27] was used, to aid acceptance in practical applications, although it can easily be modified for use with any MoM program. Coupling between the scatterer and source regions is calculated using the Reaction Theorem, with appropriate modifications to the MoM program to take account of the fact that the basis and weighting functions are not the same. Progress has been made in solving this problem by development of a formulation that allows current flow across the domain boundary, as will be discussed in the following section with some basic examples that illustrate the objectives of this novel technique.

9.2.1. Theory of Modified Hybrid MoM-FDTD Method

Figure 9.1(a) shows two regions, one enclosing the source, the other enclosing the scatterer. The source region is further subdivided into two sub-regions, *A* and *B* as shown in Figure 9.1(b). In general there is no physical attachment between source region *B* and the scatterer region, but the scatterer can be very close to, or attached to, the source region A. In Figure 9.1(b) the surface S_c encloses the entire source region. The fields due to the induced currents from both source sub-regions A and B can be computed in order to evaluate the surface currents J_{si} and M_{si} over the modified closed surface S_{c1}:

$$M_{si} = (E(J_A, M_A) + E(J_B, M_B)) \times \hat{n} \tag{9.1}$$

$$J_{si} = (H(J_A, M_A) + H(J_B, M_B)) \times \hat{n} \tag{9.2}$$

where J_A, J_B, M_A and M_B are the electric and magnetic surface currents of source regions A and B; $\hat{n}$ is the unit vector normal to the surface S_{c1}.

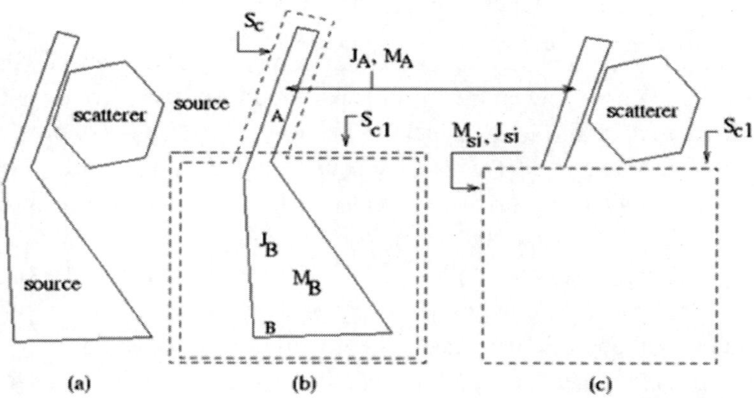

Figure 9.1. The basic geometry of the problem.

Now let the part of the surface S_c surrounding sub-region A be coincident with the surface of the conducting source structure within A. The currents on S_c then become the surface currents on the conductor which can be transferred to an FDTD model by treating them as impressed currents. The whole source region is then modelled using the FDTD method with impressed currents replacing sub-region A and a surface of equivalent currents replacing sub-region B. The hard (impressed) sources for a perfectly conducting surface ($M_A = 0$) considered in region A can be implemented in FDTD as follows:

$$E = E_{FDTD} - \frac{J_A}{(\varepsilon / \Delta t + \sigma / 2)} \tag{9.3}$$

where E_{FDTD} represents the normal electric field finite-difference updating equations with a total/scattered field formulation [22], J_A is the electric surface current calculated using MoM, Δt is the time step, ε and σ are the permittivity and the conductivity of the surrounding medium, respectively. For region *B*, which is surrounded by the closed surface Sc1, the difference equations at the surface for the FDTD method can be stated as follows:

$$E = E_{FDTD} + \frac{J_{si\ Sc1}}{(\varepsilon / \Delta t + \sigma / 2)} \tag{9.4}$$

$$H = H_{FDTD} - M_{si\ Sc1} \frac{\Delta t}{\mu} \tag{9.5}$$

where J_{si} and M_{si} are the surface electric and magnetic currents, and with the suffix S_{c1} are only those defined at the surface $S_{c1.}$ μ is the permeability of the medium. The voltage induced on the source region *A* due to the back-scattered fields from the scatterer is:

$$V_A = (E_{free} - E_{scat}) \cdot \Delta s \tag{9.6}$$

where E_{free} and E_{scat} are the induced electric fields in the region *A*, calculated in the FDTD domain, without and with the scatterer, and Δs is the FDTD cell size. The voltage induced in region *B* (the scattered field region) is found using MoM by employing the Reaction Theorem:

$$V_B = \int_{S_B} J_{ts} \cdot E_{ib} ds_B \tag{9.7}$$

$$V_B = \int_{S_{c1}^-} \left(E_{ts} \cdot J_{ib} - H_{ts} \cdot M_{ib} \right) ds_{c1}^- \tag{9.8}$$

where:

$$E_{iB} = -j\omega A(r) - \nabla V(r) - \frac{1}{\varepsilon} \nabla \times F(r) \tag{9.9}$$

$$A(r)=\mu\int_{S_{c1}^{-}} J_{ib}g(r,r')ds_{c1}^{-} \tag{9.10}$$

$$V(r)=\frac{-j}{\omega\varepsilon}\int_{S_{c1}^{-}} \nabla'_{s}\cdot J_{ib}g(r,r')ds_{c1}^{-} \tag{9.11}$$

$$F(r)=\varepsilon\int_{S_{c1}^{-}} M_{ib}g(r,r')ds_{c1}^{-} \tag{9.12}$$

The free space Green's function is given by:

$$g(r,r')=\frac{e^{-jk|r-r'|}}{4\pi|r-r'|} \tag{9.13}$$

The vectors r and r' apply to the source and observation points respectively. S_B is the conducting surface area of the structure within region B. J_{ts} is the electric test function used on the wire or the patch. It is usually a uniform pulse to account for the excitation voltage at the centre of the wire segment and the patch for the MoM, if using NEC [27]. E_{ts} and H_{ts} are the electric and magnetic fields respectively for the test function J_{ts}. It should be noted that, in applying (7), there is no magnetic test function specified in NEC [27]. Since E_{ts} and H_{ts} can be obtained easily using NEC, after ignoring the sinusoidal basis functions, and assuming that the cell meshing used in FDTD is very small compared to the operating wavelength, equation (9.8) can be reduced by ignoring the surface integral and evaluating the voltage back-scattered corresponding to the centre of the cell surface, as shown below:

$$V_b=\sum_{n_{S_{c1}^{-}}}(E_{ts}(r,r')\cdot J_{ibn}-H_{ts}(r,r')\cdot M_{ibn})a_n \tag{9.14}$$

where r_n' is the position vector of the centre of the cell surface and a_n is the surface area of the cell. Therefore J_{ibn} and M_{ibn} are considered to be the equivalent surface currents at the centre of the surface cell n. At the boundary of the hard source and the equivalent surface (S_{c1}) FDTD requires at least two or three cells of additional margin to recalculate the voltage induced in region

B, as seen from Figure 9.1(b). When the voltages have been obtained, the new currents (J_A, J_{si} and M_{si}) can be computed and the method can be repeated until the steady state is reached.

9.3. Simulation and Results

Near-fields and input impedance are taken to be convenient but rigorous validation results for the following examples:

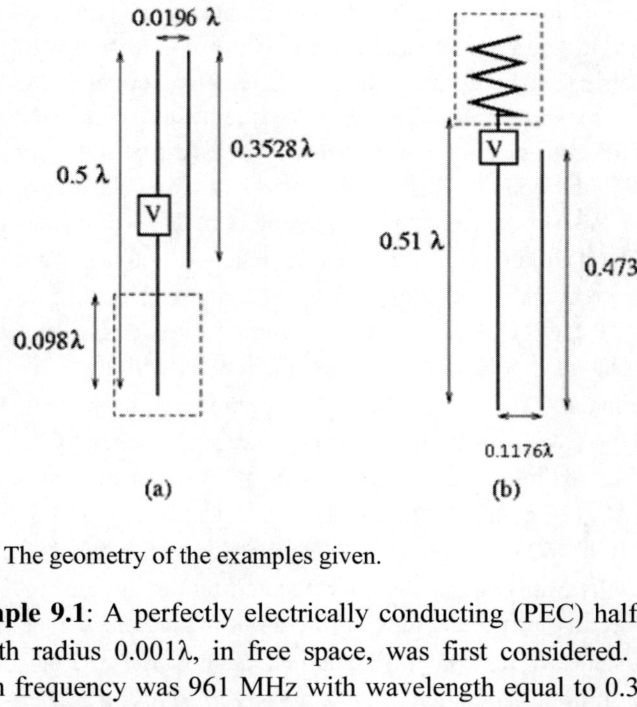

Figure 9.2. The geometry of the examples given.

Example 9.1: A perfectly electrically conducting (PEC) half-wavelength dipole with radius 0.001λ, in free space, was first considered. The chosen simulation frequency was 961 MHz with wavelength equal to 0.31212 m and the FDTD cell size equal to 0.00306 m. Thus, the dipole was considered as 51 cells in the *z*-direction. The equivalent surface treatment was arbitrarily chosen to enclose 70.4% (0.352λ) of the length of the dipole to be modelled using MoM, whereas the rest of the dipole was replaced by the impressed currents inside the FDTD problem space (as in Figure 9.2(a), without the scatterer). The free space electric field on a line parallel to the dipole at a transverse distance of 0.0346λ was examined, as shown in Figure 9.3. The results of the proposed current crossing technique are in good agreement with those from the hybrid MoM/FDTD [25] method (with a Huygens surface surrounding the full dipole) and the pure MoM NEC programs [27].

Example 9.2: A wire scatterer was introduced adjacent to the half-wavelength dipole of the previous example, in the form of a PEC rod of length 0.3528λ and radius 0.001λ, parallel to the source dipole at a distance 0.0196λ, equivalent to two FDTD cells, see Figure 9.2(a). The scatterer was located in the region of the impressed current source. The dipole here was modelled using the current crossing technique, using Huygens surface to enclose 19.6% (0.098λ) of the length of the dipole. The impressed currents were used for the rest of the dipole inside the FDTD part of the program. The input impedance of the source dipole versus the number of iterations is shown in Figure 9.4. The results are in good agreement with those from a uniform MoM treatment [27]. The electric field contours in the *y-z* plane, which contains the source and the scatterer, are shown in Figure 9.5 for the cases with and without the scatterer. This figure clarifies the methodology used and the technique of the current crossing the boundary using the equivalent surface principle.

Example 9.3: In this example, illustrated in Figure 9.2(b), the source region contains a dipole with total conductor length 1.0λ and radius 0.001λ. The working frequency was doubled to become 1923 MHz which means wavelength of 0.156 m. Similar to the previous example, cell size of 0.00306 m was used. One of dipole arms is coiled into a helix with three turns of radius 0.12λ and pitch 0.004λ. This simulates a canonical form of a transmitter cabinet directly feeding a helical antenna adjacent to a scatterer, in a case where the cabinet is intended to be simulated by FDTD. The scatterer is a wire of length 0.473λ (24 cells) and radius 0.001λ. The scatterer and the source are separated by 0.01182λ (6 cells). The equivalent surface encloses the helix region (treated by MoM) while the impressed current treatment (FDTD region) is used for the other arm of the dipole. The input impedance versus the number of iterations is shown in Figure 9.6. The results converge to $Z = 420 - j190\ \Omega$ which is also in good agreement with a non-hybrid NEC treatment $Z = 462 - j220\ \Omega$ [27]. The accuracy of this proposed modified hybrid method is demonstrated in Figure 9.7. This shows the computed near-field magnitude distribution contours for the example given in Figure 9.2, when conduction currents are crossing domain boundaries for both the dipole and the helix-dipole in free space and with a wire scatterer. The plane displayed contains the scatterer and the second arm of the source dipole that is represented by the straight wire. The near-field distribution shows the important advantage of the method over the conventional hybrid technique [24, 25, 28-30] which requires a more substantial separation distance between the source and the scatterer.

In the free space simulation of Figure 9.7(a) the symmetric patterns of the half-wavelength dipole in the total field region outside the Huygens surface can be observed. Figure 9.7(a) shows a reasonable pattern with expected asymmetry for the helix-dipole combination of length 1.0λ, with a minimum near-field value around the middle of the straight half of the dipole. In both Figures 9.5(a) and 9.7(a), the near-field values inside the Huygens surface are close to zero. Also, it is noticed that the boundary area of connection between the wire and the virtual surface shows very little disruption to the fields in the scattered region, hence indicating that the proposed current crossing treatment is implemented successfully. Figures 9.6(b) and 9.7(b) illustrate near-field contours in the presence of scatterers. It can be noted that more back-scattered field is observed inside the Huygens surface for the helix-dipole as compared with the dipole. This is because of the closer proximity of the scatterer to the crossing boundary in the first case.

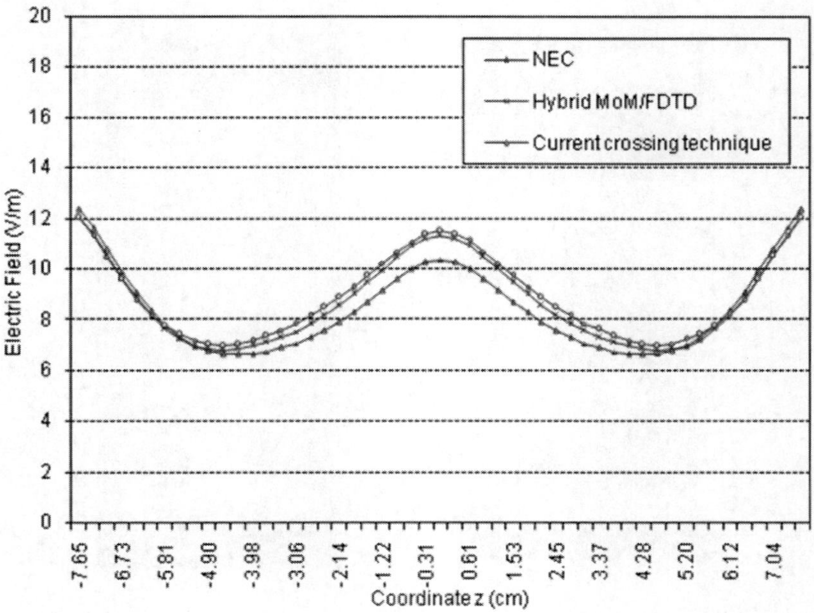

Figure 9.3. Comparison between the computed near electric fields of a dipole in free space (Example 9.1).

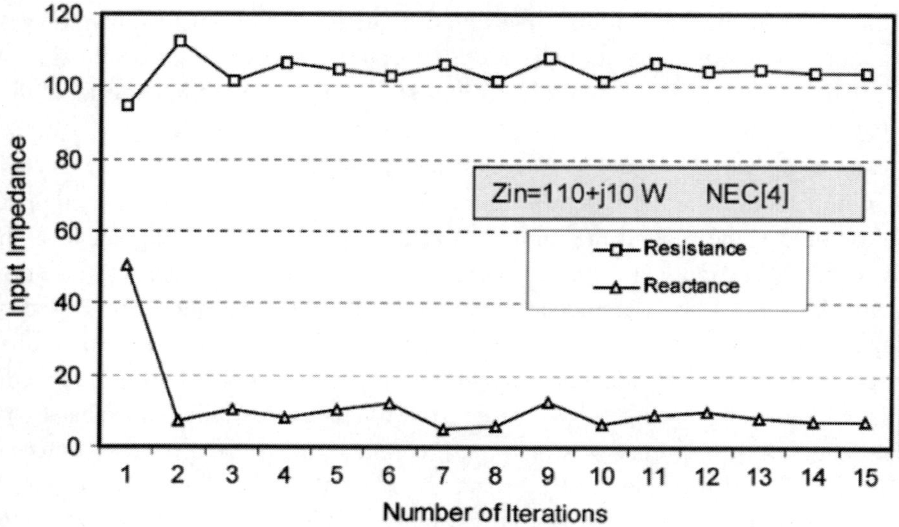

Figure 9.4. Input impedance of the source dipole with scatterer of Example 9.2, versus number of iterations, for current-crossing hybrid treatment.

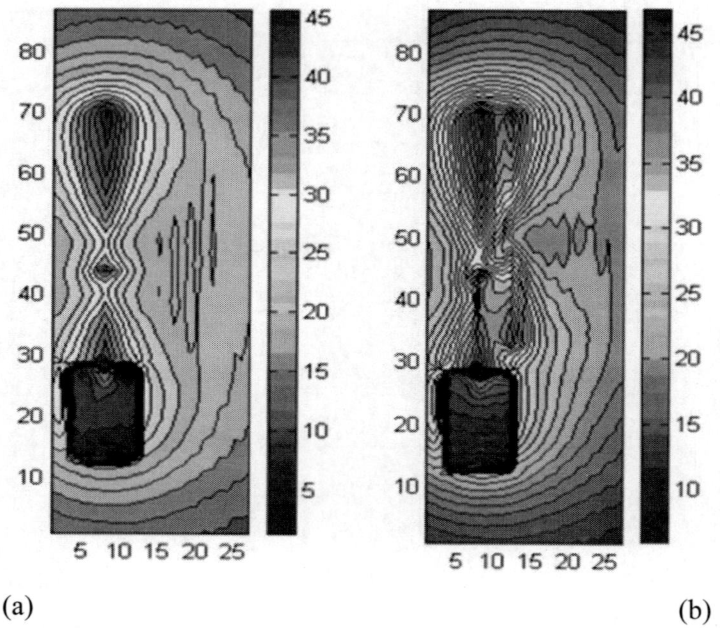

Figure 9.5. Vertical slice cut for computed near electric field magnitude distribution contours (in dB) for the Example 9.2: (a) without the scatterer; (b) with the scatterer.

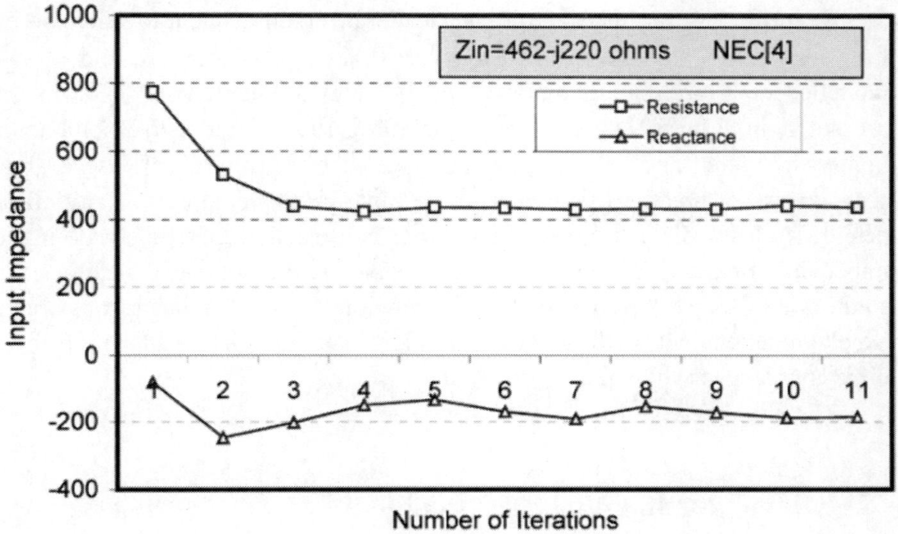

Figure 9.6. Input impedance of the semi-helical dipole with scatterer of Example 9.3, versus number of iterations, for hybrid current-crossing treatments.

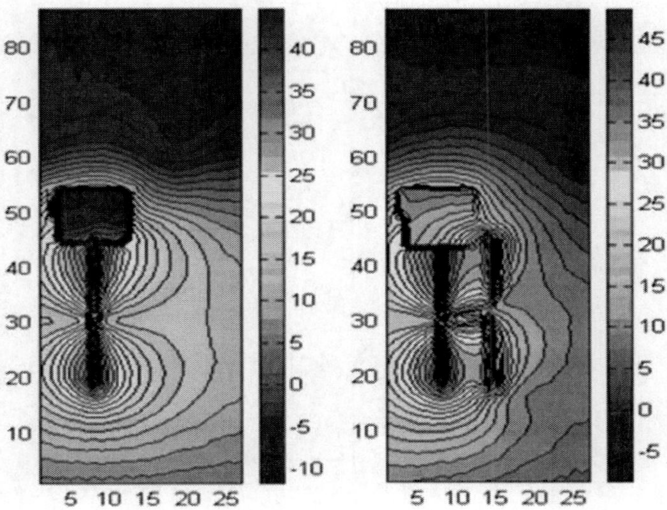

Figure 9.7. Vertical slice cut for computed near-field magnitude distribution contours (in dB) for the Example 9.3: (a) without the scatterer; (b) with the scatterer.

Example 9.4: In this example, the scatterer region is taken to be a slab of lossy dielectric. The sheet dimensions are 0.125λ×0.0625λ×0.5λ and it was modelled with a relative permittivity of 41.3 and conductivity of 1.3 S/m, as presented in Figure 9.8. These material properties would correspond to a highly hydrated and conducting material such as biological tissue, but seawater is not greatly dissimilar and the same principle applies to any other penetrable lossy dielectric, such as earth or construction materials. The input impedance versus the number of iterations required to reach steady state solution, for three different distances, is shown in Figure 9.9. The results are in excellent agreement with the data in [31]. The far field pattern for two different planes is shown in Figure 9.10.

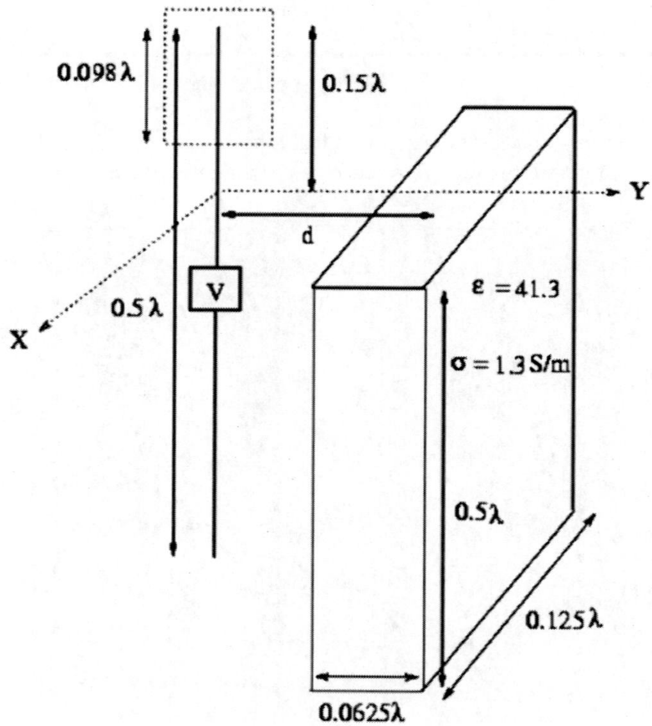

Figure 9.8. The geometry for Example 9.4.

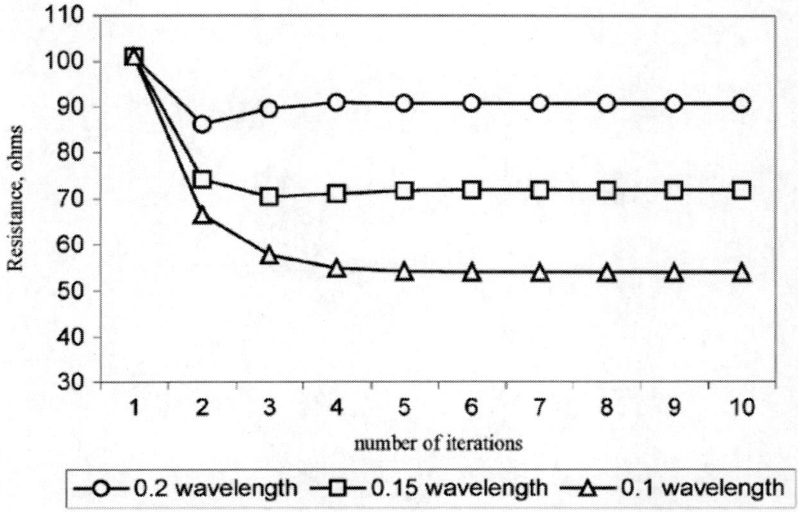

(a)

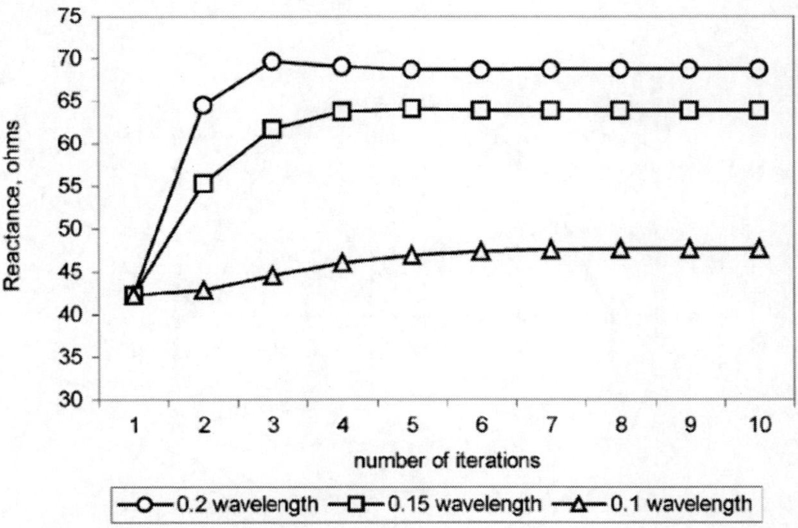

(b)

Figure 9.9. The input resistance and reactance of the source dipole in Example 9.4, for three different distances from the dielectric.

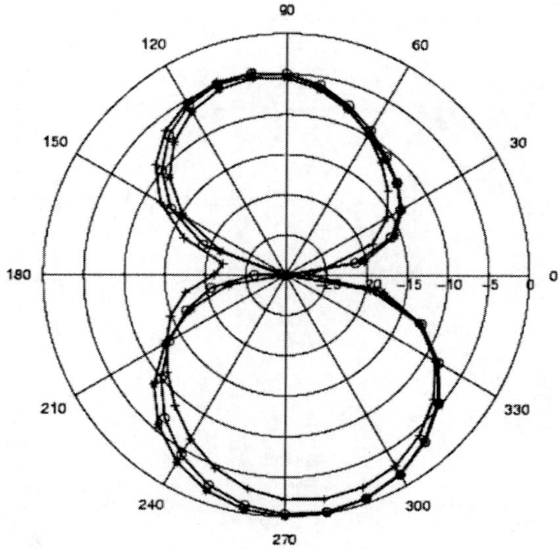

(a)

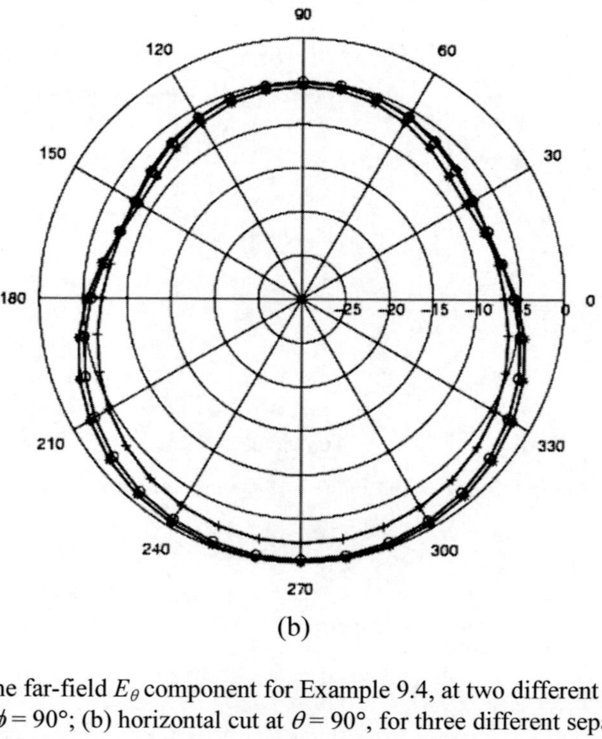

(b)

Figure 9.10. The far-field E_θ component for Example 9.4, at two different cuts: (a) vertical cut at $\phi = 90°$; (b) horizontal cut at $\theta = 90°$, for three different separation distances from the dielectric. (— + — 0.1λ, — o — 0.15λ, — * — 0.2λ).

Example 9.5: A typical HF broad-band duo-conical monopole antenna over an infinite perfectly conducting ground plane is modelled as an example of a broadcasting antenna and to insure the ability of the method to account for the ground plane effect in the presence of the current-crossing boundary. The antenna, shown in Figure 9.11, has the following properties: 34.3 m high, 45.0 m across, 110° apex angle and 2.2° sag angle. The antenna is chosen to be simulated as operating within the short wave band at 3 MHz, with a VSWR of 1.75 at the input port. The input port is placed in the segment of the antenna attaching it to the ground plane. The antenna and ground models inside the FDTD problem space are shown in Figure 9.12. The bottom set of PML layers is removed and both PML and problem space updating equations are modified to account for the effect of the infinite ground plane. The surface currents on the Huygens surface are calculated using MoM, considering the infinite conducting ground plane. The problem space is set to 72×72×67 cells of size 1 m^3 and the equivalence-principle surface is 50×50×40 cells. The antenna currents outside this surface are represented by the impressed currents, as described in the theory of the method. The near-field magnitude of the dominant vertically-polarised electric field component, observed at 27 m radial distance from the origin, as a function of height (i.e. distance in the *z*-direction) is shown in Figure 9.13. The results are in excellent agreement with those from NEC [27]. The hybrid method, however, has much greater flexibility than pure MoM, such as NEC, to handle differing types of ground and arbitrarily-shaped volumes of lossy dielectric near the antenna.

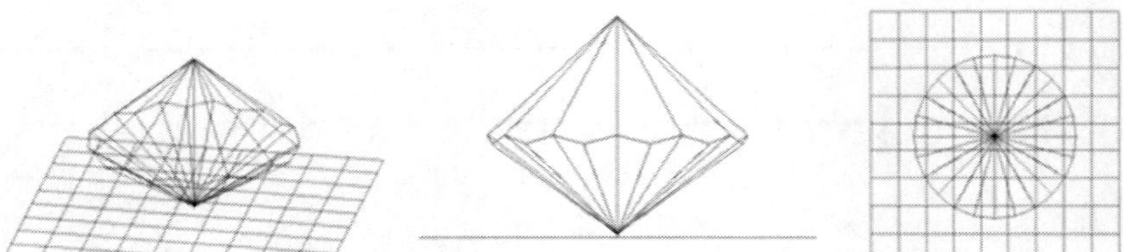

(a) 3-D view (b) Front view (c) Top view

Figure 9.11. Different views of duo-conical monopole antenna on perfect ground plane (infinite).

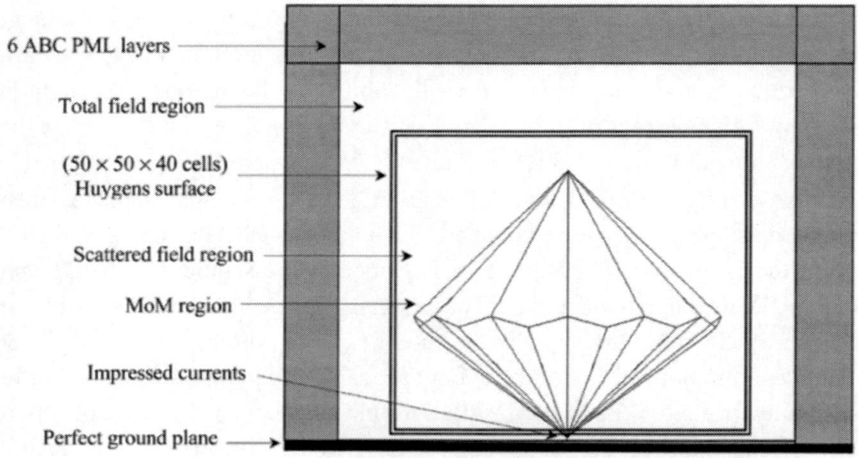

Figure 9.12. FDTD problem space details of Example 9.4.

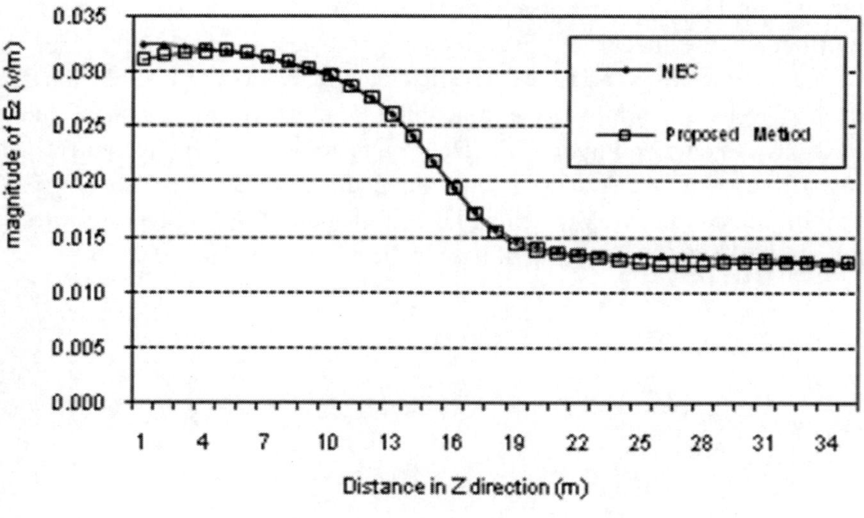

Figure 9.13. Electric field magnitude (E_z) versus distance parallel to z-axis at 27 m radial distance from the origin.

CONCLUSION

A novel hybrid method has been presented which permits conduction currents to cross the boundary between different computational domains. The objects can be physically connected or separated. The method uses the FDTD method in the scatterer and part of the source region and uses the standard frequency-domain MoM program in the remaining part of the source region to give the complete solution of the fields in the two regions. Several validation cases are examined and the results compared with available data. Also a short wave broadcasting monopole over ground plane has been successfully implemented using this modified technique. The hybrid method will have great advantage over alternative methods in the following electromagnetics applications such as cellular telephone dosimetry investigations close to real biological tissues, complex satellite quadrifilar antennas, SAR reduction using phased array antennas, base station safety assessment, subsurface radar, and in general, interaction between complex source structures and inhomogeneous dielectric materials.

REFERENCES

[1] K. Tap, P. Pathak and R. Burkholder, "Complex source beam - moment method procedure for accelerating numerical integral equation solutions of radiation and scattering problems," *IEEE Transactions on Antennas and Propagation*, vol. 62, no. 4, pp. 2052 - 2062, 2014.

[2] D. J. Ludick, R. Maaskant, D. B. Davidson, U. Jakobus, R. Mittra and D. De Villiers, "Efficient analysis of large aperiodic antenna arrays using the domain Green's function method," *IEEE Transactions on Antennas and Propagation*, vol. 62, no. 4, part 1, pp. 1579 - 1588, 2014.

[3] S. R. Zang and J. R. Bergmann, "Analysis of omnidirectional dual-reflector antenna and feeding horn using method of moments," *IEEE Transactions on Antennas and Propagation*, vol. 62, no. 3, pp. 1534-1538, 2014.

[4] X. -M. Pan and X. -Q. Sheng, "Improved algebraic preconditioning for MoM solutions of large-scale electromagnetic problems," *IEEE Antennas and Wireless Propagation Letters*, vol. 13, pp. 106-109, 2014.

[5] M. A. Echeverri Bautista, M. A. Francavilla, F. Vipiana and G. Vecchi, "A hierarchical fast solver for EFIE-MoM analysis of multiscale structures at very low frequencies," *IEEE Transactions on Antennas and Propagation*, vol. 62, no. 3, pp. 1523-1528, 2014.

[6] M. Amir, S. Bedra, S. Benkouda and T. Fortaki, "Bacterial foraging optimisation and method of moments for modelling and optimisation of

microstrip antennas," *IET Microwaves, Antennas and Propagation*, vol. 8, no. 4, pp. 295-300, 2014.

[7] M. Shafieipour, I. Jeffrey, J. Aronsson and V. I. Okhmatovski, "On the equivalence of RWG method of moments and the locally corrected Nyström method for solving the electric field integral equation," *IEEE Transactions on Antennas and Propagation*, vol. 62, no. 2, pp. 772-782, 2014.

[8] A. Schroder, G. A. Rasek, H. -D. Bruns, Z. Reznicek, J. Kucera, S. E. Loos and C. Schuster, "Analysis of high intensity radiated field coupling into aircraft using the method of moments," *IEEE Transactions on Electromagnetic Compatibility*, vol. 56, no. 1, pp. 113-122, 2014.

[9] N. Hendijani, J. Cheng and R. J. Adams, "Combined field integral equation using a constraint-based Helmholtz decomposition," *IEEE Transactions on Antennas and Propagation*, vol. 62, no. 3, pp. 1500-1503, 2014.

[10] J. Alvarez, L. D. Angulo, M.R. Cabello, A. Rubio Bretones and S. G. Garcia, "An analysis of the leap-frog discontinuous Galerkin method for Maxwell's equations," *IEEE Transactions on Microwave Theory and Techniques*, vol. 62, no. 2, pp. 197-207, 2014.

[11] R. Xiong, B. Chen, Y. Yi, H. -L. Chen and H. -F. Zhao, "Two-step method for the FDTD analysis of long apertures having depth," *IEEE Antennas and Wireless Propagation Letters*, vol. 12, pp. 39-42, 2013.

[12] S. K. Kamepally, B. P. Kumar and C. S. Paidimarry, "FDTD estimation for accurate specific absorption rate in a tumor," *Annual International Conference on Emerging Research Areas and International Conference on Microelectronics, Communications and Renewable Energy (AICERA/ICMiCR)*, pp. 1-5, 2013.

[13] W. G. Whittow, C. C. Njoku and Y.C. Vardaxoglou, "Fine scale simulations of patch antennas with heterogeneous substrates," *USNC-URSI Radio Science Meeting (Joint with AP-S Symposium)*, pp. 223, 2013.

[14] A. C. Lesina, A. Vaccari and A. Bozzoli, "A modified RC-FDTD algorithm for plasmonics in Drude dispersive media nanostructures," *6th European Conference on Antennas and Propagation (EUCAP)*, pp. 687-690, 2012.

[15] G. M. Noetscher, Y. Xu and S. N. Makarov, "Accuracy of point source models with coincident phase centers in a cubic FDTD grid for arbitrary source orientation," *IEEE Antennas and Propagation Society International Symposium (APSURSI)*, pp. 1-2, 2012.

[16] V. Nayyeri, M. Soleimani and M. Dehmollaian, "Implementation of a PEMC boundary condition in the 2-D FDTD technique," *IEEE Antennas and Propagation Society International Symposium (APSURSI)*, pp. 1-2, 2012.

[17] D. -C. Liu and H. -C. Chang, "The second-order condition of FDTD method at sloped dielectric interfaces by averaging the permittivities along surface normal," *IEEE Transactions on Antennas and Propagation*, vol. 60, no. 11, pp. 5259-5267, 2012.

[18] M. Weiwei, S. Dong and W. Xianliang, "UPML-FDTD parallel computing on GPU," *International Conference on Microwave and Millimeter Wave Technology (ICMMT)*, vol. 4, pp. 1-4, 2012.

[19] C. Guiffaut, A. Reineix and B. Pecqueux, "New oblique thin wire formalism in the FDTD method with multiwire junctions," *IEEE Transactions on Antennas and Propagation*, vol. 60, no. 3, pp. 1458-1466, 2012.

[20] D. M. Shyroki, "Modeling of sloped interfaces on a Yee grid," *IEEE Transactions on Antennas and Propagation*, vol. 59, no. 9, pp. 3290-3295, 2011.

[21] E. K. Miller, L. Medgyesi-Mitschang, and E. H. Newman, Eds., Computational electromagnetics: Frequency-domain method of moments, New York: IEEE Press, 1992.

[22] A. Taflove and S. C. Hagness, Computational electrodynamics: The finite-difference time-domain method, 3rd ed., Boston, MA: Artech House, 2005.

[23] R. A. Abd-Alhameed, P. S. Excell, J. A. Vaul, and M. A. Mangoud, "Hybrid treatment for electromagnetic field computation in multiple regions," *Electronic Letters*, vol. 34, pp. 1925-1926, 1998.

[24] G. Cerri, P. Russo, A. Schiavoni, G. Tribellini, and P. Bielli, "MoM-FDTD hybrid technique for analysing scattering problems," *Electronic Letters*, vol. 34, pp. 433-440, 1998.

[25] R. A. Abd-Alhameed, P. S. Excell, J. A. Vaul, and M. A. Mangoud, "Computation of radiated and scattered fields using separate frequency domain moment-method regions and frequency-domain MoM-FDTD hybrid methods," *in IEE Conference Antennas & Propagation, York, UK*, pp. 53-57, 1999.

[26] M. A. Mangoud, R. A. Abd-Alhameed, and P. S. Excell, "Simulation of human interaction with mobile telephones using hybrid techniques over coupled domains," *IEEE Transaction on Microwave Theory and Techniques*, vol. 48, pp. 2014-2021, Nov. 2000.

[27] G. H. Burke and A. J. Poggio, "Numerical electromagnetics code (NEC): Method of moments," *US Naval Ocean Systems Center, San Diego, CA*, Rep. no. TD116, 1981.

[28] R. A. Bretones, R. Mittra, and G. Martin, "Hybrid NEC/FDTD approach for analysing electrically short thin-wire antennas located in proximity of inhomogeneous scatterers," *Electronic Letters*, vol. 35, pp. 1594-1596, 1999.

[29] N. Rogier, F. Olyslager, and D. De Zutter, "A new hybrid FDTD-BIE approach to model electromagnetic scattering problems," *IEEE Microwave and Guided Wave Letter*, vol. 3, pp. 138-140, 1998.

[30] A. Monorchio and R. Mittra, "A hybrid finite-element-difference time-domain (FE/FDTD) technique for solving complex electromagnetic problems," *IEEE Microwave and Guided Wave Letters*, vol. 8, pp. 93-95, 1998.

[31] R. A. Abd-Alhameed and P. S. Excell, "Analysis of dielectrically-loaded wire, strip and patch antennas, using method of moments," *in Proceeding IEE Conference Computation in Electromagnetics*, pp. 306-311, 1996.

About the Editors

Khairan N. Ramli

Faculty of Electrical and Electronics Engineering
Universiti Tun Hussein Onn Malaysia
Parit Raja, Batu Pahat, Johor, Malaysia
Email: khairun@uthm.edu.my

Raed A. Abd-Alhameed

Mobile and Satellite Communications Research Centre
Bradford University
Bradford, UK
Email: r.a.a.abd@bradford.ac.uk

Peter S. Excell

Institute for Arts, Science and Technology
Glyndwr University
Wrexham, UK

INDEX

#

A

B

C

D

E

I

K

L

M

N

O

P

Q

R

S

T

U

V

W